Current Topics in Microbiology and Immunology

Volume 430

The review series *Current Topics in Microbiology and Immunology* provides a synthesis of the latest research findings in the areas of molecular immunology, bacteriology and virology. Each timely volume contains a wealth of information on the featured subject. This review series is designed to provide access to up-to-date, often previously unpublished information.

2019 Impact Factor: 3.095., 5-Year Impact Factor: 3.895

2019 Eigenfaktor Score: 0.00081, Article Influence Score: 1.363

2019 Cite Score: 6.0, SNIP: 1.023, h5-Index: 43

More information about this series at http://www.springer.com/series/82

Fabio Bagnoli · Rino Rappuoli
Editors

Three Dimensional Human Organotypic Models for Biomedical Research

Responsible series editor: Rino Rappuoli

Springer

Editors
Fabio Bagnoli
GSK Vaccines Research
Siena, Italy

Rino Rappuoli
GSK Vaccines Research
Siena, Italy

ISSN 0070-217X ISSN 2196-9965 (electronic)
Current Topics in Microbiology and Immunology
ISBN 978-3-030-62451-4 ISBN 978-3-030-62452-1 (eBook)
https://doi.org/10.1007/978-3-030-62452-1

This Springer imprint is published by the registered company Springer Nature Switzerland AG
The registered company address is: Gewerbestrasse 11, 6330 Cham, Switzerland

Preface

Animal models are currently extensively used for studying host-pathogen interactions, mechanism of action and efficacy of vaccines and drugs, cancer formation, as well as for deciphering the pathogenesis of many human diseases. On the basis of such studies many invaluable discoveries have been done and new medical interventions developed. However, animal models have several limitations that can significantly affect their reliability for translating the observations to humans. Indeed, besides the many examples of successful outcome of research performed using animal models, unfortunately there are also notable cases of failure. Furthermore, genome and immunology studies recently performed by several research teams such as the one led by Mark Davis, are showing differences between humans and mice that explain limitations of mouse models in predicting clinical outcomes and stress the need of performing research on human samples.

Therefore, models that can better reproduce human tissues are urgently needed. Three dimensional organotypic human models are informative and tractable experimental systems and they represent a valid alternative to animal models. Efforts aimed at establishing such models is receiving a lot of attention from most areas of biomedical research and many human tissues can today be cultivated in vitro and used for several research objectives. Examples of such models include reconstruction of human epithelia (skin, intestine, stomach, lung) from cells cultivated in culture. The key element of these models are the epithelial cells, but other cell types (e.g., immune cells) can be added. To accomplish this goal researchers are using cutting-edge technologies such as induced pluripotent stem cells, which can be used to generate virtually any sort of cell including macrophages and neutrophils. These models are amenable for the dissecting the role of particular cell types in a certain phenomenon. A different approach relies on the use of tissue explants obtained from donors undergoing surgery operations. Certain explants can be kept in culture for quite a long time allowing many kinds of experiments. They contain all the resident cells of the tissue and they can reflect quite closely certain processes that happen in the patients. Although most of these models still have the important limitation of lacking blood circulation, they certainly represent an important step forward for biomedical research and combined with powerful analytical tools such as RT-PCR, DNA-microarrays, confocal and two-photon microscopy as well as quantitative and *in situ* mass spectrometry they are able to shed light on many yet unknown physiological, immunological and pathological mechanisms. Observations done using these models can complement and sometimes replace animal models allowing to reduce the number of animals used in experimental research, which is another very important achievement.

This book attempts to provide a general overview, including detailed scientific and technical aspects, of on a nascent but already very active and fruitful area of research. It covers major areas of biomedical research conducted using these models such as infectious diseases, cancer, allergic diseases as well as drug/vaccine discovery and toxicology studies. It provides an unprecedented overview of recent developments and knowledge on very diverse human organotypic models and their applications.

We hope it will help many basic scientists, physicians and applied scientists in designing their next experiments for understanding pathological or medicinal mechanisms and better predict the same phenomena in patients.

Conflict of Interest

Rino Rappuoli and Fabio Bagnoli are employees of the GSK group of companies. RR and FB hold shares in the GSK group of companies. FB holds pending and issued patents on *S. aureus* vaccine formulations.

Siena, Italy

Fabio Bagnoli
Rino Rappuoli

Contents

Part I
Three Dimensional Human Organotypic Models for Studying Infectious Diseases

Human Three-Dimensional Models for Studying Skin Pathogens

Elena Boero, Malgorzata Ewa Mnich, Andrea Guido Oreste Manetti, Elisabetta Soldaini, Luca Grimaldi, and Fabio Bagnoli

Contents

Abstract Skin is the most exposed surface of the human body, separating the microbe-rich external environment, from the sterile inner part. When skin is breached or its homeostasis is perturbed, bacterial, fungal and viral pathogens can cause local infections or use the skin as an entry site to spread to other organs. In the last decades, it has become clear that skin provides niches for permanent microbial colonization, and it actively interacts with microorganisms. This crosstalk promotes skin homeostasis and immune maturation, preventing expansion of harmful organisms. Skin commensals, however, are often found to be skin most prevalent and dangerous pathogens. Despite the medical interest, mechanisms of colonization and invasion for most skin pathogens are poorly understood. This limitation is due to the lack of reliable skin models. Indeed, animal models do not adequately mimic neither the anatomy nor the immune response of human skin. Human 3D skin

Elena Boero and Malgorzata Ewa Mnich—These authors contributed equally to the work.

E. Boero · M. E. Mnich · A. G. O. Manetti · E. Soldaini · F. Bagnoli (✉)
GSK, Siena, Italy
e-mail: fabio.x.bagnoli@gsk.com

L. Grimaldi
Department of Medicine, Surgery and Neuroscience, University of Siena, Siena, Italy

Current Topics in Microbiology and Immunology (2021) 430: 3–27
https://doi.org/10.1007/82_2020_219

Published Online: 30 June 2020

models overcome these limitations and can provide new insights into the molecular mechanisms of microbial pathogenesis. Herein, we address the strengths and weaknesses of different types of human skin models and we review the main findings obtained using these models to study skin pathogens.

List of Abbreviations

AD	Atopic dermatitis
APC	Antigen-presenting cell
cDC	Conventional DC
CHIPS	Chemotaxis inhibitory protein of *Staphylococcus aureus*
CLA	Cutaneous lymphocyte antigen
ClfA	Clumping factor A
ClfB	Clumping factor B
DC	Dendritic cell
GAS	Group A Streptococcus
GFP	Green fluorescent protein
GM-CSF	Granulocyte-macrophage colony-stimulating factor
GvH	Graft-versus-host
GvHD	Graft-versus-host disease
HIV	Human immunodeficiency virus
HPV	Human papillomavirus
HSE	Human skin equivalent
HSV	Herpes simplex virus
IFN	Interferon
IL	Interleukin
LC	Langerhans cell
MCPyV	Merkel cell polyomavirus
moDC	Monocyte-derived DC
NSTI	Necrotizing soft tissue infection
PBMCs	Peripheral blood mononuclear cells
PIA	Polysaccharide intracellular adhesin
PSMα	Phenol-soluble modulins α
PVL	Panton–Valentine leukocidin
ROS	Reactive oxygen species
SCIN	Staphylococcal complement inhibitor
SEA	Staphylococcal enterotoxin A
SEB	Staphylococcal enterotoxin B
SLS	Streptolysin S
SLO	Streptolysin O
SSI	Surgical site infection
SSS	Scalded skin syndrome
SSTI	Skin and soft tissues infection
T_{Eff}	Effector T cell
T_{EM}	Effector memory T cell

T_{RM}	Tissue-resident memory T cells
TLR2	Toll-like receptor 2
TNFα	Tumour necrosis factor α
VZV	Varicella zoster virus
WGS	Whole genome shotgun

1 Introduction

The skin is the most extended outer surface of the human body and constitutes a harsh, nutrient-poor, acidic and dry landscape. Its outermost layer, the epidermis, is made of overlapping strata of keratinocytes, of which the apical one, the stratum corneum, is made of terminally differentiated, enucleated cells, which are chemically cross-linked to fortify the skin barrier. Below the epidermis, there is the dermis, a loose, irrorated and innervated connective matrix, made mainly by fibroblasts. Interrupting the continuity of the stratum corneum and digging into the dermis, there are skin appendages, such as sweat glands and hair follicles, exerting protective and homeostatic functions. They are unevenly distributed in the skin depending on the area of the body, further increasing the heterogeneity of skin landscapes. Both the epidermal and dermal compartments host the skin immune system, a network of immune cells that guard the interface with the exterior world. In particular, the epidermis is exposed to the external non-sterile environment, and it is colonized itself by a heterogeneous community of microbes—the skin microbiota. The characterization of the diversity of the skin microbiota has progressed immensely in the last decade, thanks to the whole-genome shotgun (WGS) metagenomic sequencing technology, which allowed to skip the culturing step of microbes. We now know that the composition of the skin microbiota differs widely between different parts of the body; in fact, depending on the pH, humidity, as well as salt and oxygen content, each area of the epidermis constitutes a multitude of ecological niches in a single host (Grice and Segre 2011; Erin Chen et al. 2018).

Skin is a common entry point for pathogens; bacteria, fungi, viruses and parasites can enter the body when the cutaneous barrier is breached, for example during minor abrasions, an injury or a surgical incision. The subsequent skin infections can lead to diseases of various severities, ranging from localized and uncomplicated to systemic and life-threatening. Skin and soft tissue infections (SSTIs) are a heterogeneous group of clinical conditions affecting the skin and the underlying tissues. SSTIs are very frequent in the population (Ray et al. 2013). Incidence of SSTIs is even higher in HIV and other immunocompromised patients (Burke and Lopez 2017; Poulakou et al. 2019). Skin symptoms can be a manifestation of infections that started at other sites of the body (e.g. measles virus, herpesviruses, varicella virus, etc.).

Despite the high burden of skin infections in the population, the molecular mechanisms mediating the entry of most pathogens in the skin are still poorly understood. Studies in this regard have been mainly conducted in animal models, especially mice, but results have not been entirely translatable to humans due to inherent differences between the skins of the two species. These studies are especially unreliable when studying anthropophilic pathogens, like *Staphylococcus aureus* or some dermatophytes that are specialized to colonize and infect humans (Parker 2017). New human skin models are hence needed to unravel the dynamics of infections.

Human 3D skin models are commonly used for pharmacological, toxicological and allergological studies (Abd et al. 2016), and could become useful tools to study colonization, infection and immune response towards skin-related pathogens. The advantages of human 3D skin models are that they are entirely human-derived, they closely resemble the skin complexity and they have the potential to mimic more faithfully what is happening in a real infection, thus offering an exciting and possibly cost-effective alternative to animal models. We can distinguish two types of human 3D skin models: human skin equivalents (HSEs)—artificial models produced in the laboratory from either commercially available cell lines or cells isolated from skin biopsies; and skin explants—full-thickness pieces of skin from either surgical waste or cadavers.

HSEs offer good reproducibility: standardized models are commercially available avoiding the problem of donor-to-donor variation; however, HSEs lack key features of real skin, such as skin appendages. In fact, sweat glands and hair follicles, which are currently impossible to reproduce in HSEs, have been shown to provide entry sites to the host inner skin layers and unique niches for microbial colonization. Consequently, sweat glands and hair follicles are tightly connected with the skin-resident immune system, representing unique immunologically relevant regions (Kabashima et al. 2019).

Skin explants, on the contrary, preserve the skin structure and cellular heterogeneity of skin in vivo. Skin explants can be kept in culture for weeks, maintaining their structural and immunological properties (Mason et al. 2016; Hendriks et al. 2018). However, long culture could lead to the loss of specific cell types; for example, antigen-presenting Langerhans cells crawl out from the explant within the first days of culture (Mathes et al. 2014).

Nevertheless, both human skin equivalents and human skin explants present a major limitation compared to animal models: the lack of circulation. In fact, during an inflammatory process, some components of the cellular and humoral immune response, like complement factors, antibodies or professional phagocytes, extravasate from the blood into the skin to clear the pathogens. At the same time, antigen-presenting cells (APCs), loaded with pathogen antigens, migrate from the skin to the lymph nodes to mount the adaptive immune response. The high structural complexity of immune tissues and their dynamic structure has not been fully recapitulated in vitro yet, despite the significant efforts invested in the development of microfluidic organ-on-a-chip models for tracking immune cells (reviewed by Sun et al. 2019). Connection of such an in vitro immune system to skin models represents the greatest frontier in infective research in human skin models.

2 Skin Microbiota: Between Colonization and Infection

Pathogens involved in skin infections have been extensively described in the literature; we gave a summary of the most relevant ones in Table 1. Interestingly, some of the major skin pathogens are also part of the skin microbiota, meaning that they are both able to live in close contact with the host without causing harm, perhaps even providing benefits, but they can also cause skin disorders under certain conditions. Limited evidence attributes the transition of these pathogens' phenotype from symbiotic to pathogenic, to a major transcriptional reprogramming, triggered by mechanisms that are still poorly understood. The reasons are probably attributable to alterations in the skin ecosystem leading to changes in the skin microbiota, or dysbiosis (Erin Chen et al. 2018). This perturbation would be dependent on both host-dependent factors (e.g. genetic predisposition, immune system status, the presence of abrasions or wounds, antibiotics usage/misuse) and microbe-dependent factors (microbe localization and microbe–microbe interactions) (Bliven and Maurelli 2016; Erin Chen et al. 2018).

The most abundant bacterial skin commensals belong to three phyla: Firmicutes, Actinobacteria and Proteobacteria (Wang et al. 2017). Major skin bacterial pathogens belong to Firmicutes or Actinobacteria. For example, *S. aureus,* which colonizes transiently or persistently the anterior nares of 80% of healthy subjects asymptomatically (Sakr et al. 2018), belongs to Firmicutes. However, *S. aureus* is also one of the most common causes of SSTIs, with the resident strain of *S. aureus* often causing the infections to its host (Davis et al. 2004; Albrecht et al. 2015).

Streptococcus pyogenes, or group A streptococcus (GAS), also belonging to the Firmicutes phylum, is a human commensal bearing potential benefits for the host since in vitro studies have shown that some of its virulence factors can improve wound healing (Stevens and Bryant 2016). Nevertheless, *S. pyogenes* is a common cause of necrotizing soft tissue infections (NSTIs). As for *S. aureus*, *S. pyogenes* main virulence factors are toxins, such as the pore-forming toxins streptolysins S and O (SLS and SLO).

Finally, *Staphylococcus epidermidis,* one of the most abundant bacteria of the human skin flora, belongs to Firmicutes. As described previously for other commensals, also *S. epidermidis* has been demonstrated to have beneficial effects for the host. For example, *S. epidermidis* can selectively inhibit the invasiveness of *S. aureus* through its quorum-sensing system. In particular, the autoinducing peptide produced by *S. epidermidis* can inhibit the agr-controlled expression of *S. aureus* toxins, like phenol soluble modulins α (PSMα) and alpha-toxin (Williams et al. 2019).

Although *S. epidermidis* is much more abundant than *S. aureus* in the healthy human skin microbiome, it causes disease much less frequently. This is since *S. aureus* has a much broader arsenal of virulence factors than *S. epidermidis,* which confers a better capacity to invade the tissues and to evade the immune system. Only in rare cases, *S. epidermidis* can become an opportunistic pathogen, causing nosocomial sepsis, especially via biofilm formation on catheters and medical devices (Otto 2009; Nguyen et al. 2017).

Table 1 Microbes associated with skin infections, most important virulence factors, pathogenic mechanisms and examples of related pathological outcomes

Skin pathogen	Main virulence factors and mechanisms of skin pathogenesis	Skin-related conditions
Bacteria		
Staphylococcus aureus	Adhesins (e.g. fibrinogen-binding proteins: ClfA and ClfB), proteases, exotoxins (e.g. pore-forming toxins like haemolysins, leukocidins and phenol soluble modulins, PSMs), superantigens (e.g. staphylococcal enterotoxins A and B, SEA and SEB), complement inhibitory proteins (e.g. CHIPS and SCIN) (Shumba et al. 2019)	SSTI: e.g. abscess, cellulitis, impetigo, folliculitis, furuncles, carbuncles, scalded skin syndrome (SSS), wound infections, including surgical site infections (SSIs)
Staphylococcus epidermidis	Biofilm formation, extracellular proteases (e.g. SepA), adhesins and toxins (Otto 2014)	SSTI: e.g. cellulitis, cysts formation, abscesses, folliculitis and furuncle, SSI (Rupp 2014; Natsis and Cohen 2018)
Staphylococcus lugdunensis	Adhesins like polysaccharide intracellular adhesin (PIA) leading to biofilm formation, slime and toxin production (Lambe et al. 1990; Frank et al. 2008)	SSTI: wound infections including SSI, abscesses, cellulitis (Heldt Manica and Cohen 2017)
Group A streptococcus (GAS) *Streptococcus pyogenes*	Toxins (e.g. pore-forming toxins, streptolysin S (SLS), proteases, adhesins (Shumba et al. 2019)	SSTI: e.g. including necrotizing fasciitis, impetigo, cellulitis (Walker et al. 2014; Shumba et al. 2019)
Cutibacterium acnes	Pore-forming toxins, adhesins, porphyrins, enzymes (e.g. hyaluronate lyase), biofilm formation (Dréno et al. 2018)	Acne vulgaris
Acinetobacter baumannii	Biofilm formation, expression of surface polysaccharides, degradation of host cell membranes by phospholipase, porins (Lee et al. 2017; Harding et al. 2018)	SSTI: e.g. cellulitis, wound infections, burn site infections (Guerrero et al. 2010; McConnell et al. 2013)
Pseudomonas aeruginosa	Biofilm formation, pore-forming toxins, adhesins (Crousilles et al. 2015)	SSTI: e.g. chronic wound infections and ulcer infections (Serra et al. 2015)
Fungi		
Candida albicans	Adherence proteins (i.e. Gpm1), extracellular matrix production, hydrolases (Mayer et al. 2013; Kühbacher et al. 2017a)	Superficial skin infections with thickening of the skin, hyperkeratosis and erythema, especially among atopic dermatitis psoriatic or immunocompromised patients (Kühbacher et al. 2017a)

(continued)

Table 1 (continued)

Skin pathogen	Main virulence factors and mechanisms of skin pathogenesis	Skin-related conditions
Malassezia spp.	Secretion of skin degrading enzymes (i.e. proteases and lipases), reactive oxygen species (ROS), pigment production (Hort and Mayser 2011)	Skin infections leading to tinea versicolor, folliculitis and seborrhoeic dermatitis (Gaitanis et al. 2012; Prohic et al. 2016)
Exophiala dermatitidis	Pigment production, extracellular polysaccharides, adhesins, biofilm formation (Poyntner et al. 2016; Kirchhoff et al. 2019)	Cutaneous and subcutaneous abscess formations (Zeng et al. 2007)
Trichophyton rubrum	Production of keratinases, biofilm formation, toxins, adhesins (Toussaint and Sticherling 2019)	Dermatophytosis and abscesses (Seebacher et al. 2008; Lee and Park 2015; Toussaint and Sticherling 2019)
Viruses		
Human papillomavirus (HPV)	Stimulation of the epidermal cell proliferation and interruption of physiological cell cycle (Doorbar et al. 2015)	Various types of skin warts, e.g. common warts (back of the hands and knees), plantar warts (soles of the feet), flat warts (back of the hands and face). Skin cancer (e.g. squamous cell carcinoma) and epidermoid cysts. (Bacaj and Burch 2018)
Herpes simplex virus (HSV)	Immune evasion mechanisms, downregulation of antiviral responses, i.e. modulation of interferon production, complement activation and cellular responses (Tognarelli et al. 2019)	Skin lesion formation after HSV infection, eczema among AD patients (Toussaint and Sticherling 2019; Damour et al. 2019)
Molluscum contagiosum virus (MCV)	Disruption of epidermal cell differentiation and enhancement of cell proliferation (Chen et al. 2013; Shisler 2015)	Skin lesions with higher prevalence in AD and immunocompromised patients (Chen et al. 2013)

In the phylum Actinobacteria, *Cutibacterium acnes* (formerly known as *Propionibacterium acnes)* is a skin commensal that resides mostly in hair follicles and sebaceous glands. *C. acnes* could have a protective role for the host, since its ability to ferment skin glycerol into short-chain fatty acids can inhibit common pathogenic strains of *S. aureus* (Shu et al. 2013). Nevertheless, *C. acnes* can cause common skin diseases such as folliculitis and acne vulgaris (Yang et al. 2018). Although the specific triggering factors behind these diseases are not known, there is evidence indicating that greater *C. acnes* abundance on the skin correlates with an excessive production of filaggrin by the keratinocytes, a protein with a role in the differentiation of the stratum corneum, already linked to acne and other skin conditions (Jarrousse et al. 2007).

As with bacterial infections, also SSTIs of fungal origin can be caused by species belonging to the human skin mycobiome. The most abundant fungal species on human skin belong to the genus *Malassezia*, which represents 50–80% of total skin fungi, except for the foot web spaces, where a higher diversity is present including Ascomycetes such as *Arthrodermataceae* or *Aspergillus* (White et al. 2014; Theelen et al. 2018). *Malassezia* spp. are mostly associated with mild skin disorders such as dandruff or tinea versicolor; however, they have also been associated with atopic dermatitis and psoriasis (Rudramurthy et al. 2014; Valencia-Herrera et al. 2018). These species are lipid-dependent since they lack a fatty acid synthase, and therefore they can quickly adapt and benefit from the lipid-rich skin environment (Wu et al. 2015). Their main virulence factors contributing to skin colonization or infection are proteases, which can degrade components of the skin barrier such as keratin.

The genus *Candida* also comprises a few commensal species that can cause skin infections, among which *C. albicans* is the most prevalent (Kühbacher et al. 2017a). One of the first steps allowing fungi to disseminate into the skin is their ability to bind and attach via fungal proteins. Among them, the virulence factor Gpm1 has been described as a ligand for vitronectin, a glycoprotein expressed by human endothelial cells and keratinocytes, whose binding results in evasion of complement and other immune mechanisms (Lopez et al. 2014). Colonizing fungi have been also associated with the impairment of skin pigmentation. Indeed, it has been described that *C. albicans* can suppress the transcription of the enzymes responsible for the melanogenesis (Kippenberger et al. 1997).

Finally, a wide range of clinically relevant skin infections is caused by dermatophytes (genera *Microsporum*, *Epidermophyton* and *Trichophyton*), which are not skin commensals but comprise anthropophilic and zoophilic species that evolved to colonize specifically the stratum corneum of human skin from which the infection starts. They cause a range of diseases, the most important of which are foot mycoses (in particular those caused by *Trichophyton rubrum*) (White et al. 2014).

Compared to the bacterial and fungal microbiome, the virobiome of healthy skin has been less investigated, partly because only DNA viruses are detected by WGS metagenomic sequencing. This technique has shown that the highest fraction of the virobiome is composed of bacteriophages infecting the common bacterial skin commensals, like *S. aureus*. Bacteriophages do not have a direct effect on the skin, but may affect the composition of its microbiome conferring antibiotic resistance to their bacterial hosts via transduction of antibiotic resistance and virulence factors genes, thus influencing the skin metabolism and predisposition to infections (Modi et al. 2013; Hannigan et al. 2015).

Eukaryotic viruses that maintain a latent state within the nucleus of the host cells are also present in the human skin virome. Evidence about the contribution of viruses in maintaining the equilibrium of the skin microbiota has not been adequately investigated yet; nevertheless, recent studies conducted on gut microbiome in macaques after antiretroviral treatment transiently perturbed the gut microbiota acting on endogenous retroviruses (Ortiz et al. 2019). This evidence suggests that endogenous retroviruses could have a similar balancing role also for skin microbiota.

Papillomaviridae, *Polyomaviridae* and *Circoviridae* families have been identified as resident in healthy skin (Foulongne et al. 2012; Ma et al. 2014; Hannigan et al. 2015). Interestingly, these groups include virus strains associated with dermatological infections and/or oncogenesis; however, it is not clear whether these viruses are responsible for the exacerbation of the associated pathological conditions (Duerkop and Hooper 2013).

The analysis of the family heterogeneity of *Papillomaviridae* revealed the presence of a complex virobiome, with more than 80 different types of human papillomaviruses (HPVs) in healthy skin samples. The majority of these types are known to cause human diseases and pre-malignant skin conditions in both immunocompromised and immunocompetent patients; therefore, it has been hypothesized that they could represent a risk for the host. For instance, patients with a rare hereditary defect are more susceptible to HPV types 5 and 8 and develop the disease epidermodysplasia verruciformis. Furthermore, HPV 50 can cause squamous cell carcinoma (Foulongne et al. 2012). Among the HPV types found in the skin, there are also three types that are high-risk factors for cervical cancer (Ma et al. 2014).

Similarly, in the *Polyomaviridae* family, the Merkel cell polyomavirus (MCPyV) was found in healthy individuals, although it is known to be associated with Merkel cell carcinoma, an aggressive neoplastic disease affecting neuroendocrine cells of the skin (Foulongne et al. 2012). Other types of polyomaviruses have been mentioned to reside in the skin, but their relationship with non-melanoma cancers is still not clear.

In some pathological conditions, such as atopic dermatitis, psoriasis or during immunosuppression, the microbiota of the host can differ greatly compared to the healthy condition. These situations of dysbiosis are thought to be driven by both host-dependent factors, like immune system deregulation and alteration of body homeostasis, and environmental factors, such as exposure to irritants, pathogens or climate factors (Sanmiguel and Grice 2015; Kantor and Silverberg 2017).

For example, it has been observed that patients with primary immunodeficiencies display a higher heterogeneity of fungal species on the skin (Oh et al. 2013). From the viral point of view, one typical example of virome dysregulation involves herpesviruses, which are usually not considered part of the skin virobiome. In fact, healthy people get cutaneous infections by herpes simplex virus (HSV) only during primary infection or relapse episodes. However, HSV becomes a common opportunistic infectious agent of the skin of patients with atopic dermatitis and, together with HPV and molluscum contagiosum virus, has been reported to worsen the development of the disease. HSV-1 and HSV-2 can cause a serious infection called eczema herpeticum in 3% of atopic patients, but not in healthy patients (Saintive et al. 2017). Another hallmark of atopic dermatitis patient's microbiota is a higher ratio of colonization and infection with *S. aureus*. One of the mechanisms thought to be associated with the exacerbation of atopic dermatitis is the production of virulence factors that disrupt the skin barrier. For instance, the release of proteases, alpha-toxin and PSMα is thought to facilitate penetration of HSV. However, coagulase-negative staphylococci have been shown to counteract this injury mechanism by exploiting quorum sensing between bacterial species, demonstrating how important the composition of the microbiome is for the onset and development of such a complex disease (Geoghegan et al. 2018; Williams et al. 2019).

3 Skin Colonization and Infection Studies in Human Skin Models

Several skin colonization studies aimed at understanding how micro-organisms interact with the skin and its most abundant cells, keratinocytes and fibroblasts. Research in this field can provide valuable information regarding the mechanisms of adhesion and persistent colonization of microbes on the skin, the latter occasionally linked to a higher chance of developing the correspondent infectious disease (Davis et al. 2004; Albrecht et al. 2015). Moreover, these studies provide insights on how microorganisms influence the local immune response stimulating skin cells to secrete cytokines, chemokines and antimicrobial peptides.

Human skin equivalents (HSEs) are affordable and reproducible models to study the interaction of pathogens with the skin. Developed mostly in the fields of regenerative medicine or drug screening, HSEs have been also employed to study bacterial colonization, mainly analysing the gene expression of skin cells (Holland et al. 2008). HSEs comprise a heterogeneous group of artificial models with a different complexity, ranging from simple multi-layered epidermis to dermal equivalents. Hence, the proper HSE model must be chosen to answer each specific research question.

The simplest HSE model is the so-called reconstructed human epidermis, obtained by culturing keratinocytes at the air–liquid interface (Prunieras et al. 1983). In this model, stratified skin-derived keratinocytes are seeded on filter inserts and eventually the upper layer differentiates to resemble the stratum corneum. This reconstructed human epidermis was used to study the persistence and replication of *Acinetobacter baumannii* on human skin, in order to understand how it transmits between hospitalized patients. The authors did not evidence any major changes in the expression profile of keratinocytes during colonization, but demonstrated the importance of disinfection to eradicate the bacteria from the skin (De Breij et al. 2012). Using a similar epidermal model, researchers investigated the response of the epithelial barrier when exposed to multiple bacterial species. When *C. acnes*, *S. epidermidis* and *S. aureus* were applied on the top layer of keratinocytes, no transcriptional changes in the cells were observed, suggesting that in physiological conditions these bacteria harmlessly colonize skin. However, when bacteria were interacting with keratinocytes from the bottom layer, a situation more like a wound infection, keratinocytes showed significant changes in their transcriptional profiles. Remarkably, while the contact with *C. acnes* and *S. epidermidis* induced a repair-type of response, the contact with *S. aureus* induced a proinflammatory response, with the release of cytokines and induction of cell death (Duckney et al. 2013).

More complex HSE models were developed with keratinocytes seeded on top of a dermal equivalent made of fibroblasts, immersed in an extracellular matrix of collagen or fibrin (Carlson et al. 2008). The fibroblast layer promotes keratinocyte division and optimal differentiation, making the epidermal layer anatomically and functionally more similar to skin structure (El-Ghalbzouri et al. 2002). Colonization of *S. epidermidis* and *S. aureus* was investigated in such a dermal equivalent model, and the gene expression profile of keratinocytes was evaluated by microarray analysis. *S. aureus*, but not *S. epidermidis*, was found to stimulate transcription of genes associated with the innate-immune response (Holland et al. 2009). The same group also studied the colonization of *C. acnes* and *Malassezia furfur* using the previously described model (Holland et al. 2008). The colonization of *S. aureus* was also investigated in another study using a dermal equivalent model. The authors found that the growth and survival of *S. aureus* microcolonies in the epidermis were dependent on the interactions of bacteria with the keratinocytes, rather than on the skin culture medium (Popov et al. 2014).

Overall, HSE models have been proven useful to unravel some mechanisms of bacterial colonization and persistence in the skin. However, they miss crucial components such as skin appendages, sweat glands and hair follicles, that indeed represent privileged sites for microorganism accumulation (Lange-Asschenfeldt et al. 2011), and weak spots in the skin barrier through which pathogens can start an infection (Byrd et al. 2018). Additionally, the presence of skin appendages is needed to maintain skin physiological conditions, such as hydration and acidification. For example, *P. acnes* in sweat glands releases short-chain fatty acids that help maintaining the epidermis at pH 5 (Christensen and Brüggemann 2014). Hence, the lack of skin appendages in HSEs not only makes it difficult to study bacteria that have their physiological niche there, but can have also other consequences, such as an imbalanced pH. Therefore, efforts have been made to reproduce skin appendages in HSEs. Indeed, a recent study reported the development of the first 3D model of skin sweat glands, using primary sweat gland cells extracted from skin biopsies. When these cells were grown in a special hanging-drop culture system with a dedicated medium containing epidermal growth factor, they formed 3D spheroids that reflected real sweat gland polarity (Klaka et al. 2017). Despite resembling native sweat glands more closely than a 2D culture, sweat gland spheroids are scaffold-free and have not been embedded into a skin equivalent model yet. The only attempt to transfer 3D sweat gland models to skin so far was performed in mice. Murine cells immersed in Matrigel and allowed to develop in vitro were subsequently transplanted in mice, where they formed tubular structures resembling sweat glands in terms of polarity, protein expression and functionality (Li et al. 2015, 2016). Unfortunately, none of these models has been used to study sweat glands as a niche of bacterial infection.

Similar attempts were made to develop 3D models of human hair follicles and to implement them in HSEs. Crucial for this process is the understanding of the mechanism of hair growth. The most important components of hair follicles that are necessary for this process are mesenchymal cells from dermal papillae. A dermal papillae model based on spheroid formation in Matrigel showed similar gene and protein expression as native hair follicles, in contrast to monolayer cultures (Miao et al. 2014). Moreover, hair fibre-like structures could be detected, suggesting that the reconstructed hair follicle could undertake its physiological functions. Even though first attempts to develop sweat gland and hair follicle 3D models were made, there is still a long way for their implementation into HSE models. Therefore, full-thickness skin explants represent the most physiologically relevant model to conduct colonization or infection studies.

For example, the use of skin explants was necessary to study the molecular basis of skin tropism for varicella zoster virus (VZV). Previous studies were conducted on human 3D organotypic models with an artificial dermis, but these lacked hair follicles, which have been identified as the primary site of VZV transfer to the underlying skin layers, via infected T cells expressing skin-homing markers. The model was therefore considered unreliable, and the infection studies were transferred to human skin explants. By using foetal skin cultures, the authors established a relevant model resembling VZV infection in humans that could give insights on the pathogenesis of varicella, as well as on the genes required for the infection in each part of differentiated skin. Moreover, this model provides a new drug screening tool, as well as a prototype to start human infection models of other human viruses (Taylor and Moffat 2005).

In the field of mycotic infections, skin explants were used to compare the initial adherence of *Candida* species *C. albicans*, *C. tropicalis* and *C. parapsilosis*, to the stratum corneum of human thigh skin explants. This study evidenced that *C. albicans* proliferated better than other species on human skin, in concordance with previous studies conducted in rodents (Bendel 2003). In addition, using the human skin explant model the authors evidenced a previously unobserved production of biofilm-related material also in *C. tropicalis* and *C. parapsilosis.* This study therefore suggested a link between biofilm formation and pathogenic potential of these species (Raz-Pasteur et al. 2011).

In some infection studies, the barrier function of the skin is disrupted by removing the stratum corneum or by injuring the surface of the skin explant to favour the bacterial or fungal invasion. Usually, skin explants are wounded, and the infections are carried on for a few weeks, followed by immunohistochemical imaging or transcriptional analysis, for instance of cytokines. Studies on *S. aureus* and *P. aeruginosa* evidenced a production of inflammatory cytokines, in particular IFN-α, IFN-β and TNFα, which was consistent with previous studies on wound infections in animal models (Steinstraesser et al. 2010; Schaudinn et al. 2017). A scarce interleukin response, and specifically the suppression of the IL-6 response in infected vs. uninfected wounds, was observed. Another traumatic infection model was set up on breast skin explants for *Exophiala dermatitidis* (Poyntner et al. 2016) and *Trichophyton rubrum* (Peres et al. 2016; Corzo-León et al. 2019). In these studies, the pathogen transcriptome was analysed during the infection, evidencing the expression of new protein-coding and non-coding genes, while the expression of virulence factor genes and metabolic pathways was not altered. Moreover, these studies demonstrated that gene expression can be significantly modulated depending on the infection model of choice.

Human skin explants with damaged barrier function were also used to investigate the impact of individual virulence factors in skin pathology. Upon disruption of the stratum corneum via tape-stripping, skin explants from abdominal tissue were infected with *S. aureus* and used to study the effects of staphylococcal alpha-toxin and Panton–Valentine leukocidin (PVL). Alpha-toxin showed the most pronounced effect, determining high tissue toxicity and compromising skin epithelial integrity. On the other hand, PVL was associated with milder toxicity and skin structural integrity was not affected. Besides, a cooperative mechanism of the two toxins in inducing tissue pathology was also found (Olaniyi et al. 2018). This study demonstrated the relevance of human skin explants to study pathogenic mechanisms and local immune responses triggered by bacterial toxins.

Recently, human skin explants have also been used to compare the movements of wild-type and radiation-attenuated *Plasmodium falciparum* malaria sporozoites. GFP-expressing sporozoites were injected intradermally in abdominal skin explants, and their movements were tracked using confocal microscopy movies. This study indicated that attenuation via radiation may alter sporozoite motility (Winkel et al. 2019).

4 Immune Response to Skin Infections

The skin immune system is composed by a dynamic network of resident and recirculating cells and proteins. Under physiological conditions, the skin immune system guards the tissue and surveys the surface without eliciting inflammation (Kabashima et al. 2019). The skin microbiota cross-talks with the immune system to keep skin homeostasis and induce immune tolerance towards innocuous commensals (Cogen et al. 2008). When the skin barrier is perturbed and microbes invade the host, an inflammatory response is triggered. This process implies not only the activation of skin-resident cells, but also the recruitment of effector cells and proteins from the bloodstream to mediate bacterial clearance, and the migration of APCs loaded with microbial antigens to the lymph nodes, to initiate the development of the adaptive immunity (Figs. 1 and 2).

Skin DCs comprise a heterogeneous family of APCs differing for localization and function. Epidermal DCs are called Langerhans cells (LCs) and are essential first-line APCs surveying the epidermis. The dermal layers, instead, host other subtypes of DCs, the most abundant of which are conventional DCs (cDC1 and cDC2) (Kashem et al. 2017). Several attempts were made so far to include immune cells into HSE models. In particular, LCs, other DCs, macrophages and T cells, either as cell lines or derived from the blood, have been used (Pupovac et al. 2018).

The first attempts to include LCs in HSE models were made in the 1990s (Nakazawa et al. 1997). First, umbilical cord blood-derived $CD34^{+}$ hematopoietic progenitor cells were included in the HSE model and the cells were induced to differentiate into LCs with GM-CSF and TNFα (Régnier et al. 1998). Studies performed on HSEs showed that melanocytes and keratinocytes created an

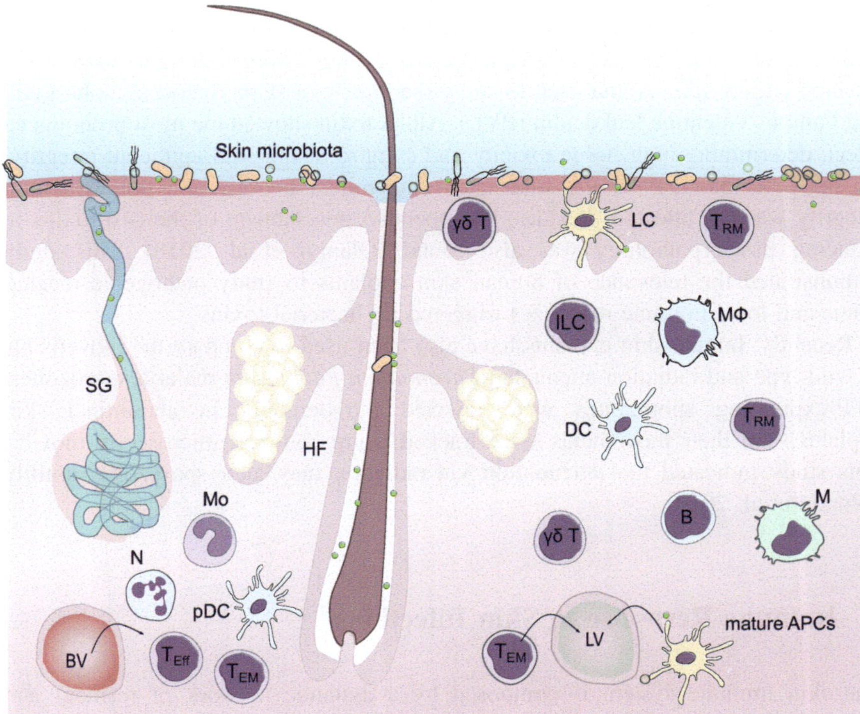

Fig. 1 Overview of human skin and immune cells. Human skin is composed by the epidermis (in purple) layered over the dermis (in pink). Both layers are crossed by skin appendages, hair follicles (HF) and sweat glands (SG), and home tissue-resident and blood-derived immune cells. Skin-resident immune cells comprise: T cells, which can be either $CD4^+$ or $CD8^+$ tissue-resident memory T cells (T_{RM}) and $\gamma\delta$ T cells ($\gamma\delta$ T); professional antigen-presenting cells (APCs), like Langerhans cells (LCs), different kinds of dendritic cells (DCs) such as plasmacytoid DCs, and macrophages (MΦ); and other cell types, like mast cells (M), innate lymphoid cells (ILC), including natural killer cells (NK) and B cells (B). APCs that have taken up and processed the antigen (Ag) migrate out of the skin to draining lymph nodes where they present the Ag to T cells. Ag-specific T cells, either naïve or memory, become activated and can extravasate from high venules and enter the skin as effector T cells (T_{Eff}) and effector memory T cells (T_{EM}), respectively. Other immune cells, like neutrophils (N) and monocytes (Mo), can be recruited from the circulation as a consequence of the inflammation caused by the infection

environment enough for CD34+ cell differentiation into LCs. Another study focused on the incorporation of DCs into a three-layer model consisting of fibroblasts, monocyte-derived DCs (moDCs) and keratinocytes. MoDCs from blood of healthy donors were inserted in between a fibroblast and a keratinocyte layer and proved to be responding to a skin sensitizer applied topically (Chau et al. 2013). Although these models would be useful to evaluate the first events in the host–pathogen interaction, no infection studies have been performed so far.

Experimental evidence suggests that healthy human skin contains 1 million T cells per cm^2 and therefore a total of $\approx 2 \times 10^{10}$ T cells, roughly twice that of the

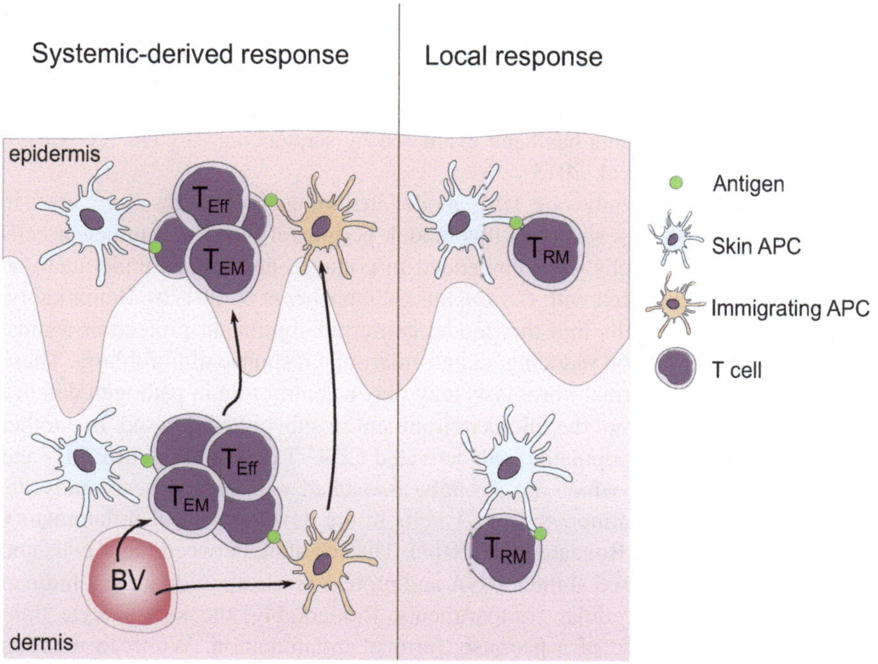

Fig. 2 Antigen presentation to T cells in human skin. *Local response*—In a local response, tissue-resident memory T cells (T_{RM}) interact with resident antigen-presenting cells (APCs). This type of interaction can occur in human skin explants. *Circulation-derived response*—During an infection, antigen-specific effector T cells (T_{Eff}), and effector memory T cells (T_{EM}) in case the same infection occurred in the past, are activated in the skin-draining lymph nodes and recruited to the skin from the circulation. Monocytes are also attracted from the blood to the site of infection where they behave as APCs, thus increasing the efficacy of the response. Circulation-derived response occurs only *in vivo* and, so far, can be reproduced only implanting human skin grafts in NGS mice

total blood T cell population (Clark et al. 2006). Several types of T cells can be found in the human skin. Tissue-resident memory T cells (T_{RM}) are the most abundant (Watanabe et al. 2015). These cells are memory T cells that encountered the antigen in the skin and remained there after the T cell contraction phase and skin homeostasis re-establishment. T_{RM} were initially defined as permanently residing in tissues, while more recent data have shown that they can recirculate, although they are in significant disequilibrium with the circulation (Klicznik et al. 2019). T_{RM} are characterized by the expression of the skin-homing marker cutaneous lymphocyte antigen (CLA), the tissue-retention marker CD69, while part of them express also the integrin CD103 (Watanabe et al. 2015; Klicznik et al. 2019). Skin-resident memory T cells comprise several subsets such as $CD4^+$ TH1, $CD4^+$ TH2, $CD4^+$ TH17, $CD8^+$ Tc1, $CD8^+$ Tc17 and regulatory T cells (Treg). In addition, non-conventional highly plastic commensal-specific T cells have been found in mouse skin. If present in humans, such cellular plasticity would allow commensal-specific type 17 cells to promote IL-17A-mediated antimicrobial defence under homeostatic conditions, as well as tissue repair in an IL-13-dependent manner in the context of tissue injury (Harrison et al. 2019). Moreover, evidence on the importance of the skin-resident antigen-specific $CD8^+$ T cells in the HSV-2 patients has been evaluated by sequencing of $CD8^+$ cells from genital biopsies (Zhu et al. 2013).

To our knowledge, only one HSE model including T cells has been used in infectious disease studies so far. In this model, peripheral blood mononuclear cells (PBMCs) or $CD4^+$ T cells were embedded in collagen and placed underneath an HSE that was then infected with *C. albicans* (Kühbacher et al. 2017b). Remarkably, inclusion of $CD4^+$ T cells into this model conferred significant protection against deep *C. albicans* infection inducing an anti-microbial response of fibroblasts. These results indicated that dermal fibroblasts may play a central role in pathogen defence by integrating signals from the microenvironment in subepithelial tissue. In another study, HSEs have been populated with activated $CD4^+$ T cells and in vitro Th1- and Th17-polarized T cells, which are strongly associated with psoriasis, to study the interaction between keratinocytes and T cells in the development of inflammatory skin diseases (Van Den Bogaard et al. 2014). The crosstalk between the T cells and keratinocytes was analysed at the mRNA and protein levels upon direct and indirect contact between both cellular compartments. Remarkably, the keratinocyte transcriptome was indicative of a psoriasis form of inflammation. Worth to mention here is the skin explant assay developed to predict the occurrence of graft-versus-host disease (GvHD) in human leucocyte antigen (HLA)-matched bone marrow transplant patients. This assay is based on co-culturing mixed lymphocytes from donor and recipient with skin biopsy taken from recipient prior to transplantation. Then, the degree of GvH reaction is evaluated by histopathological analysis of the skin sections after 3 days (Dickinson et al. 1999; Qesari et al. 2016). More recently, the same group developed a human in vitro skin test for predicting low molecular weight drug hypersensitivity reactions. Full-thickness skin biopsies and immune cells derived from the blood of the same individual were used to perform this test. Following exposure of moDCs to the drug, cells were co-cultured with autologous T cells. Then, skin biopsies obtained from the same donor were co-cultured with the primed T cells and evaluated for histopathology. This skin explant test showed promising results as a potential predictive assessment tool for drug hypersensitivity reactions (Ahmed et al. 2019).

In summary, even though HSEs with APCs and T cells could provide a valuable tool to unravel some aspects of skin immune response to pathogens, skin explants remain the preferred model for this type of investigations. Indeed, thanks to the presence of the resident immune cells embedded in their physiological anatomical organization, skin samples offer valuable opportunities to study the activation of resident APCs following the contact with the bacterium.

Alternatively, immune cells were isolated from the skin by enzymatic digestion and their response was evaluated ex vivo (Watanabe et al. 2015). Several protocols to isolate and characterize T cells from skin explants have been developed (Watanabe et al. 2015; Klicznik et al. 2019). This approach has been used to show that IL-17-producing $CD4^+$ T_{RM} specific for *C. albicans* are present in human skin explants from healthy subjects (Park et al. 2018).

Instead of using enzymatic digestion, DCs and T cells can be isolated by exploiting their migratory abilities. In fact, activated DCs in vivo normally travel out of the skin to the lymph nodes to present the antigen to naïve T cells. Therefore, DCs stimulated in the experimental setting spontaneously crawl out of the skin explant and can be harvested. DCs migration was exploited in multiple models of viral infection: for example, epithelial tissue explants from human thighs were infected with two different types of HIV-1 isolates to compare their fitness in the skin. HIV-infected LCs that had crawled out after 3–4 days were harvested, and their gene expression profile was analysed (Kawamura et al. 2000; Ball et al. 2003). Isolated DCs were also studied in a dengue virus infection model. After superficially injecting cadaveric skin explants to imitate the mosquito bites, migrating DCs were harvested and analysed to investigate the mechanisms of transmission of the virus (Wu et al. 2000). Finally, a skin explant model was used to study the infection by *Borrelia burgdorferi*, the aetiological agent of Lyme borreliosis. Upon intradermal injection of the bacterium, crawled-out DCs were harvested and the results evidenced a role of TLR2 in DCs migration. Simultaneously, the authors characterized the maturation of LCs and dermal DCs upon challenge with the pathogen (Mason et al. 2016). Finally, infected skin explants have been used to study the interaction of immune cells and pathogens in situ. For example, the interaction of LCs and DCs was observed in a human model of genital skin, in order to study the establishment of the immunity against HSV-1 and HSV-2 in primary herpetic lesions. Immunohistochemistry revealed the interplay between LCs and dermal DCs, both in normal foreskin tissues obtained from infants and young children undergoing circumcision and in biopsies of infected tissue (Kim et al. 2015). A similar model of cutaneous infection was developed to study the local immune response upon encounter of dengue virus. By infecting healthy human skin, the authors observed that the virus would impede the migration of epidermal LCs, therefore preventing antigen presentation to T cells (Limon-Flores et al. 2005).

5 Bloodstream-Derived Elements

During skin infections, immune elements are recalled from the bloodstream to aid the resident immune response with acute phase proteins, antibodies and effector cells. Among these, neutrophils are considered the first responders of skin infections and have been shown to play a crucial role, for example, in the clearance of *S. aureus* (Spaan et al. 2013). Evidence shows that neutrophils not only kill bacteria, but also skew the adaptive response towards a Th17-type, as well as playing an important role in wound healing (Vono et al. 2017). Overall, this points to the importance of supplying human skin models to be used in infection studies with the crucial humoral and cellular elements recalled from the bloodstream during microbial invasion.

In the context of human skin models, the absence of blood supply has been overcome grafting the skin on a recipient animal that provides it with its own circulation. Several xenograft skin models have been established, mainly in mice and especially for tumour or wound healing studies, demonstrating that the grafted skin can be vascularized. These models use immunodeficient strains of mice that are unable to reject the skin transplant (Salgado et al. 2017). Obviously, this approach introduces several biases since human skin cells are gradually replaced by mouse cells. The usage of allogenic skin grafts also presents major technical difficulties and drawbacks including manual ability for the insertion, ethical concerns and costs (Benichou et al. 2011).

Recently, a xenograft humanized model was developed in mice to investigate the colonization of pathogenic *S. aureus* USA300 on human skin. In this study, the authors evidenced an IL-8 response of keratinocytes that induced the recruitment of mouse neutrophils to the epidermis, possibly to control the bacterial burden on the surface (Schulz et al. 2019). This work suggests that the containment of pathogenic *S. aureus* on the skin may require the involvement of cellular elements that otherwise would have been overlooked.

The only way to avoid the use of skin grafts would be to develop new technologies combining skin explants with their autologous blood elements. This was achieved in a recently published paper in which the authors present a novel model that combines microfluidics to allow neutrophils, present in a very small volume of fresh whole blood, to migrate towards skin micro-biopsies, inserted into a skin-on-a-chip model, in response to *S. aureus*. Using this model, the number of neutrophils migrating towards bacteria on the skin micro-biopsies was assessed, suggesting that neutrophils could be a potential diagnostic biomarker for SSTIs. Furthermore, this model could be used to directly collect skin and blood samples in very small amounts from patients, making it appealing for clinical research (Kim et al. 2019).

6 Conclusions and Future Perspectives

Skin is the portal of entry of many relevant pathogens, but the pathophysiology of these infections is often not sufficiently understood. The line between colonization and infection remains unclear, especially for commensals. Unravelling what predisposes microbes to shift from commensalism to pathogenic phenotypes and understanding what favours certain niches for causing infections is a very timely topic that could lead to important discoveries, possibly translating to new medical interventions able to reduce the high burden of skin infections.

SSTIs are among the most important and common infections afflicting humans. They vary from relatively mild disease manifestations, such as cellulitis and abscesses, to severe and life-threatening diseases, like surgical site infections. The first ones are very frequent in the community setting, while surgical infections obviously occur in hospitals. Antibiotics are partially effective against bacterial

SSTIs, antimicrobial resistance is increasing, and therapeutic treatment of infections is associated with high healthcare cost-burden. Therefore, alternative medical interventions such as vaccines or passive immunotherapies against pathogens associated with SSTIs are needed. Unfortunately, there are no vaccines available on the market against the major skin pathogens yet. Research and development of vaccines against these pathogens has been hampered by failures and difficulties in identifying the right antigens. Research on skin infections can contribute in understanding pathogenesis and immune response associated with SSTIs, and thus to identify critical antigens against skin pathogens.

Most of the research on skin infections has been conducted using animal models, especially mice; however, these present several limitations in resembling human infections. Indeed, most pathogens have a specific tropism for human tissues, and there are several intrinsic differences between murine and human skin in terms of structure, immune system and microbial flora. The use of human skin explants or equivalents can provide a significant contribution to the research in this field. In addition, this approach would not present ethical concerns implying no animal use, while it would give the possibility to exploit valuable surgical waste from hospitals and clinics.

In this review article, we presented the current view and state of the art on research conducted using human skin equivalents and skin explants. Both models have their advantages and weaknesses, which make them suitable only for specific purposes.

Human models have been used extensively in pharmacological and allergological studies, but also in the cosmetic industry. Surprisingly, human skin equivalents and explants are not yet commonly used in infection studies, although in the last decade, a greater number of authors have started working in this area.

Human skin models are more reliable than cell cultures since they better reproduce the multi-component, complex three-dimensional structure that represents the physiological settings of colonization and infection. However, in order to make human skin models fully representative of the in vivo situation they should be supplied with the microbial flora and blood components from the homologous donor. It is anticipated that skin models will be a great platform to study the immune response towards skin-related pathogens (Lacey et al. 2016).

Competing Interests Elena Boero (EB) and Malgorzata Ewa Mnich are participating in a postgraduate studentship programme (DISSection) at GSK. Andrea Guido Oreste Manetti, Elisabetta Soldaini and Fabio Bagnoli are employees of GSK. Fabio Bagnoli reports ownership of GSK stocks and owns patents on *S. aureus* vaccine candidates. Luca Grimaldi is an employee of the University of Siena, Italy, and previously was involved in a collaboration between GSK and the Azienda Ospedaliera Universitaria Senese, Siena, Italy, providing human samples for research work conducted in GSK. The study was approved by local ethical committees and conducted according to good clinical practice in accordance with the declaration of Helsinki. Patients had given their written consent to the study. The authors have no other relevant affiliations or financial involvement with any organization or entity with a financial interest in or financial conflict with the subject matter or materials discussed in the manuscript, apart from those disclosed.

Funding This work was undertaken at the request of and sponsored by GlaxoSmithKline Biologicals SA, and it was also supported by the European Union's Horizon 2020 research H2020-MSCA-ITN (No. 675106 coordinated by Dr. Fabio Bagnoli, GSK, Siena, Italy).

Bibliography

Abd E, Yousef SA, Pastore MN et al (2016) Skin models for the testing of transdermal drugs. Clin Pharmacol Adv Appl 8:163–176

Ahmed SS, Whritenour J, Ahmed MM, et al (2019) Evaluation of a human in vitro skin test for predicting drug hypersensitivity reactions. Toxicol Appl Pharmacol 369:39–48

Albrecht VS, Limbago BM, Moran GJ et al (2015) *Staphylococcus aureus* colonization and strain type at various body sites among patients with a closed abscess and uninfected controls at U.S. emergency departments. J Clin Microbiol 53:3478–3484

Bacaj P, Burch D (2018) Human papillomavirus infection of the skin. Arch Pathol Lab Med 142:700–705

Ball SC, Abraha A, Collins KR et al (2003) Comparing the ex vivo fitness of CCR5-tropic human immunodeficiency virus type 1 isolates of subtypes B and C. J Virol 77:1021–1038

Bendel CM (2003) Colonization and epithelial adhesion in the pathogenesis of Neonatal candidiasis. Semin Perinatol 27:357–364

Benichou G, Yamada Y, Yun SH et al (2011) Immune recognition and rejection of allogeneic skin grafts. Immunotherapy 3:757–770

Bliven KA, Maurelli AT (2016) Evolution of bacterial pathogens within the human host. In: Virulence mechanisms of bacterial pathogens, 5th edn. NIH Public Access, pp 3–13

Burke VE, Lopez FA (2017) Approach to skin and soft tissue infections in non-HIV immunocompromised hosts. Curr Opin Infect Dis 30:354–363

Byrd AL, Belkaid Y, Segre JA (2018) The human skin microbiome. Nat Rev Microbiol 16:143–155

Carlson MW, Alt-Holland A, Egles C, Garlick JA (2008) Three-dimensional tissue models of normal and diseased skin. Curr Protoc, Cell Biol

Chau DYS, Johnson C, MacNeil S et al (2013) The development of a 3D immuno competent model of human skin. Biofabrication 5:35011

Chen X, Anstey AV, Bugert JJ (2013) Molluscum contagiosum virus infection. Lancet Infect Dis 13:877–888

Christensen GJM, Brüggemann H (2014) Bacterial skin commensals and their role as host guardians. Benef Microbes 5:201–215

Clark RA, Chong B, Mirchandani N et al (2006) The vast majority of CLA + T cells are resident in normal skin. J Immunol 176:4431–4439

Cogen AL, Nizet V, Gallo RL (2008) Skin microbiota: a source of disease or defence? Br J Dermatol 158:442–455

Corzo-León DE, Munro CA, MacCallum DM (2019) An ex vivo human skin model to study superficial fungal infections. Front Microbiol 10

Crousilles A, Maunders E, Bartlett S et al (2015) Which microbial factors really are important in *Pseudomonas aeruginosa* infections? Future Microbiol 10:1825–1836

Damour A, Garcia M, Seneschal J et al (2019) Eczema herpeticum: clinical and pathophysiological aspects. Clin Rev Allergy Immunol. https://doi.org/10.1007/s12016-019-08768-3

Davis KA, Stewart JJ, Crouch HK et al (2004) Methicillin-resistant *Staphylococcus aureus* (MRSA) nares colonization at hospital admission and its effect on subsequent MRSA infection. Clin Infect Dis 39:776–782

De Breij A, Haisma EM, Rietveld M et al (2012) Three-dimensional human skin equivalent as a tool to study *Acinetobacter baumannii* colonization. Antimicrob Agents Chemother 56:2459–2464

Dickinson AM, Hromadníková I, Sviland L et al (1999) Use of a skin explant model for predicting GVHD in HLA-matched bone marrow transplants: Effect of GVHD prophylaxis. Bone Marrow Transplant 24:857–863

Doorbar J, Egawa N, Griffin H et al (2015) Human papillomavirus molecular biology and disease association. Rev Med Virol 25:2–23

Dréno B, Pécastaings S, Corvec S et al (2018) Cutibacterium acnes (Propionibacterium acnes) and acne vulgaris: a brief look at the latest updates. J Eur Acad Dermatology Venereol 32:5–14

Duckney P, Wong HK, Serrano J et al (2013) The role of the skin barrier in modulating the effects of common skin microbial species on the inflammation, differentiation and proliferation status of epidermal keratinocytes. BMC Res Notes 6

Duerkop BA, Hooper LV (2013) Resident viruses and their interactions with the immune system. Nat Immunol 14:654–659

El-Ghalbzouri A, Gibbs S, Lamme E et al (2002) Effect of fibroblasts on epidermal regeneration. Br J Dermatol 147:230–243

Erin Chen Y, Fischbach MA, Belkaid Y (2018) Skin microbiota-host interactions. Nature 553:427–436

Foulongne V, Sauvage V, Hebert C et al (2012) Human skin Microbiota: high diversity of DNA viruses identified on the human skin by high throughput sequencing. PLoS One 7

Frank KL, del Pozo JL, Patel R (2008) From clinical microbiology to infection pathogenesis: how daring to be different works for *Staphylococcus lugdunensis*. Clin Microbiol Rev 21:111–133

Gaitanis G, Magiatis P, Hantschke M et al (2012) The Malassezia genus in skin and systemic diseases. Clin Microbiol Rev 25:106–141

Geoghegan JA, Irvine AD, Foster TJ (2018) *Staphylococcus aureus* and atopic dermatitis: a complex and evolving relationship. Trends Microbiol 26:484–497

Grice EA, Segre JA (2011) The skin microbiome. Nat Rev Microbiol 9:244–253

Guerrero DM, Perez F, Conger NG et al (2010) *Acinetobacter baumannii*—associated skin and soft tissue infections: recognizing a broadening spectrum of disease. Surg Infect (Larchmt) 11:49–57

Hannigan GD, Meisel JS, Tyldsley AS et al (2015) The human skin double-stranded DNA virome: topographical and temporal diversity, genetic enrichment, and dynamic associations with the host microbiome. MBio 6

Harding CM, Hennon SW, Feldman MF (2018) Uncovering the mechanisms of *Acinetobacter baumannii* virulence. Nat Rev Microbiol 16:91–102

Harrison OJ, Linehan JL, Shih H-Y et al (2019) Commensal-specific T cell plasticity promotes rapid tissue adaptation to injury. Science (80)363:eaat6280

Heldt Manica LA, Cohen PR (2017) *Staphylococcus lugdunensis* infections of the skin and soft tissue: a case series and review. Dermatol Ther (Heidelb) 7:555–562

Hendriks A, Cruz AR, Soldaini E et al (2018) Human organotypic models for anti-infective research. Springer, Berlin, Heidelberg, pp 1–22

Holland DB, Bojar RA, Farrar MD, Holland KT (2009) Differential innate immune responses of a living skin equivalent model colonized by *Staphylococcus epidermidis* or *Staphylococcus aureus*. FEMS Microbiol Lett 290:149–155

Holland DB, Bojar RA, Jeremy AHT et al (2008) Microbial colonization of an in vitro model of a tissue engineered human skin equivalent—a novel approach. FEMS Microbiol Lett 279:110–115

Hort W, Mayser P (2011) Malassezia virulence determinants. Curr Opin Infect Dis 24:100–105

Jarrousse V, Castex-Rizzi N, Khammari A et al (2007) Modulation of integrins and filaggrin expression by Propionibacterium acnes extracts on keratinocytes. Arch Dermatol Res 299:441–447

Kabashima K, Honda T, Ginhoux F, Egawa G (2019) The immunological anatomy of the skin. Nat Rev Immunol 19:19–30

Kantor R, Silverberg JI (2017) Environmental risk factors and their role in the management of atopic dermatitis. Expert Rev Clin Immunol 13:15–26

Kashem SW, Haniffa M, Kaplan DH (2017) Antigen-presenting cells in the skin. Annu Rev Immunol 35:469–499

Kawamura T, Cohen SS, Borris DL et al (2000) Candidate microbicides block HIV-1 infection of human immature Langerhans cells within epithelial tissue explants. J Exp Med 192:1491–1500

Kim JJ, Ellett F, Thomas CN et al (2019) A microscale, full-thickness, human skin on a chip assay simulating neutrophil responses to skin infection and antibiotic treatments. Lab Chip. https://doi.org/10.1039/C9LC00399A

Kim M, Truong NR, James V et al (2015) Relay of herpes simplex virus between Langerhans cells and dermal dendritic cells in human skin. PLoS Pathog 11:e1004812

Kippenberger S, Bernd A, Menzel I et al (1997) *Candida albicans* suppresses transcription of melanogenesis enzymes in cultured melanocytes. Mycoses 40:373–375

Kirchhoff L, Olsowski M, Rath P-M, Steinmann J (2019) *Exophiala dermatitidis*: key issues of an opportunistic fungal pathogen. Virulence 10:984–998

Klaka P, Grüdl S, Banowski B et al (2017) A novel organotypic 3D sweat gland model with physiological functionality. PLoS ONE 12:e0182752

Klicznik MM, Morawski PA, Höllbacher B et al (2019) Human CD4+ CD103+ cutaneous resident memory T cells are found in the circulation of healthy individuals. Sci Immunol 4: eaav8995

Kühbacher A, Burger-Kentischer A, Rupp S (2017a) Interaction of *Candida* species with the skin. Microorganisms 5:32

Kühbacher A, Henkel H, Stevens P et al (2017b) Central role for dermal fibroblasts in skin model protection against *Candida albicans*. J Infect Dis 215:1742–1752

Lacey KA, Geoghegan JA, McLoughlin RM (2016) The role of *Staphylococcus aureus* virulence factors in skin infection and their potential as vaccine antigens. Pathogens 5

Lambe DW, Ferguson KP, Keplinger JL, Gemmell CG (1990) Pathogenicity of *Staphylococcus lugdunensis*, *Staphylococcus schleiferi*, and three other coagulase-negative staphylococci in a mouse model and possible virulence factors. Can J Microbiol 455–63

Lange-Asschenfeldt B, Marenbach D, Lang C et al (2011) Distribution of bacteria in the epidermal layers and hair follicles of the human skin. Skin Pharmacol Physiol 24:305–311

Lee C-R, Lee JH, Park M et al (2017) Biology of *Acinetobacter baumannii*: pathogenesis, antibiotic resistance mechanisms, and prospective treatment options. Front Cell Infect Microbiol 7

Lee WS, Park SY (2015) Deep dermatophytosis induced by *Trichophyton rubrum*. Korean J Dermatology 53:666–667

Li H, Chen L, Zeng S et al (2015) Matrigel basement membrane matrix induces eccrine sweat gland cells to reconstitute sweat gland-like structures in nude mice. Exp Cell Res 332:67–77

Li H, Li X, Zhang B et al (2016) Changes in keratins and alpha-smooth muscle actin during three-dimensional reconstitution of eccrine sweat glands. Cell Tissue Res 113–122

Limon-Flores AY, Perez-Tapia M, Estrada-Garcia I et al (2005) Dengue virus inoculation to human skin explants: An effective approach to assess in situ the early infection and the effects on cutaneous dendritic cells. Int J Exp Pathol. https://doi.org/10.1111/j.0959-9673.2005.00445.x

Lopez CM, Wallich R, Riesbeck K et al (2014) *Candida albicans* uses the surface protein Gpm1 to attach to human endothelial cells and to Keratinocytes via the adhesive protein vitronectin. PLOS ONE 9

Ma Y, Madupu R, Karaoz U et al (2014) Human Papillomavirus community in healthy persons, defined by metagenomics analysis of human microbiome project shotgun sequencing data sets. J Virol 88:4786–4797

Mason LMKK, Wagemakers A, Van't Veer C et al (2016) Borrelia burgdorferi induces TLR2-mediated migration of activated dendritic cells in an ex vivo human skin model. PLoS ONE 11:439–449

Mathes SH, Ruffner H, Graf-Hausner U (2014) The use of skin models in drug development. Adv Drug Deliv, Rev

Mayer FL, Wilson D, Hube B (2013) *Candida albicans* pathogenicity mechanisms. Virulence 4:119–128

McConnell MJ, Actis L, Pachón J (2013) *Acinetobacter baumannii*: human infections, factors contributing to pathogenesis and animal models. FEMS Microbiol Rev 37:130–155
Miao Y, Sun Y Bin, Liu BC (2014) Controllable production of transplantable adult human high-passage dermal papilla spheroids. Tissue Eng Part A 20:2329–2338
Modi SR, Lee HH, Spina CS, Collins JJ (2013) Antibiotic treatment expands the resistance reservoir and ecological network of the phage metagenome. Nature 499:219–222
Nakazawa K, Nakazawa H, Sahuc F et al (1997) Pigmented human skin equivalent: new method of reconstitution by grafting an epithelial sheet onto a non-contractile dermal equivalent. Pigment Cell Res 10:382–390
Natsis NE, Cohen PR (2018) Coagulase-negative Staphylococcus skin and soft tissue infections. Am J Clin Dermatol 19:671–677
Nguyen TH, Park MD, Otto M (2017) Host response to *Staphylococcus epidermidis* colonization and infections. Front Cell Infect Microbiol
Oh J, Freeman AF, Park M et al (2013) The altered landscape of the human skin microbiome in patients with primary immunodeficiencies. Genome Res 23:2103–2114
Olaniyi RO, Pancotto L, Grimaldi L, Bagnoli F (2018) Deciphering the pathological role of staphylococcal α-Toxin and panton-valentine leukocidin using a novel ex vivo human skin model. Front Immunol 9:951
Ortiz AM, Flynn JK, DiNapoli SR et al (2019) Antiretroviral therapy administration in healthy rhesus macaques is associated with transient shifts in intestinal bacterial diversity and modest immunological perturbations. J Virol 93
Otto M (2009) *Staphylococcus epidermidis*—the "accidental" pathogen. Nat Rev Microbiol 7:555–567
Otto M (2014) *Staphylococcus epidermidis* pathogenesis. Humana Press, Totowa, NJ. pp 17–31
Park CO, Fu X, Jiang X et al (2018) Staged development of long-lived T-cell receptor αβ T H 17 resident memory T-cell population to *Candida albicans* after skin infection. J Allergy Clin Immunol 142:647–662
Parker D (2017) Humanized mouse models of *Staphylococcus aureus* infection. Front Immunol 8:1–6
Peres NTDA, da Silva LG, da Santos RS et al (2016) In vitro and ex vivo infection models help assess the molecular aspects of the interaction of *Trichophyton rubrum* with the host milieu. Med Mycol 54:420–427
Popov L, Kovalski J, Grandi G et al (2014) Three-dimensional human skin models to understand *Staphylococcus aureus* skin colonization and infection. Front Immunol 5:41
Poulakou G, Lagou S, Tsiodras S (2019) What's new in the epidemiology of skin and soft tissue infections in 2018? Curr Opin Infect Dis 32:77–86
Poyntner C, Blasi B, Arcalis E et al (2016) The transcriptome of *Exophiala dermatitidis* during ex-vivo skin model infection. Front Cell Infect Microbiol 6:136
Prohic A, Jovovic Sadikovic T, Krupalija-Fazlic M, Kuskunovic-Vlahovljak S (2016) Malassezia species in healthy skin and in dermatological conditions. Int J Dermatol 55:494–504
Prunieras M, Regnier M, Woodley D (1983) Methods for cultivation of keratinocytes with an air-liquid interface. J Invest Dermatol 81:S28–S33
Pupovac A, Senturk B, Griffoni C et al (2018) Toward immuno competent 3D skin models. Adv Healthc Mater 7:1–11
Qesari M, Richter A, Ogonek J et al (2016) Cytomegalovirus-specific T cells isolated by IFN-γ secretion assay do not induce significant graft-versus-host reactions in vitro. Transplantation 100:2352–2361
Ray GT, Suaya JA, Baxter R (2013) Incidence, microbiology, and patient characteristics of skin and soft-tissue infections in a U.S. population: a retrospective population-based study. BMC Infect Dis 13:1
Raz-Pasteur A, Ullmann Y, Berdicevsky I (2011) The pathogenesis of *Candida* infections in a human skin model: scanning electron microscope observations. ISRN Dermatol 2011:1–6

Régnier M, Patwardhan A, Scheynius A, Schmidt R (1998) Reconstructed human epidermis composed of keratinocytes, melanocytes and Langerhans cells. Med Biol Eng Comput 36:821–824

Rudramurthy SM, Honnavar P, Chakrabarti A et al (2014) Association of Malassezia species with psoriatic lesions. Mycoses 57:483–488

Rupp ME (2014) Clinical characteristics of infections in humans due to *Staphylococcus epidermidis*. Methods Mol Biol 1–16

Saintive S, Abad E, Ferreira DDC et al (2017) What is the role of *Staphylococcus aureus* and herpes virus infections in the pathogenesis of atopic dermatitis? Future Microbiol 12:1327–1334

Sakr A, Brégeon F, Mège JL,et al (2018) *Staphylococcus aureus* nasal colonization: an update on mechanisms, epidemiology, risk factors, and subsequent infections. Front Microbiol 9

Salgado G, Ng YZ, Koh LF et al (2017) Human reconstructed skin xenografts on mice to model skin physiology. Differentiation 98:14–24

Sanmiguel A, Grice EA (2015) Interactions between host factors and the skin microbiome. Cell Mol Life Sci 72:1499–1515

Schaudinn C, Dittmann C, Jurisch J et al (2017) Development, standardization and testing of a bacterial wound infection model based on ex vivo human skin. 12:e0186946

Schulz A, Jiang L, de Vor L et al (2019) Neutrophil recruitment to noninvasive MRSA at the stratum corneum of human skin mediates transient colonization. Cell Rep. https://doi.org/10.1016/j.celrep.2019.09.055

Seebacher C, Bouchara JP, Mignon B (2008) Updates on the epidemiology of dermatophyte infections. Mycopathologia 166:335–352

Serra R, Grande R, Butrico L et al (2015) Chronic wound infections: the role of *Pseudomonas aeruginosa* and *Staphylococcus aureus*. Expert Rev Anti Infect Ther 13:605–613

Shisler JL (2015) Immune evasion strategies of *Molluscum contagiosum* virus. In: Advances in virus research, 1st edn. Elsevier Inc., pp 201–252

Shu M, Wang Y, Yu J et al (2013) Fermentation of Propionibacterium acnes, a Commensal Bacterium in the human skin microbiome, as skin probiotics against methicillin-resistant *Staphylococcus aureus*. PLoS ONE 8:e55380

Shumba P, Shambat SM, Siemens N (2019) The role of streptococcal and staphylococcal exotoxins and proteases in human necrotizing soft tissue infections. Toxins (Basel) 11

Spaan AN, Surewaard BGJ, Nijland R, van Strijp JAG (2013) Neutrophils versus *Staphylococcus aureus*: a biological tug of war. Annu Rev Microbiol 67:629–650

Steinstraesser L, Niederbichler MSAD, Stupka MBJ et al (2010) A novel human skin chamber model to study wound infection ex vivo. 357–365

Stevens DL, Bryant AE (2016) Impetigo, erysipelas and cellulitis. In: Ferretti JJ, Stevens DL, Fischetti VA (eds) Streptococcus pyogenes: basic biology to clinical manifestations. Oklahoma City: University of Oklahoma Health Sciences Center. Available at: https://www.ncbi.nlm.nih.gov/books/NBK333408/

Sun W, Luo Z, Lee J et al (2019) Organ-on-a-chip for cancer and immune organs modeling. Adv Healthc Mater 8

Taylor SL, Moffat JF (2005) Replication of varicella-zoster virus in human skin organ culture. J Virol 79:11501–11506

Theelen B, Cafarchia C, Gaitanis G et al (2018) Malassezia ecology, pathophysiology, and treatment. Med Mycol 56:S10–S25

Tognarelli EI, Palomino TF, Corrales N et al (2019) Herpes simplex virus evasion of early host antiviral responses. Front Cell Infect Microbiol 9

Toussaint F, Sticherling M (2019) Multiple dermal abscesses by *Trichophyton rubrum* in an immunocompromised patient. Front Med 6:1–4

Valencia-Herrera A, Toledo-Bahena M, Ramírez-Cortés E et al (2018) Association of Malassezia to atopic dermatitis. Curr Fungal Infect Rep 12:201–206

Van Den Bogaard EH, Tjabringa GS, Joosten I et al (2014) Crosstalk between keratinocytes and T Cells in a 3D microenvironment: a model to study inflammatory skin diseases. J Invest Dermatol 134:719–727

Vono M, Lin A, Norrby-Teglund A et al (2017) Neutrophils acquire the capacity for antigen presentation to memory CD4+ T cells in vitro and ex vivo. Blood 129:1991–2001

Walker MJ, Barnett TC, McArthur JD et al (2014) Disease manifestations and pathogenic mechanisms of group A Streptococcus. Clin Microbiol Rev 27:264–301

Wang L, Ding Y, Lin Z et al (2017) Tailoring the binding properties of SpA Ig binding domains by in vitro molecular evolution. Int J Clin Exp Med 10:14244–14255

Watanabe R, Gehad A, Yang C et al (2015) Human skin is protected by four functionally and phenotypically discrete populations of resident and recirculating memory T cells. Sci Transl Med 7

White TC, Findley K, Dawson TL et al (2014) Fungi on the skin: dermatophytes and Malassezia. Cold Spring Harb Perspect Med 4

Williams MR, Costa SK, Zaramela LS et al (2019) Quorum sensing between bacterial species on the skin protects against epidermal injury in atopic dermatitis. Sci Transl Med 11

Winkel BMF, de Korne CM, van Oosterom MN et al (2019) A tracer-based method enables tracking of *Plasmodium falciparum* malaria parasites during human skin infection. Theranostics 9:2768–2778

Wu G, Zhao H, Li C et al (2015) Genus-wide comparative genomics of Malassezia delineates its phylogeny, physiology, and niche adaptation on human skin. PLoS Genet 11:1–26

Wu SJL, Grouard-Vogel G, Sun W et al (2000) Human skin Langerhans cells are targets of dengue virus infection. Nat Med 6:816–820

Yang J-J, Chang T-W, Jiang Y et al (2018) Commensal *Staphylococcus aureus* provokes immunity to protect against skin infection of methicillin-resistant *Staphylococcus aureus*. Int J Mol Sci 19

Zeng JS, Sutton DA, Fothergill AW et al (2007) Spectrum of clinically relevant Exophiala species in the United States. J Clin Microbiol 45:3713–3720

Zhu J, Peng T, Johnston C et al (2013) Immune surveillance by CD8αα+ skin-resident T cells in human herpes virus infection. Nature 497:494–497

Human Organotypic Respiratory Models

Mattias Svensson and Puran Chen

Contents

Abstract Biomedical research aiming to understand the molecular basis of human lung tissue development, homeostasis and disease, or to develop new therapies for human respiratory diseases, requires models that faithfully recapitulate the human condition. This has stimulated biologists and engineers to develop in vitro organotypic models mimicking human respiratory tissues. In this chapter, we provide examples of different types of model systems ranging from simple unicellular cultures to more complex multicellular systems. The models contain, in varying degree, cell types present in real tissue in combination with different extracellular matrix components that can critically affect cell phenotype and function. We also

Current topics in Microbiology and Immunology,
"Three-dimensional human organotypic models for biomedical research"
Eds Rino Rappuoli and Fabio Bagnoli
Human organotypic respiratory models.

M. Svensson (✉) · P. Chen
F59, Department of Medicine, Center for Infectious Medicine, Karolinska Institutet, Karolinska University Hospital, Huddinge, 141 86, Stockholm, Sweden
e-mail: mattias.svensson@ki.se

Current Topics in Microbiology and Immunology (2021) 430: 29–54
https://doi.org/10.1007/82_2018_91

Published Online: 29 May 2018

describe how organotypic respiratory models can be combined with human innate immune cells, to better recapitulate tissue inflammation, a key component in, for example, infectious diseases. These models have the potential to provide new insights into lung physiology, tissue infection and inflammation, disease mechanisms, as well as provide a platform for identification of novel targets and screening of candidate drugs in human lung disorders.

1 Introduction

Over the past five to ten years, there have been exciting advances in our understanding of relative roles of specialized cells and molecules important for proper lung function. This has largely been accomplished using advanced experimental animal models, particularly genetically engineered mouse models. However, these models often times fail to accurately recapitulate human lung in health and disease. Many promising drugs that perform well in preclinical animal studies have failed in humans due to toxicity and/or poor efficacy, suggesting that the current preclinical testing focusing on murine in vivo models is not always adequate. For example, asthma represents a disease where few new drugs make it to the clinic. Today, the majority of asthma research is carried out in non-human models, which poorly recapitulates human asthma making it difficult to translate findings from these models to efficacious therapies (Holmes et al. 2011). Also, the host tropism of pathogens for mice or humans as well as differences in immune cell composition between species is likely to affect the host immune response. For example, important human pathogens (e.g., *Mycobacterium tuberculosis and Staphylococcus aureus*) induce species-specific inflammatory responses which cannot be recapitulated in mice (Birkness et al. 1999; Loffler et al. 2010). When studying tissue inflammation, it is also important to take into consideration the effect that immune cells as wells as non-hematopoietic cells, such as fibroblasts and epithelial cells, have on the inflammatory response. The development of complex human tissue models is therefore needed.

Due to histological differences between the proximal and distal part of the respiratory tree, different types of models will be required depending on what scientific question is being addressed. The epithelium has been the main focus when it comes to developing organotypic models of lung tissue. A hallmark of the respiratory epithelium is the pseudostratified, columnar ciliary epithelial cells, which are found from the distal part of the larynx to the terminal bronchioles (Fig. 1). The cilia perform synchronized waves to transport unwanted, inhaled particles from the distal part of the respiratory tree to the pharynx. The particles are transported in a thin layer of mucus produced mainly by goblet cells, which also have important immunological functions (Fahy and Dickey 2010). The lamina propria is characterized by a dense layer of collagenous fibers that supports the epithelium. Underneath is a highly cellular, but loose connective tissue containing immune cells, fibroblasts, blood vessels, and lymphatic tissue. The trachea branches into the bronchi which are similar in structure as the trachea. As the respiratory tree branches and the diameter of the airway decreases, the epithelium gets thinner and the prominent cartilage structure gets thinner and scarcer. The substantial branching of the airways results in filtration of the air, as particles larger than five μm are too large to reach the respiratory bronchioles and alveoli. The respiratory bronchioles are characterized by a non-ciliated, cuboidal epithelium with no cartilage structures beneath the submucosa. Although gas exchange is allowed here, the alveoli are the primary site of gas exchange. Smaller particles (<1 μm) reaching the alveoli are removed by the numerous alveolar macrophages. The alveolar macrophages perform their function in the interphase of the alveolar space and interstitial space. Interstitial macrophages in the lung are distinct from the alveolar macrophages, and their role in the immunological response is likely very different (Cai et al. 2014). The epithelium of the alveoli is primarily composed of type I and type II alveolar cells. The type II alveolar cells are dome-shaped and secrete surfactant and other immunological active compounds from their lamellar bodies. The type I alveolar cells are exceptionally thin squamous cells that are similar in numbers as the type II alveolar cells but cover most of the surface of the alveoli. Type II alveolar cells proliferate during lung injury and replenish both type I and type II alveolar cells. The majority of the gas exchange occurs in the alveolar septum and is referred to the air–blood barrier. This structure is extremely thin, where the squamous epithelial cells and capillary endothelial cells share the basal lamina. Areas where connective tissue and fibers widen the distance between the alveolar lumen and capillary lumen are referred to the "thick portion" of the alveolar septum.

When choosing an organotypic model system to model health and disease in human lung, it is important to take into consideration the features associated with distinct parts of the lung to ensure experimental data are interpreted correctly and can be translated to in vivo situations. For example, studying gas exchange in the alveolar septum, and the function of alveolar epithelium is vastly different from investigating cancer formation, which often occurs in more proximal areas of the lung at the site of branching (Ishizumi et al. 2010). As many respiratory diseases involve the immune system, choosing a model system with immune cells, here referred to as immunocompetent model systems, may be highly relevant. Tissue infection and inflammation also involve the crosstalk between immune cells and epithelial cells, as well as other tissue constituent cells such as fibroblasts and endothelial cells, which implies the need for relatively complex organotypic model systems to mimic infectious and inflammatory processes of human lung (Fig. 2).

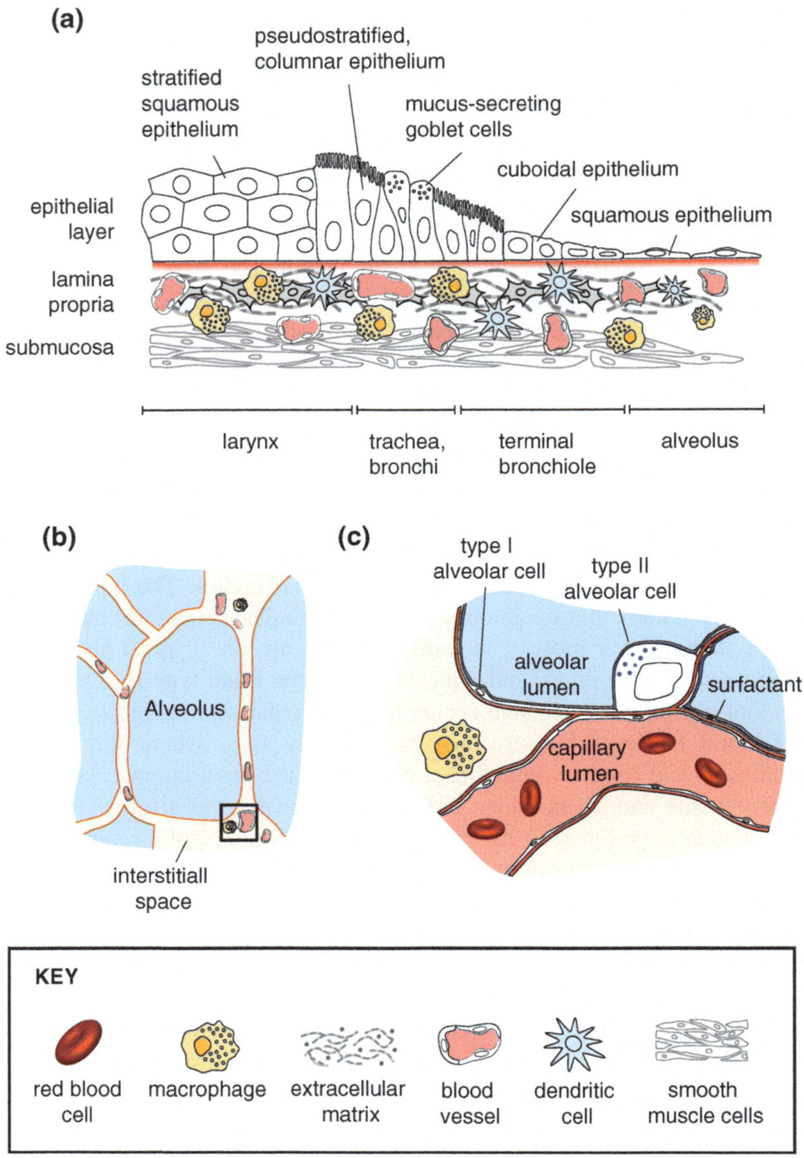
(a)
pseudostratified, columnar epithelium
stratified squamous epithelium
mucus-secreting goblet cells
cuboidal epithelium
squamous epithelium
epithelial layer
lamina propria
submucosa
larynx
trachea, bronchi
terminal bronchiole
alveolus
(b)
Alveolus
interstitiall space
(c)
type I alveolar cell
type II alveolar cell
alveolar lumen
surfactant
capillary lumen
KEY
red blood cell
macrophage
extracellular matrix
blood vessel
dendritic cell
smooth muscle cells

◂**Fig. 1** Illustration of airway epithelium. Schematic representation of the major cell types and structures in the different regions of the human lung. **a** As the airways progresses from the upper airways to the gas-exchanging region, the cell types transform to accommodate the functions and defense mechanisms active in each region of the respiratory system. The conducting airway consists of trachea, bronchi, and terminal bronchioles. When air passes through these structures, no gas exchange takes place. The respiratory zone consists of respiratory bronchioles, the alveolar ducts, and alveolar sac. Specialized cells, such as goblet cells are interspersed throughout the whole respiratory epithelium. **b** The alveolus is the main part where gas exchange takes place between the alveolar lumen and pulmonary capillaries. Flexible walls separate the alveolar sacs and contain the capillaries where gas exchange occur. **c** Higher magnification of the selected region in (**b**), illustrating the alveolar type I and type II epithelial cells, and the shared basal lamina between alveolar epithelial cells and endothelial cells. Alveolar macrophages are localized in the intersection between the interstitial space and the alveolar space

2 Organotypic Respiratory Models

Modeling of lung tissue spans the spectrum from traditional two-dimensional (2D) cultures to the intricate multicellular three-dimensional (3D) cultures. All model systems have their respective strengths and weaknesses depending on the scientific question asked. The 2D culture systems are easy to use and highly reproducible, but the overall representation of intact tissue is poor (Drewitz et al. 2011; Ho et al. 2010; Kunz-Schughart et al. 2004; Nam et al. 2015). Although such systems are very straightforward, cheap, and suitable for high-throughput screening of candidate drugs, they are not always ideal for assessing clinically relevant parameters such tissue integrity. For example, cells grown in two-dimensional culture tend to lose their original phenotype and functionality. This includes genetic alterations and epigenetic drift as well as alterations in cell–cell and cell–extracellular matrix interactions (Guo et al. 2014). These 2D monolayer systems are also limited by the lack of polarized cell phenotype and large number of cell–cell contacts, which affects their function and response to external stimuli (Roskelley et al. 1994). In contrast, organotypic respiratory models can include epithelium alone or in combination with extracellular matrix proteins and/or other cells such as fibroblasts which can critically affect cell phenotype and function (Black et al. 2005; Choe et al. 2006a, b; Cukierman et al. 2001; Mishra et al. 2012). There are various aspects that need to be taken into account when choosing appropriate organotypic model. This include the anatomical location to be modeled, and what specific processes which will be investigated. To date, highly reproducible 3D organotypic respiratory models have been developed by using primary, tissue-specific cells or cell lines (Table 1). More recently, progress has also been made using inducible pluripotent stem cells (iPSC) to generate equivalents of lung tissue, so-called organoids, and new applications of microtechnology have been demonstrated useful to create a human lung-on-a-chip.

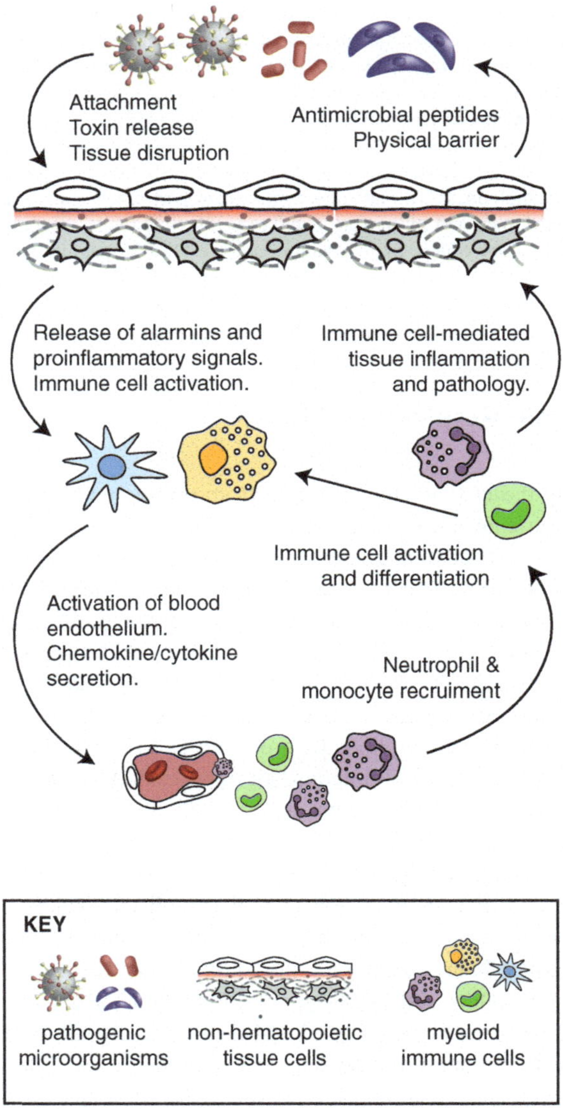

Fig. 2 Illustration of cellular contributors shaping the early host response against pathogenic microorganisms in tissue. Schematic drawing of the interactions of pathogenic microorganisms with their host. Early events of host–pathogen interactions includes mechanisms that pathogens use for invasion and innate mechanisms that the host uses to control pathogens

Table 1 Examples of respiratory organotypic models and their use in biomedical research

Model systems	Cell types	Area of use	References
Unicellular/monolayer systems			
Epithelial	NHBE, Calu-3	Transcriptional comparison	(Pezzulo et al. 2011)
	NHBE	Smoke-related toxicology	(Balharry et al. 2008)
		Pharmacodynamics	(Lin et al. 2007)
Fibroblast	MRC-5	NF-kB-signaling	(Htwe et al. 2015)
Layer-by-layer 3D systems			
w/o immune cells	16HBE14o-, MRC-5	Staphylococcal pneumonia	(Mairpady Shambat et al. 2015)
		Hantavirus infection	(Sundstrom et al. 2016)
	A549, HULEC	Mycobacterial infection	(Birkness et al. 1999)
	NHBE	Rhinovirus infection	(Sharma et al. 2009)
Immunocompetent	Calu-3, Mo-DCs, MRC-5	Asthma and allergen exposure	(Harrington et al. 2014)
	NHBE, SAE, Mo-DCs, Mo-MΦ	Aspergillus fumigatus	(Chandorkar et al. 2017)
	16HBE14o-, Mo-DCs, MRC-5	Tissue imprinting and cell migration	(Nguyen Hoang et al. 2014, 2012)
	16HBE14o-, Mo-MΦ, MRC-5	Mycobacterial infection	(Parasa et al. 2014)
Lung-on-a-chip			
	NPF (CCL-210), LMVEC, NHBE, neutrophils	Fungal infection	(Barkal et al. 2017)
	AEC, LMVEC	COPD	(Benam et al. 2016)
	SPCA-1, HFL1z	Drug screening in lung cancer	(Xu et al. 2013)
	LMVEC, AEC (NCI-H441)	Pulmonary edema	(Huh et al. 2010)
Spheroids			
	Resected nasal polyps	Bacterial infections and antibiotics	(Ulrich and Doring, 2004)
		Pseudomonas infection in cystic fibrosis	(Worlitzsch et al. 2002)
	NHBE	COPD	(Deslee et al. 2007)
	NHBE, A549	lung cancer	(Fessart et al. 2013)
	A549	Pseudomonas infection	(Carterson et al. 2005)

(continued)

Table 1 (continued)

Model systems	Cell types	Area of use	References
Organoids			
	NAEC	CRISPR-Cas9, MUC18-mediated inflammation	(Chu et al. 2015)
	Embryonic stem cells	iPSC, establishment of type II alveolar cells	(Yamamoto et al. 2017)
	hPSC	Syncytial virus infection	(Chen et al. 2017)
Explants			
	–	LPS-induced inflammation	(Switalla et al. 2010)
	–	Adenovirus infection	(Booth et al. 2004)
	–	Influenza infection	(Nicholas et al. 2015)

Abbreviations: *NHBE* normal human bronchial epithelial cells, *HULEC* human lung endothelial cells, *SAE* small airway epithelial cells, *Mo-DCs* monocyte-derived dendritic cells, *Mo-MΦ* monocyte-derived macrophages, *COPD* chronic obstructive pulmonary disorder, *NAEC* nasal airway epithelial cells, *NPF* normal pulmonary fibroblast, *AEC* airway epithelial cells, *LMVEC* lung microvascular endothelial cells, *hPSC* human pluripotent stem cells

2.1 *Unicellular 3D Models*

The most straightforward way of creating a 3D tissue-like model is by allowing a single cell type to grow in, or on top of, layers of either organic matrices or synthetic polymer scaffolds (Fig. 3a). The structures have in common that they allow deposition of extracellular matrix (ECM), and transportation of nutrients, signaling molecules and gases. In addition, the matrix can be remodeled by the cells and achieve a more physiologically relevant environment compared to traditional polystyrene-based culture flasks.

2.1.1 Epithelial Models

For the purpose of studying disease-specific differences of human lung epithelial cells, or studying effects of direct exposure to inhaled environmental stressors, such as pathogens, allergens, or air pollutants, human lung epithelial cells are cultured to form functional 3D epithelial barriers in vitro. Human lung epithelial cells of choice can be purchased commercially or isolated from various lung tissue specimens depending which part of the lung is desired to be modeled. The lung epithelial cells can be grown on non-biological porous supports, such as Transwell® inserts. Prior to seeding the epithelial cells, the inserts are often pre-coated with an ECM protein. The insert and cell culture are kept submerged in media, and once epithelial cells reach confluence, media is replaced and only added back to the basolateral compartment, rendering the apical side of the epithelial cells air-exposed (Fig. 3b). By mimicking air-exposed conditions, the epithelial cells differentiate and polarization of the epithelial barrier is established. At this stage, the method is referred to as an air–liquid interface (ALI) culture (Pezzulo et al. 2011) and has frequently been used as an in vitro model for airway drug transport studies as well as the evaluation of inflammatory and cytotoxic responses to aerosol content. Both bronchial epithelial cell lines and primary bronchial epithelial cells have been used frequently in ALI cultures. However, it may be advantageous to use primary epithelial cells rather than immortalized epithelial cell lines as cell lines often fail to differentiate into heterogeneous cell populations composed of polarized ciliated and mucus-producing cells (similar to what is observed in vivo (Figs. 1a and 3a). Three-dimensional cultures of primary epithelial cells in the air–liquid interface have been used to study host–pathogen interactions, and such studies have, for example, revealed that epithelial cells actively engage in the uptake of Haemophilus influenza. The surface of epithelial cells showed evidence of actin polymerization and lamellipodia surrounding bacteria, and bacteria were found within vacuoles inside of the epithelial cells (Ketterer et al. 1999). ALI cultures have also been used in combination with aerosol delivery systems, and this has proven to provide great predictive power with respect to toxicology, as well as drug delivery efficiency and efficacy (Andreoli et al. 2003; Balharry et al. 2008; Lin et al. 2007; Tralau and Luch 2012).

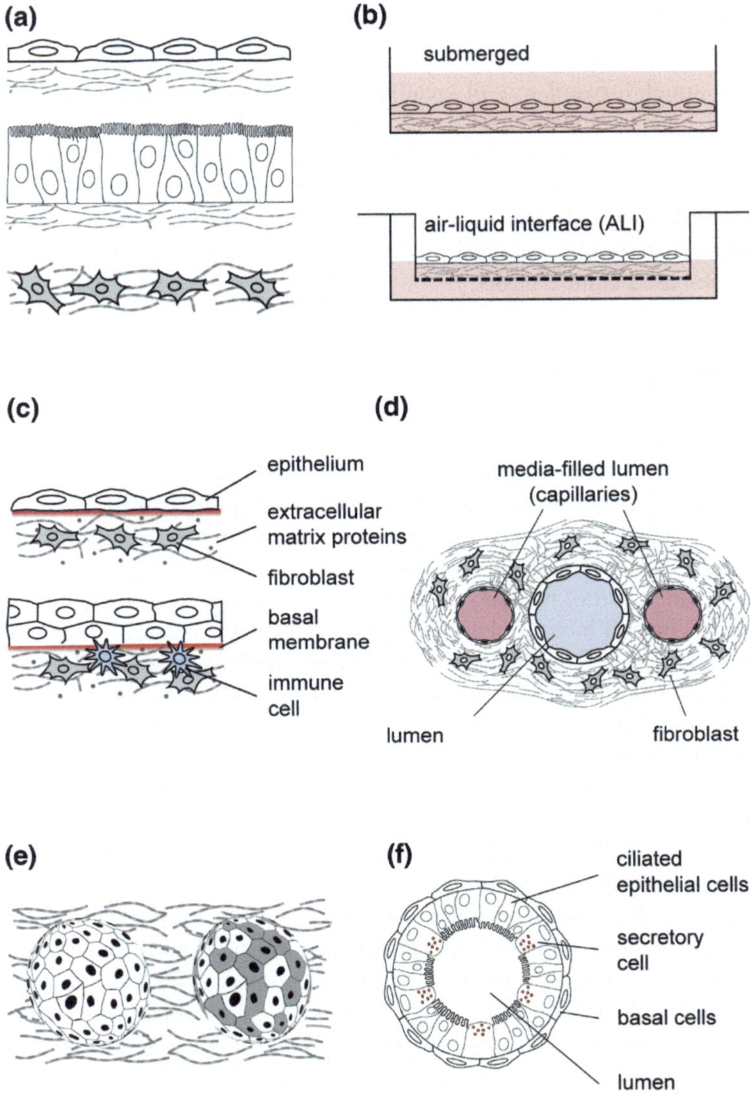
(a)
(b)
submerged
air-liquid interface (ALI)
(c)
(d)
epithelium
extracellular matrix proteins
fibroblast
basal membrane
immune cell
media-filled lumen (capillaries)
lumen
fibroblast
(e)
(f)
ciliated epithelial cells
secretory cell
basal cells
lumen

◂**Fig. 3** Illustration of respiratory organotypic models. Schematic representations of the different organotypic models described in this chapter. **a** Representation of unicellular 3D models cultured together with an extracellular matrix protein. Model tissue with pseudostratified airway epithelial cells that are non-ciliated (upper) or ciliated (lower), as well as model tissue with lung fibroblasts (lower), is shown. **b** Representation of submerged versus air–liquid interface cultures. **c** Representation of multicellular 3D models built layer by layer. Composition of tissue models with epithelial cells and fibroblasts (upper), or with epithelial cells, fibroblasts, and immune cells (lower). The formation of a basal membrane (red) is seen in the boundary between the epithelial layer and the layer of fibroblasts and extracellular matrix proteins. **d** Representation of a lung-on-a-chip. Microfabricated compartmentalized microchannels are used to form an alveolar–capillary barrier, where alveolar epithelial cells and blood endothelial cells are seeded on each side of a thin, porous flexible membrane coated with ECM proteins. Fibroblast is embedded in the ECM. **e** Representation of spheroids, which are self-assembling structures made from differentiated epithelial cells. Spheroids can be made with single epithelial cell lines (left) or with a combination of cell types as well as with epithelial sheets acquired from bronchial brushings (right). **f** Representation of organoids, which are self-assembling structures originating from epithelial progenitor cells, stem cells, or inducible pluripotent stem cells

2.1.2 Fibroblast Models

Fibroblasts provide support and structure to other cells present in lung tissue. In addition, fibroblasts play an important role in lung inflammatory responses. Thus, there is a need for model systems that can provide insight into questions related to fibroblast function in lung tissue. While fibroblasts grown in planar dishes are maintained in a 2D orientation, fibroblasts within the normal 3D microenvironment of the lung extend membrane protrusions in all directions. For example, fibroblasts extend to contact endothelial cells and type I cells in numerous alveolar spaces (White 2015). The 3D architecture of lung tissue is further organized by the deposition and remodeling of ECM proteins, processes predominantly performed by fibroblasts, and which in turn play an important role in shaping fibroblast behavior both in health and disease. Thus, the strong link between the ECM and fibroblasts makes it important to create the right prerequisites for studying fibroblast tissue-organizing functions in as in vivo tissue-like conditions as possible. In vitro cultures with fibroblasts should therefore contain some form of scaffold that mimic normal tissue conditions (Fig. 3a). The selection of scaffolds influences cell functionality, at least initially, before the cells themselves begin to produce ECM proteins on their own and build structures essential for tissue function. In this context, both biological and non-biological compounds are used to initiate the cultures. Biological material includes various types of collagen type I and type IV, or, Matrigel®, which is a complex protein mixture. More recently, non-biological scaffolds based on nanotechnology and the synthesis of fiber structures using polymers have been developed (Kim et al. 2010). The material properties of the scaffold may be vital in determining cellular fate and responses, such that the ability of pore size and porosity of scaffolds to direct cellular responses (Loh and Choong 2013). As an example, the non-biological polyethylene terephthalate was used to engineer scaffolds with defined thickness, diameter of fibers, and pore size to resemble the geometry of human lung ECM and to culture lung fibroblasts in (Htwe et al. 2015). Under these conditions, researchers found that fibroblasts grown in the 3D scaffolds responded much weaker to TNF as compared to fibroblasts grown in simpler 2D cultures. It was also found that fibroblasts in 3D cultures remained responsive to TNF-α at higher concentrations. The different responses seen in fibroblast were concluded to be a consequence of the fact that the fibroblasts in 3D obtain a more in vivo-like TNF receptor I expression. Thus, highlighting the importance of mimicking the in vivo situation where the cells can interact and behave according to the mechanical cues acquired from the neighboring 3D environment.

Although unicellular 3D lung tissue models of normal airway mucosa have proven to be particularly informative for analyzing the behavior of epithelial cells (Birkness et al. 1999; Choe et al. 2006b; Cozens et al. 1994), 3D multicellular cultures that recapitulate not only the epithelium layer, but aims to take into consideration the interaction between different cell types, is needed (Pampaloni et al. 2007).

2.2 Models Built Layer by Layer

The more complex 3D multicellular cultures consist of at least two distinct cell populations within the same culture environment and were initially utilized to study cell–cell communication. Building layer-by-layer multicellular model systems are usually accomplished by using a cellular stroma layer as the foundation for the 3D multicellular culture. Fibroblasts mixed with an ECM protein, such as collagen, are seeded on a non-biological scaffold. Once the stroma layer polymerizes, epithelial cells are seeded on top to form the epithelial lining (Fig. 3c). This approach takes advantage of paracrine signaling and/or cell–cell contacts to direct cell phenotype and control cellular migration and organization, thereby creating a more physiologically relevant in vitro model. For optimal lung epithelial cell function, it is necessary that there is supportive tissue, a scaffold or stroma, to anchor to. Stroma consists principally of fibroblasts and ECM proteins. Although, ECM proteins are predominantly produced by fibroblasts, it is only when fibroblast is cultured together with other cells such as epithelial cells, endothelial cells, smooth muscle cells, and some immune cells, that the full potential to make organ structures is achieved. One such example is the basement membrane, which is a structure that serves to delineate boundaries and to compartmentalize tissues. It acts as a scaffold for epithelial cell anchoring, tissue repair, and cell trafficking. Fibroblasts also promote cell growth and survival, remodeling of the ECM, and maintenance of elasticity to lung tissue (Black et al. 2005; Choe et al. 2006a). By actively interacting with the adjacent epithelial layer, fibroblasts have a key role in inflammation and repair of a damaged epithelium (Tomasek et al. 2002). Several respiratory models with a combination of fibroblast matrix layer and an epithelial layer that better recapitulates human tissue have been developed (Choe et al. 2006b; Harrington et al. 2014; Nguyen Hoang et al. 2012). Under these conditions, fibroblasts proliferate slower and their production of extracellular matrix protein and growth factors contributes to basal membrane formation and differentiation of epithelial cells. This is in contrast to when fibroblasts are grown on artificial gels or synthetic membranes (Choe et al. 2006a; Griffith and Swartz 2006). Fibroblasts are relatively easy to derive from human tissue biopsies and kept in culture without immortalization, but it may be more difficult to establish in vitro co-cultures with primary fibroblasts and lung epithelial cells, depending on the culture requirements. An alternative approach used by many investigators is therefore to use in vitro immortalized cells. However, it is recommended to verify that the cells used have not lost important features, and that deposition of structural framework proteins, such as E-cadherin and claudin-1, can be found in the appropriate compartment of the model (Nguyen Hoang et al. 2012). Carcinoma-derived cell lines should preferably be avoided, unless certain functionalities associated with cancer is studied, as these cells can display abnormal organ-/tissue-specific functions. For example, the bronchial 16HBE14o- (16HBE) cell line is derived from a healthy individual and forms proper epithelial barriers, while the adenocarcinoma cell line A549 forms aberrant tight junctions and epithelial barriers (Heijink et al. 2010).

To date, respiratory multicellular culture systems built layer by layer have significantly contributed to the development of various model systems for biomedical research and in particular for studies on human lung infections. Among other things, a human 3D model with a combination of human fibroblasts and bronchial epithelial cells have been used to study *Aspergillus fumigatus* (*A. fumigatus*), *Mycobacterium tuberculosis* (*M. tuberculosis*) and *Staphylococcus aureus* (*S. aureus*) infections, as well as Andes hantavirus infections (Chandorkar et al. 2017; Mairpady Shambat et al. 2015; Parasa et al. 2014; Sundstrom et al. 2016). The flexibility of these model systems to study the pathogenesis of several pathogens (viral and bacterial) implies their broad potential. Modeling staphylococcal pneumonia and exposure of the lung model to *S. aureus*-derived toxins revealed tissue-destructive events, and demonstrated the usefulness of the model to delineate mechanisms of lung epithelium destruction (Mairpady Shambat et al. 2015). This study also showed that the addition of human poly-specific intravenous immunoglobulins (IVIG) inhibited Staphylococcal toxin-mediated cytotoxicity and tissue injury. Thus, these multicellular co-culture systems may also represent a valuable tool for drug testing applications. These types of multicellular tissue models have enormous potential to improve preclinical drug testing. Depending on the scientific question, it may be of interest to add other cell types such as endothelial and airway smooth muscle cells, as well as immune cells.

2.3 Immunocompetent 3D Models

Less complex tissue models that examine the direct action of toxins/drugs that cause cell death, induce apoptosis, or modulate biomolecule production by lung epithelial cells do not take into consideration the accompanying responses by other relevant cell types, such as immune cells. The multicellular assembly provides additional secreted factors and multiple cell-to-cell communications within the tissue microenvironment that is likely to play an essential role in regulating leukocyte function. In addition, subpopulations of leukocytes, such as the myeloid immune cells, have important roles in regulating tissue homeostasis and inflammation (Condon et al. 2011). Macrophages, monocytes, and dendritic cells (DCs) are phagocytes belonging to heterogeneous populations of widely distributed myeloid immune cells, with overlapping and unique functions, and they play central roles in homeostatic events, as well as the initiation and orchestration of immune responses (Banchereau and Steinman 1998; Guilliams et al. 2013; Pluddemann et al. 2011). Within lung, DCs mainly associate with the epithelial layer and there is evidence of DC regulation by the epithelium and that epithelial dysfunction leads to overzealous immune cell activation (Hammad and Lambrecht 2008; Lambrecht and Hammad 2009). Three-dimensional models with multiple cell types could, for example, provide valuable insight into mechanisms involving lung inflammation, damage, and fibrosis in response to drugs as well as pathogens in the human lung. For example, an ALI system with normal primary human bronchial epithelial or small airway epithelial cells that was kept under perfusion condition to allow seeding of immune cells was used to study tissue interactions with the airborne pathogen *Aspergillus fumigatus* (*A. fumigatus*) (Chandorkar et al. 2017). Under perfused conditions, epithelial cells had higher ciliogenesis compared to static conditions and monocyte-derived cells appeared fully functional with respect to their migratory and phagocytic capacities in response to *A fumigatus*. Significant progress has also been made in introducing myeloid immune cells, such as monocytes and monocyte-derived DCs into 3D models with both fibroblasts and epithelial cells (Fig. 3d) (Harrington et al. 2014; Nguyen Hoang et al. 2012; Parasa et al. 2014).

A limited number of organotypic models combining tissue-specific cells and immune cells have been developed. One of the systems uses the "normal" bronchial epithelial cell line 16HBE14o- and MRC-5 lung fibroblasts, and monocyte-derived DCs (Nguyen Hoang et al. 2012). The intactness of the epithelial barrier is not affected by the immune cells, and enzymatic digestion of the tissue models reveals DC survival and preserves phenotype (CD45, HLA-DR, and CD1a) (Nguyen Hoang et al. 2014). The confocal microscopy analyses of lung models also revealed that DCs were distributed mainly in the subepithelial layer. Notably, the implanted DCs can survive for at least one month without the addition of external growth factors, suggesting that the tissue model provide necessary survival factors. Further evidence that DCs are regulated by the lung tissue microenvironment comes from the observations that DCs implanted in the lung model produces CCL18, a chemokine that is constitutively expressed in lung at steady state and is elevated in several human disorders, including various malignancies and inflammatory lung, skin, and joint diseases (Hieshima et al. 1997; Schutyser et al. 2005). Lung tissue models with DCs abundantly expressed CCL18 mRNA and protein, while CCL18 expression was undetectable in the lung tissue model without DCs (Nguyen Hoang et al. 2012). In contrast, CCL17 and CCL22, which are chemokines that are barely detectable in peripheral tissues at steady state and that are induced during inflammatory reactions and pulmonary inflammatory diseases (Godiska et al. 1997; Imai et al. 1996), were suppressed in DCs from tissue models compared to DCs cultured in GM-CSF and IL-4. These findings support the importance of studying DC biology in more physiological-relevant milieus rather than in monolayers on plastic surfaces. Thus, introducing DCs and other leukocytes to tissue models will not only provide better insights of tissue responses in health and disease, but also provide a more physiological relevant platform to study specific immunological responses.

Host responses to pathogens and inflammatory reactions in tissue are highly dynamic processes dependent on the interplay between leukocytes, tissue-specific cells, and ECM components (Kleinman et al. 2003; Svensson et al. 2004; Wrenshall 2003). Complexities of cell–cell interactions between different types of cells, as well as cell–matrix interactions, must be considered when performing studies of, for example, host–pathogen interactions and inflammatory reactions in tissue.

Studying dynamic cell-cell and cell–matrix interactions in human tissue is nearly impossible for ethical and practical reasons. Developing new tools to visualize immune cell behavior in tissue is therefore of great importance. Previous studies have shown that DC migration is largely influenced by tissue origin, cellular activation, and the 3D structure of the microenvironment (Gunzer et al. 2000). Organotypic models have been used to enable live imaging of phagocyte cells (DCs or macrophages) migratory behavior in lung epithelial tissue (Nguyen Hoang et al. 2014). The nature of the experimental setup enables transduction of fluorescent proteins of the epithelial cells while primary immune cells can be labeled with a cell tracker dye prior to implantation (Nguyen Hoang et al. 2014). In addition to defining phagocyte cell distribution relative to fibroblasts and epithelial cells, the live imaging setup also allows quantitative analysis of phagocyte cell migration by tracking individual cells in the tissue at steady state and in response to stimulation. These experimental setups provide new possibilities investigating mechanisms of tissue communication circuits underlying DC motility in the microenvironment (Acton et al. 2012) of human lung tissue. Understanding mechanisms behind changes in DC motility in response to specific stimulating agents is important for understanding mechanistic actions of, for example, respiratory infections. The presence and spatial distribution of phagocytic cells in lung tissue is particularly important when studying human tuberculosis. To model human tuberculosis, infections are usually performed with human blood cells in monolayer cultures (Kapoor et al. 2013; Puissegur et al. 2004). As an alternative approach, monocytes and monocyte-derived macrophages have been infected and implanted in respiratory organotypic models. In a study by Parasa et al. fluorescently labeled monocytes in the model system migrated toward *M. tuberculosis*-infected macrophages (Parasa et al. 2014), resembling clustering and initiation of granuloma formation. When the model was infected with non-virulent strains of *M. tuberculosis*, clustering was not observed. The ability to distinguish between immunological responses induced by virulent and non-virulent strains of *Mycobacterium tuberculosis* in the model system underlines the suitability of immunocompetent lung models for studying respiratory infections, such as human tuberculosis.

To date, immunocompetent models comprise of epithelial cells and fibroblasts combined with various innate, phagocytic cells. Yet, there are no 3D models built layer by layers in which lymphocytes, such as T cells, B cells, or NK are introduced. T cells play an important role in the formation of a mature granuloma in human tuberculosis, and cytotoxic T cells and NK cells are also of particular interest when studying the immunomodulatory events in the tumor microenvironment. Thus, introducing lymphocytes to organotypic models would broaden the application of immunocompetent models. Introducing T cells, particularly cytotoxic T cells, or NK cells is challenging due to the fact that cell haplotype has to be taken into account, to avoid unwanted cytotoxic responses. Future development of immunocompetent models should aim at recapitulating the complete course of an immunological response, from innate to adaptive immunity.

2.4 Lung-on-a-Chip

Recently, the application of microfluidic systems has enabled novel approaches to mimic human lung tissue. These microfluidic-based in vitro systems recreate specific features of lung physiology such as a continuously perfused capillary bed with an air-exposed epithelial barrier (Fig. 3d). Several microfluidic 3D lung-on-a-chip models have been produced that are specifically designed to meet the high-throughput need of drug and toxicity testing (Barkal et al. 2017; Benam et al. 2016; Huh et al. 2010; Xu et al. 2013). Huh and colleagues, originally described a microfluidic system of functional alveolar–capillary structures that could recapitulate bacterial respiratory infections and inflammatory responses in the alveolar space as seen in vivo (Huh et al. 2010). The lung-on-a-chip organotypic model was further developed and integrated with a computer-controlled vacuum to produce cyclic stretching of the tissue to mimic physiological breathing movements. It was found that mechanical strain together with silica nanoparticles synergistically activated endothelial cells as evident by increased expression of ICAM-1 and ROS-production (Huh et al. 2010). The activated endothelium also promoted firm adhesion of fluorescently labeled human neutrophils flowing in the vascular microchannel, and mechanical strain of the tissue significantly enhanced neutrophil translocation through the capillary-epithelial interface. The opportunity to introduce mechanically active organotypic models that recapitulate critical organ functions may therefore provide new avenues for understanding cell and tissue biology under next to in vivo conditions.

Barkal and colleagues, (Barkal et al. 2017), created an organotypic model of the bronchiole to study pulmonary infections with *A. fumigatus* and *Pseudomonas aeruginosa*. Infections could be established using an aerosol-based method, rather than inoculating with a suspension of pathogens, to better reflect in vivo situations. In contrast to static models that are built layer by layer, this model allows introduction of neutrophils in capillary-like structures to study neutrophil recruitment and migration toward these pathogens. The complex nature of lung-on-a-chip models, including the cell heterogeneity and relevant tissue organization, takes into consideration many of the features of a real infection.

Another interesting development uses microfluidic systems to create human disease model-on-a-chip that mimics pulmonary edema (Huh et al. 2012), as well as an in vitro model for studying human chronic obstructive pulmonary disorder (COPD) (Benam et al. 2016). In many respects, the lung-on-a-chip model is superior to many other model systems as it offers multiple integrated capabilities. However, there is potential for further development also of these model systems, such as the use of primary epithelial cells and inclusion of alveolar macrophages and fibroblasts. Further development of microscale tissue subunits holds great promise to provide novel platforms for unraveling and screening of new therapies targeting lung disorders, at a personalized level.

2.5 Spheroids and Organoids

The terms "*spheroids*" and "*organoids*" are sometimes used interchangeably. Here, we define spheroids as self-assembled 3D structures made from differentiated epithelial cells (primary or cell lines) (Fig. 3e), whereas organoids are self-assembling structures originating from epithelial progenitor cells, pluripotent embryonic stem cells, or inducible pluripotent stem cells (Fig. 3f). In contrast to spheroids, cells in the organoid structure differentiate into several different tissue-relevant cell types, such as basal cells, ciliated epithelial cells, and goblet cells. Spheroids and organoids are cultured with or without support from fibroblasts and/or an ECM.

2.5.1 Spheroids

Spheroids can be created with respiratory epithelial cell lines as well as primary epithelial cells isolated from dissociated lung tissue. The spheroids can later also be cultured in an ECM with or without supporting stromal cells (Ulrich et al. 2005; Ulrich and Doring 2004; Worlitzsch et al. 2002). Depending on the source of cells used, the cell composition of the spheroids will be more or less heterogeneous. The methods generating spheroids with single cells using normal human bronchial epithelial cells or cancer epithelial cell lines have been used to study differences in the basolateral expression of cell–cell junction proteins between normal and cancer epithelial cells (Fessart et al. 2013). Bronchial epithelial sheets obtained from bronchial washes also form distinctive bronchial epithelial spheroids when placed in culture. These spheroids, generated from non-dissociated bronchial epithelium obtained by bronchial brushing, show interesting characteristics as evident by the presence of a polarized, ciliated bronchial epithelium with tight junctions and ciliary beating, as well as functional basal cells and secretory cells. It was revealed that spheroids from patients with COPD had ciliary beat frequencies similar to that of spheroids from healthy individuals but that LPS stimulation of spheroids induced a higher release of IL-8, but not PGE2 and LTB4 in COPD bronchial spheroids

(Deslee et al. 2007), suggesting that the COPD epithelial cells may have an intrinsic capability to respond stronger to TLR-ligands.

The spheroid models are easy to generate, inexpensive, and have long-term viability (>6 weeks). These types of three-dimensional spheroids or aggregates have been used in various drug-testing settings, but also for studying bacterial infections. For example, the spheroid-epithelial response to pseudomonas infection is vastly different compared to its 2D counterpart, as indicated by increased mucoglycoprotein production, cell adhesion, and production of proinflammatory molecules (Carterson et al. 2005). Although the epithelial cells in the spheroid 3D environment show an in vivo-like phenotype, the physiological relevance may be questioned as cells do not come into contact with air that is the natural environment for the airway epithelium. It is also worth mentioning that function and behavior of epithelial cells is largely influenced by other cells, such as fibroblasts, and that the strength and nature of an inflammatory response is likely to be influenced by the presence and actions of both fibroblasts and immune cells.

2.5.2 Organoids from Progenitor Cells

Basal cells make up approximately 30% of the pseudostratified mucociliary epithelium. These cells rapidly adapt and proliferate in response to injury and quickly regenerate the epithelium and restore barrier function. Basal cells isolated and expanded from lung tissue form complex, multicellular organoids in 3D cultures in the presence of epidermal growth factor and fibroblasts (Hegab et al. 2015). Organoids derived from basal cells have been used as a model system to investigate which cytokines and other molecules that affect the ratio of ciliated and secretory cells (Danahay et al. 2015). Worth remembering, basal cells from individuals with respiratory disorders behave differently compared to basal cells isolated from healthy individuals and may inherently differentiate into secretory cells rather than ciliated cells (Shaykhiev et al. 2013). As basal cells have proliferative capacity, it is also possible to utilize CRISPR/Cas9 gene-editing prior to organoid formation and this will greatly improve the ability to identify factors important in lung physiology and function (Chu et al. 2015; Gao et al. 2015). Since basal cells exist in most parts of the airways, including the nasal tract, they may offer a feasible approach with minimal invasiveness to establish personalized organoids.

Another example of making respiratory organoid is to use alveolar type II cells. Alveolar type II cells are dome-shaped cells characterized by the production of pulmonary surfactant proteins. Human type II cells are typically isolated with antibody-based assays using antibodies specifically recognizing the type II cells (Gonzalez et al. 2010). In combination with supporting stromal cells, type II cells proliferate and differentiate into alveolar type I cells and form organoids, referred to as alveolospheres.

2.5.3 Organoids from Stem Cells

Another approach to create respiratory organoids is by using pluripotent embryonic stem cells. This allows studies of human lung organogenesis. Induced pluripotent stem cells (iPSCs) can also be used to study lung organogenesis. One of the main advantages using pluripotent stem cells is that it allows natural differentiation of progenitor cells into multiple differentiated tissue-specific cells that assemble into an organ-like structure (Clevers 2016). Another advantage of using stem cells or iPSCs to form organoids is that the CRISPR/Cas9 approach may be used. Recently, iPSCs was used to generate organoids harboring alveolar stem cells for long-term expansion (Yamamoto et al. 2017). Using this method, alveolar stem cells self-renewed, with transcriptomes and morphology consistent with those of alveolar type II cells, and were able to differentiate into alveolar epithelial type I-like cells. These advancements of generating alveolar epithelial cells can provide help to overcome the lack of cell heterogeneity of current approaches and useful in future disease modeling and regenerative medicine. Similarly, Chen and colleagues reported on the generation of lung bud organoids from human pluripotent stem cells (Chen et al. 2017). These organoids contained mesoderm and pulmonary endoderm and had the potential to develop into branching airway and early alveolar structures after xenotransplantation in Matrigel 3D cultures. These organoids were used to study respiratory syncytial virus infection and lung fibrosis. Thus, the generation of lung tissue using human pluripotent stem cells including embryonic stem cells and iPSCs have the potential to make a powerful impact on our understanding and lung physiology and pathology.

Like any model, organoids are imperfect and come with their own set of limitations (Fatehullah et al. 2016); for example, they are unvascularized and lack cell types such as immune cells. However, scientists are advancing these 3D models, and by improving the culture conditions, efforts are made to make organoids even more similar to organs. Even if organoids could develop more heterogeneous cell types, there would still be the problem of making larger models as cells inside these structures die without nutrient-conducting structures, such as the perfused capillaries in the lung-on-a-chip systems. Similar to the spheroids, the organoids lack of immune cells which limits the use of these systems studying certain aspects of tissue homeostasis and inflammation.

2.6 Tissue Explants

Human lung tissue explants represent a unique type of experimental system. The human lung tissue explants can be used to verify findings from the in vitro studies, with respect to, for example, organ structure, cell localization, and functionality. The explant is often acquired during bronchoscopy or surgery, and the piece of tissue is kept in nutrient-rich media (Nicholas et al. 2015). Although consisting of all the relevant cell types, properly organized histology and with a complex extracellular matrix, tissue explants have several limitations in comparison to more artificial in vitro systems and organotypic models described in this chapter. Traditionally, the tissue is stimulated or infected, with external stimuli added in the media or on top of the explant which means the site of exposure cannot be controlled in a precise manner. For example, it is challenging with the 3D orientation of the explant, and exposing specifically the apical part of the lung epithelium is virtually impossible, whereas the orientation of the in vitro models is easier to control. This may be of particular relevance when studying airborne pathogens and their interactions with the respiratory epithelium. The explant is also considerably more difficult to manipulate or alter prior to performing the experiment, complicating many types of mechanistic studies. Also, following dissociation from its parent tissue, the conditions for the tissue explants change considerably and a significant degree of cell death and loss of tissue integrity can be observed already after 48 h (Jager et al. 2014). Thus, lung tissue explants serve as an excellent ex vivo testing system in certain settings but are vastly different from the organotypic model systems established in vitro.

2.7 *Considerations for In Vitro/Ex Vivo Modeling*

In order to be able to do any of the above-described respiratory organotypic models, it is essential to obtain human cells and tissue samples from reliable sources. It is important to consider how to obtain the samples, and what questions will be addressed. Human lung epithelial cells and cell lines are commercially available, as are ALI cultures. Samples from patients with lung diseases, such as COPD, asthma, cystic fibrosis are also commercially available. Alternatively, lung tissue samples can be obtained from individuals willing to donate tissue for research purposes, requiring approval from institutional ethic review boards and informed consents. In such cases, it is important to obtain relevant information of the donors to interpret possible inter-donor variations. This is often times possible at hospital clinics and academic centers where lung surgery and transplantation occur. Tissue samples from bronchial brushes, bronchial biopsies, or lung resections are also possible sources. Researchers should be aware that regardless of the source of tissue and cells, there may be large donor variations, and preferably, samples from many individuals should be used and tested experimentally. Also, depending on what part of the lung the tissue was taken from the cells may behave differently, and it is always wise to be consistent using standardized sampling and isolation protocols.

3 Conclusions and Future Perspectives

Lung diseases contribute significantly to deaths and reduced quality of life in humans all over the world. Disorders include conditions such as respiratory infections, COPD, malignancies, and autoimmune diseases. Developing robust and reliable respiratory organotypic models is especially important to further advance the field of basic human lung biology and immunology, as well as the translational exploitation of such basic knowledge. Immunologists are increasingly appreciating the contribution of tissue constituent cells such as epithelial cells, fibroblasts, and endothelial cells in inflammatory responses. Improving the cell culture conditions for how we study cell–cell interactions will likely greatly improve our knowledge in basic lung biology and pathology. Comparisons of the gene expression profiles between cell cultures kept in monolayers and three-dimensional models implicate the importance of extracellular matrix in regulating cell migration, apoptosis, transcriptional regulation, and receptor expression (Bissell et al. 1982; Ghosh et al. 2005). However, it needs to be better understood what other specific conditions influence lung epithelial phenotype and function, as well as other cell types of the lung, and under what specific experimental conditions organotypic systems have to be kept to resemble the in vivo situation. Such improved knowledge will likely make organotypic models even more sophisticated and relevant to use.

Complex organotypic models may not only improve preclinical and basic research, but also provide an exciting platform for translational research. Creating "personalized" models to study diseases and use as a platform for predicting treatment outcomes is an exciting prospect and could have high impact on the healthcare system, as selecting the patients that will benefit the most from a specific drug is one of the main challenges. In lung cancer, the use of checkpoint inhibitors is currently being intensely studied and tested. The average responder exhibits very promising results, but it is problematic that relatively few patients respond, and therefore, identifying the differences between responders and non-responder patients is key for the implementation of immunomodulatory drugs. It is believed that the tissue microenvironment plays an important role in shaping the immunological response, thus warranting the use of human model systems in which those processes can be modeled and tweaked. We anticipate that the use of respiratory organotypic models will be broadened in this context in the future, as the appreciation of how tissue complexity greatly influences the outcome of multifactorial diseases have increased significantly over the last two decades.

To date, several challenges remain. Development of organotypic lung models will need to be carefully validated, as more complex models are continuously being developed. Also, many diseases are caused by, or influenced by cells of both myeloid and lymphoid origin. NK cells and different T cell subsets are lymphocytes of special interest in, for example, lung cancer and autoimmune disorders. If the immune cells and tissue constituent cells are not haplotype matched, and will be cultured together, the immune cells might cause tissue damage by attacking the tissue constituent cells. This could potentially be avoided by establishing autologous tissue models in which the leukocytes and tissue-specific cells are derived from the same donor or with a matched HLA-haplotype. Neutrophils on the other hand are myeloid cells that in response to microbial challenge are one of the first immune cells to infiltrate into inflamed tissue. Due to the fact that neutrophils easily get activated, they are difficult to keep in long-term cultures. This have so far been solved by using microfluidic systems in which neutrophils are recruited to the tissue via a conducting system in the lung on a chip system, resembling endothelial cells. The further development of model systems with microfluidics should facilitate the introduction of different types of immune cells at any given time. The diversity of respiratory organotypic models is good considering the unlimited number of scientific questions that need to be addressed. However, a future goal would be to find consensus of "golden standard frameworks" for different respiratory organotypic model protocols, each with their own set of options to create variable models. This would improve the comparison and verification of data between groups and increase the likelihood of making progress within the field of research on human lung health and disease. In conclusion, the future for organotypic respiratory models looks bright, and especially exciting is the future of the organ-on-a-chip model, which may be used to connect several different organ models to move toward whole body representation. The further development of immunocompetent model systems also looks promising, as does the future use of more complex organoid systems. The use of respiratory organotypic models is likely to be advantageous in high-throughput drug testing by the industry, and thus have societal impact in the near future, as it hopefully will increase the likelihood of candidate pharmaceuticals to reach the market.

Acknowledgements Our work is supported by grants from the Karolinska Institutet, Stockholm County Council, and the Swedish Research Council.

References

Acton SE, Astarita JL, Malhotra D, Lukacs-Kornek V, Franz B, Hess PR, Jakus Z, Kuligowski M, Fletcher AL, Elpek KG et al (2012) Podoplanin-rich stromal networks induce dendritic cell motility via activation of the C-type lectin receptor CLEC-2. Immunity 37:276–289

Andreoli C, Gigante D, Nunziata A (2003) A review of in vitro methods to assess the biological activity of tobacco smoke with the aim of reducing the toxicity of smoke. Toxicol in Vitro 17:587–594

Balharry D, Sexton K, BeruBe KA (2008) An in vitro approach to assess the toxicity of inhaled tobacco smoke components: nicotine, cadmium, formaldehyde and urethane. Toxicology 244:66–76

Banchereau J, Steinman RM (1998) Dendritic cells and the control of immunity. Nature 392:245–252

Barkal LJ, Procknow CL, Alvarez-Garcia YR, Niu M, Jimenez-Torres JA, Brockman-Schneider RA, Gern JE, Denlinger LC, Theberge AB, Keller NP et al (2017) Microbial volatile communication in human organotypic lung models. Nat Commun 8:1770

Benam KH, Villenave R, Lucchesi C, Varone A, Hubeau C, Lee HH, Alves SE, Salmon M, Ferrante TC, Weaver JC et al (2016) Small airway-on-a-chip enables analysis of human lung inflammation and drug responses in vitro. Nat Methods 13:151–157

Birkness KA, Deslauriers M, Bartlett JH, White EH, King CH, Quinn FD (1999) An in vitro tissue culture bilayer model to examine early events in Mycobacterium tuberculosis infection. Infect Immun 67:653–658

Bissell MJ, Hall HG, Parry G (1982) How does the extracellular matrix direct gene expression? J Theor Biol 99:31–68

Black AF, Bouez C, Perrier E, Schlotmann K, Chapuis F, Damour O (2005) Optimization and characterization of an engineered human skin equivalent. Tissue Eng 11:723–733

Booth JL, Coggeshall KM, Gordon BE, Metcalf JP (2004) Adenovirus type 7 induces interleukin-8 in a lung slice model and requires activation of Erk. J Virol 78:4156–4164

Cai Y, Sugimoto C, Arainga M, Alvarez X, Didier ES, Kuroda MJ (2014) In vivo characterization of alveolar and interstitial lung macrophages in rhesus macaques: implications for understanding lung disease in humans. J Immunol 192:2821–2829

Carterson AJ, Honer zu Bentrup K, Ott CM, Clarke MS, Pierson DL, Vanderburg CR, Buchanan KL, Nickerson CA, Schurr MJ (2005) A549 lung epithelial cells grown as three-dimensional aggregates: alternative tissue culture model for *Pseudomonas aeruginosa* pathogenesis. Infect Immun 73, 1129–1140

Chandorkar P, Posch W, Zaderer V, Blatzer M, Steger M, Ammann CG, Binder U, Hermann M, Hortnagl P, Lass-Florl C, Wilflingseder D (2017) Fast-track development of an in vitro 3D lung/immune cell model to study Aspergillus infections. Sci Rep 7:11644

Chen YW, Huang SX, de Carvalho A, Ho SH, Islam MN, Volpi S, Notarangelo LD, Ciancanelli M, Casanova JL, Bhattacharya J et al (2017) A three-dimensional model of human lung development and disease from pluripotent stem cells. Nat Cell Biol 19:542–549

Choe MM, Sporn PH, Swartz MA (2006a) Extracellular matrix remodeling by dynamic strain in a three-dimensional tissue-engineered human airway wall model. Am J Respir Cell Mol Biol 35:306–313

Choe MM, Tomei AA, Swartz MA (2006b) Physiological 3D tissue model of the airway wall and mucosa. Nat Protoc 1:357–362

Chu HW, Rios C, Huang C, Wesolowska-Andersen A, Burchard EG, O'Connor BP, Fingerlin TE, Nichols D, Reynolds SD, Seibold MA (2015) CRISPR-Cas9-mediated gene knockout in primary human airway epithelial cells reveals a proinflammatory role for MUC18. Gene Ther 22:822–829

Clevers H (2016) Modeling development and disease with organoids. Cell 165:1586–1597

Condon TV, Sawyer RT, Fenton MJ, Riches DW (2011) Lung dendritic cells at the innate-adaptive immune interface. J Leukoc Biol 90:883–895

Cozens AL, Yezzi MJ, Kunzelmann K, Ohrui T, Chin L, Eng K, Finkbeiner WE, Widdicombe JH, Gruenert DC (1994) CFTR expression and chloride secretion in polarized immortal human bronchial epithelial cells. Am J Respir Cell Mol Biol 10:38–47

Cukierman E, Pankov R, Stevens DR, Yamada KM (2001) Taking cell-matrix adhesions to the third dimension. Science 294:1708–1712

Danahay H, Pessotti AD, Coote J, Montgomery BE, Xia D, Wilson A, Yang H, Wang Z, Bevan L, Thomas C et al (2015) Notch2 is required for inflammatory cytokine-driven goblet cell metaplasia in the lung. Cell Rep 10:239–252

Deslee G, Dury S, Perotin JM, Al Alam D, Vitry F, Boxio R, Gangloff SC, Guenounou M, Lebargy F, Belaaouaj A (2007) Bronchial epithelial spheroids: an alternative culture model to investigate epithelium inflammation-mediated COPD. Respir Res 8:86

Drewitz M, Helbling M, Fried N, Bieri M, Moritz W, Lichtenberg J, Kelm JM (2011) Towards automated production and drug sensitivity testing using scaffold-free spherical tumor microtissues. Biotechnol J 6:1488–1496

Fahy JV, Dickey BF (2010) Airway mucus function and dysfunction. N Engl J Med 363:2233–2247

Fatehullah A, Tan SH, Barker N (2016) Organoids as an in vitro model of human development and disease. Nat Cell Biol 18:246–254

Fessart D, Begueret H, Delom F (2013) Three-dimensional culture model to distinguish normal from malignant human bronchial epithelial cells. Eur Respir J 42:1345–1356

Gao X, Bali AS, Randell SH, Hogan BL (2015) GRHL2 coordinates regeneration of a polarized mucociliary epithelium from basal stem cells. J Cell Biol 211:669–682

Ghosh S, Spagnoli GC, Martin I, Ploegert S, Demougin P, Heberer M, Reschner A (2005) Three-dimensional culture of melanoma cells profoundly affects gene expression profile: a high density oligonucleotide array study. J Cell Physiol 204:522–531

Godiska R, Chantry D, Raport CJ, Sozzani S, Allavena P, Leviten D, Mantovani A, Gray PW (1997) Human macrophage-derived chemokine (MDC), a novel chemoattractant for monocytes, monocyte-derived dendritic cells, and natural killer cells. J Exp Med 185:1595–1604

Gonzalez RF, Allen L, Gonzales L, Ballard PL, Dobbs LG (2010) HTII-280, a biomarker specific to the apical plasma membrane of human lung alveolar type II cells. J Histochem Cytochem 58:891–901

Griffith LG, Swartz MA (2006) Capturing complex 3D tissue physiology in vitro. Nat Rev Mol Cell Biol 7:211–224

Guilliams M, Lambrecht BN, Hammad H (2013) Division of labor between lung dendritic cells and macrophages in the defense against pulmonary infections. Mucosal Immunol 6:464–473

Gunzer M, Friedl P, Niggemann B, Brocker EB, Kampgen E, Zanker KS (2000) Migration of dendritic cells within 3-D collagen lattices is dependent on tissue origin, state of maturation, and matrix structure and is maintained by proinflammatory cytokines. J Leukoc Biol 67:622–629

Guo L, Zhou Y, Wang S, Wu Y (2014) Epigenetic changes of mesenchymal stem cells in three-dimensional (3D) spheroids. J Cell Mol Med 18:2009–2019

Hammad H, Lambrecht BN (2008) Dendritic cells and epithelial cells: linking innate and adaptive immunity in asthma. Nat Rev Immunol 8:193–204

Harrington H, Cato P, Salazar F, Wilkinson M, Knox A, Haycock JW, Rose F, Aylott JW, Ghaemmaghami AM (2014) Immunocompetent 3D model of human upper airway for disease modeling and in vitro drug evaluation. Mol Pharm 11:2082–2019

Hegab AE, Arai D, Gao J, Kuroda A, Yasuda H, Ishii M, Naoki K, Soejima K, Betsuyaku T (2015) Mimicking the niche of lung epithelial stem cells and characterization of several effectors of their in vitro behavior. Stem Cell Res 15:109–121

Heijink IH, Brandenburg SM, Noordhoek JA, Postma DS, Slebos DJ, van Oosterhout AJ (2010) Characterisation of cell adhesion in airway epithelial cell types using electric cell-substrate impedance sensing. The Eur Respir J 35:894–903

Hieshima K, Imai T, Baba M, Shoudai K, Ishizuka K, Nakagawa T, Tsuruta J, Takeya M, Sakaki Y, Takatsuki K et al (1997) A novel human CC chemokine PARC that is most homologous to macrophage-inflammatory protein-1 alpha/LD78 alpha and chemotactic for T lymphocytes, but not for monocytes. J Immunol 159:1140–1149

Ho WJ, Pham EA, Kim JW, Ng CW, Kim JH, Kamei DT, Wu BM (2010) Incorporation of multicellular spheroids into 3-D polymeric scaffolds provides an improved tumor model for screening anticancer drugs. Cancer Sci 101:2637–2643

Holmes AM, Solari R, Holgate ST (2011) Animal models of asthma: value, limitations and opportunities for alternative approaches. Drug Discov Today 16:659–670

Htwe SS, Harrington H, Knox A, Rose F, Aylott J, Haycock JW, Ghaemmaghami AM (2015) Investigating NF-kappaB signaling in lung fibroblasts in 2D and 3D culture systems. Respir Res 16:144

Huh D, Leslie DC, Matthews BD, Fraser JP, Jurek S, Hamilton GA, Thorneloe KS, McAlexander MA, Ingber DE (2012) A human disease model of drug toxicity-induced pulmonary edema in a lung-on-a-chip microdevice. Sci Transl Med 4:159ra147

Huh D, Matthews BD, Mammoto A, Montoya-Zavala M, Hsin HY, Ingber DE (2010) Reconstituting organ-level lung functions on a chip. Science 328:1662–1668

Imai T, Yoshida T, Baba M, Nishimura M, Kakizaki M, Yoshie O (1996) Molecular cloning of a novel T cell-directed CC chemokine expressed in thymus by signal sequence trap using Epstein-Barr virus vector. J Biol Chem 271:21514–21521

Ishizumi T, McWilliams A, MacAulay C, Gazdar A, Lam S (2010) Natural history of bronchial preinvasive lesions. Cancer Metastasis Rev 29:5–14

Jager J, Marwitz S, Tiefenau J, Rasch J, Shevchuk O, Kugler C, Goldmann T, Steinert M (2014) Human lung tissue explants reveal novel interactions during Legionella pneumophila infections. Infect Immun 82:275–285

Kapoor N, Pawar S, Sirakova TD, Deb C, Warren WL, Kolattukudy PE (2013) Human granuloma in vitro model, for TB dormancy and resuscitation. PLoS ONE 8:e53657

Ketterer MR, Shao JQ, Hornick DB, Buscher B, Bandi VK, Apicella MA (1999) Infection of primary human bronchial epithelial cells by Haemophilus influenzae: macropinocytosis as a mechanism of airway epithelial cell entry. Infect Immun 67:4161–4170

Kim DH, Lee H, Lee YK, Nam JM, Levchenko A (2010) Biomimetic nanopatterns as enabling tools for analysis and control of live cells. Adv Mater 22:4551–4566

Kleinman HK, Philp D, Hoffman MP (2003) Role of the extracellular matrix in morphogenesis. Curr Opin Biotechnol 14:526–532

Kunz-Schughart LA, Freyer JP, Hofstaedter F, Ebner R (2004) The use of 3-D cultures for high-throughput screening: the multicellular spheroid model. J Biomol Screen 9:273–285

Lambrecht BN, Hammad H (2009) Biology of lung dendritic cells at the origin of asthma. Immunity 31:412–424

Lin H, Li H, Cho HJ, Bian S, Roh HJ, Lee MK, Kim JS, Chung SJ, Shim CK, Kim DD (2007) Air-liquid interface (ALI) culture of human bronchial epithelial cell monolayers as an in vitro model for airway drug transport studies. J Pharm Sci 96:341–350

Loffler B, Hussain M, Grundmeier M, Bruck M, Holzinger D, Varga G, Roth J, Kahl BC, Proctor RA, Peters G (2010) *Staphylococcus aureus* panton-valentine leukocidin is a very potent cytotoxic factor for human neutrophils. PLoS pathogens 6:e1000715

Loh QL, Choong C (2013) Three-dimensional scaffolds for tissue engineering applications: role of porosity and pore size. Tissue Eng Part B Rev 19:485–502

Mairpady Shambat S, Chen P, Nguyen Hoang AT, Bergsten H, Vandenesch F, Siemens N, Lina G, Monk IR, Foster TJ, Arakere G et al (2015) Modelling staphylococcal pneumonia in a human 3D lung tissue model system delineates toxin-mediated pathology. Dis Model Mech 8:1413–1425

Mishra DK, Sakamoto JH, Thrall MJ, Baird BN, Blackmon SH, Ferrari M, Kurie JM, Kim MP (2012) Human lung cancer cells grown in an ex vivo 3D lung model produce matrix metalloproteinases not produced in 2D culture. PLoS ONE 7:e45308

Nam KH, Smith AS, Lone S, Kwon S, Kim DH (2015) Biomimetic 3D tissue models for advanced high-throughput drug screening. J Lab Autom 20:201–215

Nguyen Hoang AT, Chen P, Bjornfot S, Hogstrand K, Lock JG, Grandien A, Coles M, Svensson M (2014) Technical Advance: live-imaging analysis of human dendritic cell migrating behavior under the influence of immune-stimulating reagents in an organotypic model of lung. J Leukoc Biol 96:481–489

Nguyen Hoang AT, Chen P, Juaréz J, Sachamitr P, Billing B, Bosnjak L, Dahlén B, Coles M, Svensson M (2012) Dendritic cell functional properties in a three-dimensional tissue model of human lung mucosa. Am J Physiol—Lung Cell Mol Physiol 302:226–237

Nicholas B, Staples KJ, Moese S, Meldrum E, Ward J, Dennison P, Havelock T, Hinks TS, Amer K, Woo E et al (2015) A novel lung explant model for the ex vivo study of efficacy and mechanisms of anti-influenza drugs. J Immunol 194:6144–6154

Pampaloni F, Reynaud EG, Stelzer EH (2007) The third dimension bridges the gap between cell culture and live tissue. Nat Rev Mol Cell Biol 8:839–845

Parasa VR, Rahman MJ, Ngyuen Hoang AT, Svensson M, Brighenti S, Lerm M (2014) Modeling *Mycobacterium tuberculosis* early granuloma formation in experimental human lung tissue. Dis Model Mech 7:281–288

Pezzulo AA, Starner TD, Scheetz TE, Traver GL, Tilley AE, Harvey BG, Crystal RG, McCray PB Jr, Zabner J (2011) The air-liquid interface and use of primary cell cultures are important to recapitulate the transcriptional profile of in vivo airway epithelia. Am J Physiol Lung Cell Mol Physiol 300:L25–L31

Pluddemann A, Mukhopadhyay S, Gordon S (2011) Innate immunity to intracellular pathogens: macrophage receptors and responses to microbial entry. Immunol Rev 240:11–24

Puissegur MP, Botanch C, Duteyrat JL, Delsol G, Caratero C, Altare F (2004) An in vitro dual model of mycobacterial granulomas to investigate the molecular interactions between mycobacteria and human host cells. Cell Microbiol 6:423–433

Roskelley CD, Desprez PY, Bissell MJ (1994) Extracellular matrix-dependent tissue-specific gene expression in mammary epithelial cells requires both physical and biochemical signal transduction. Proc Natl Acad Sci USA 91:12378–12382
Schutyser E, Richmond A, Van Damme J (2005) Involvement of CC chemokine ligand 18 (CCL18) in normal and pathological processes. J Leukoc Biol 78:14–26
Sharma M, Anderson SA, Schoop R, Hudson JB (2009) Induction of multiple pro-inflammatory cytokines by respiratory viruses and reversal by standardized Echinacea, a potent antiviral herbal extract. Antiviral Res 83:165–170
Shaykhiev R, Zuo WL, Chao I, Fukui T, Witover B, Brekman A, Crystal RG (2013) EGF shifts human airway basal cell fate toward a smoking-associated airway epithelial phenotype. Proc Natl Acad Sci U S A 110:12102–12107
Sundstrom KB, Nguyen Hoang AT, Gupta S, Ahlm C, Svensson M, Klingstrom J (2016) Andes hantavirus-infection of a 3D human lung tissue model reveals a late peak in progeny virus production followed by increased levels of proinflammatory cytokines and VEGF-A. PLoS ONE 11:e0149354
Svensson M, Maroof A, Ato M, Kaye PM (2004) Stromal cells direct local differentiation of regulatory dendritic cells. Immunity 21:805–816
Switalla S, Lauenstein L, Prenzler F, Knothe S, Forster C, Fieguth HG, Pfennig O, Schaumann F, Martin C, Guzman CA et al (2010) Natural innate cytokine response to immunomodulators and adjuvants in human precision-cut lung slices. Toxicol Appl Pharmacol 246:107–115
Tomasek JJ, Gabbiani G, Hinz B, Chaponnier C, Brown RA (2002) Myofibroblasts and mechano-regulation of connective tissue remodelling. Nat Rev Mol Cell Biol 3:349–363
Tralau T, Luch A (2012) Drug-mediated toxicity: illuminating the 'bad' in the test tube by means of cellular assays? Trends Pharmacol Sci 33:353–364
Ulrich M, Berger J, Moller JG, Doring G (2005) Moxifloxacin and ciprofloxacin protect human respiratory epithelial cells against Streptococcus pneumoniae, Staphylococcus aureus, Pseudomonas aeruginosa, and Haemophilus influenzae in vitro. Infection 33(Suppl 2):50–54
Ulrich M, Doring G (2004) Three-dimensional human airway epithelial cell cultures. J Cyst Fibros 3(Suppl 2):55–57
White ES (2015) Lung extracellular matrix and fibroblast function. Ann Am Thorac Soc 12(Suppl 1): S30–S33
Worlitzsch D, Tarran R, Ulrich M, Schwab U, Cekici A, Meyer KC, Birrer P, Bellon G, Berger J, Weiss T et al (2002) Effects of reduced mucus oxygen concentration in airway Pseudomonas infections of cystic fibrosis patients. J Clin Invest 109:317–325
Wrenshall L (2003) Role of the microenvironment in immune responses to transplantation. Springer Semin Immunopathol 25:199–213
Xu Z, Gao Y, Hao Y, Li E, Wang Y, Zhang J, Wang W, Gao Z, Wang Q (2013) Application of a microfluidic chip-based 3D co-culture to test drug sensitivity for individualized treatment of lung cancer. Biomaterials 34:4109–4117
Yamamoto Y, Gotoh S, Korogi Y, Seki M, Konishi S, Ikeo S, Sone N, Nagasaki T, Matsumoto H, Muro S et al (2017) Long-term expansion of alveolar stem cells derived from human iPS cells in organoids. Nat Methods 14:1097–1106

Human Gastrointestinal Organoid Models for Studying Microbial Disease and Cancer

Jayati Chakrabarti, Martha B. Dua-Awereh, Loryn Holokai, and Yana Zavros

Contents

Jayati Chakrabarti, Martha B. Dua-Awereh—Equal contribution.

J. Chakrabarti · Y. Zavros (✉)
Department of Cellular and Molecular Medicine, University of Arizona College of Medicine-Tucson, 1333 N. Martin Avenue, LSN 412, Tucson AZ 85719-0000, AZ, USA
e-mail: yzavros@arizona.edu

J. Chakrabarti
e-mail: chakraj@email.arizona.edu

M. B. Dua-Awereh
Department of Pharmacology and Systems Physiology, University of Cincinnati College of Medicine, 231 Albert Sabin Way, MSB 4202, Cincinnati OH 45267-0000, OH, USA
e-mail: duaawemb@mail.uc.edu

L. Holokai
Merck Pharmaceuticals, San Francisco, CA, USA

Current Topics in Microbiology and Immunology (2021) 430: 55–75
https://doi.org/10.1007/82_2020_223

Published Online: 6 September 2020

Abstract One of the major discoveries in stem cell research in the past decade embraces the development of "organs in a dish," also known as "organoids." Organoids are three-dimensional cellular structures derived from primary stem cells of different organ-specific cell types which are capable of self-renewal and maintenance of the parental lineages. Researchers have developed in vitro organoid models to mimic in vivo host–microbial interactions and disease. In this review, we focus on the use of gastrointestinal organoids as models of microbial disease and cancer.

Abbreviation

3D	Three dimensional
iPSC	Induced pluripotent stem cell
ESC	Embryonic stem cell
GI	Gastrointestinal
Lgr5	Leucine-rich repeat-containing G-protein-coupled receptor 5
HuNoV	Human norovirus
ZIKV	Zika virus
SPEM	Spasmolytic polypeptide expressing metaplasia
PBMCs	Peripheral blood mononuclear cells
DCs	Dendritic cells
TCL	Tumor cell lysate
CD	Cluster of differentiation
MHC	Major histocompatibility complex
IEL	Intraepithelial lymphocytes
CTLA4	Cytotoxic T lymphocyte-associated protein 4
PD-1	Program death 1
PD-L1	Program death ligand 1
RCC	Renal cell carcinoma
ATB	Ampicillin, colistin, and streptomycin
TME	Tumor microenvironment
CagA	Cytotoxin-associated gene A
TFF2	Trefoil factor 2
PMN-MDSCs	Polymorphonuclear myeloid-derived suppressor cells
Slfn	Schlafen
VEGF	Vascular endothelial growth factor
TNFα	Tumor necrosis factor α
IL-1β	Interleukin 1β
Treg	Regulatory T cell
IL-2	Interleukin 2
ALI	Air–liquid interface
Th1/2	T helper cell ½

1 Introduction

"Organotypic" is a term that has often been used to describe tissue that is explanted from an organ that continues to function as it would in vivo (Randall et al. 2011; Strangeways and Canti 1926; Dedhia et al. 2016). The term "organotypic" has been changed with recent advances of three-dimensional cultures. We typically use the term "organotypic" or "organoid" interchangeably to define long-term in vitro culture systems that recapitulate the complex cellular diversity and function of the native organ. Such long-term organoid cultures include tissue or epithelial-derived organoids (Barker et al. 2010; DeWard et al. 2014; Greggio et al. 2013; Huch et al. 2013; Jung et al. 2011; Kalabis et al. 2012; Sato et al. 2009; Stange et al. 2013; Sato et al. 2011a, b; Bertaux-Skeirik et al. 2017; Engevik et al. 2016; Steele et al. 2019) and pluripotent stem cell (PSC)-derived organoids (McCracken et al. 2014; Spence et al. 2011) (Table 1). The current review will focus on the use of gastrointestinal organoids for studies of microbial–host interactions and cancer.

The robust and diverse nature of gastrointestinal (GI) organoids has made this organoid culture system an important tool for studies of organ development, physiology, and preclinical models for personalized medicine. Long-term organoid cultures can be derived from pluripotent stem cells (PSCs) or adult primary tissue-resident cells (stem or differentiated cells) (reviewed in (Huch and Koo 2015; Dedhia et al. 2016). Table 1 summarizes the major differences in cellular complexity, morphology, applications, and limitations of use of induced pluripotent

Table 1 Complexity, morphology and applications of organotype (organoid) cultures

	Organotype Culture	
	iPSC-Derived	**Tissue-Derived**
Characteristics/Complexity	• Epithelium • Mesenchyme/Stroma	• Epithelium • No Mesenchyme
Morphology	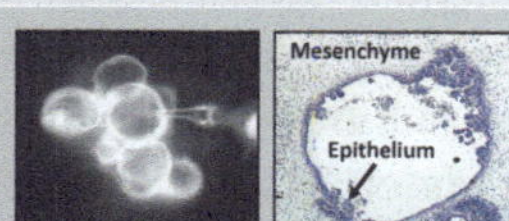	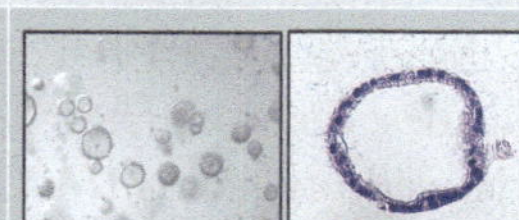
Advantages/Applications	• Studies of mesenchymal-epithelial interactions • Developmental biology • Host-pathogen interactions • Physiology	• Rapid/robust generation of cultures • Physiology • Host-pathogen interactions • Preclinical models for cancer, patient response, targeted therapies
Limitations	• Immature phenotype of differentiated cells	• Loss of resident immune cells with passage • Loss of differentiated cell types with passage

stem cell (iPSC)-derived and tissue-derived gastric organoids. It is important to consider the strengths and limitations of each culture when designing experiments. For example, tissue-derived epithelial organoids can be rapidly generated in less than 2 weeks, whereas PSC-derived organoids require up to 4–5 weeks to establish. Both cultures have been used for in vitro studies of organ physiology and host–pathogen interactions, whereas PSC-derived organoids are more suitable for elucidating mechanisms of organ development.

The ability to grow organotypic/organoid culture long-term in vitro was made possible by the development of experimental conditions that allow for the support and growth of stem cells (Ootani et al. 2009; Sato et al. 2009). The long-term growth of tissue-derived organoids requires media containing growth factors that stimulate Wnt signaling and support the growth and differentiation of the stem cell niche. In 2009, Sato et al. were the first group to isolate intestinal crypts from mice that gave rise to organoids comprising epithelium when cultured with laminin-rich Matrigel (BD Biosciences, San Jose, CA) and LGR5-positive stem cell-supporting growth factors (Sato et al. 2009). These organoid cultures were the first to reflect the intestinal niche in vivo (Sato et al. 2009). Following this technological advancement, Ootani et al. then developed an alternate strategy whereby epithelial–mesenchymal organoids were cultured using minced small intestinal or colonic fragments in a collagen matrix exposed to an air–liquid interface (Ootani et al. 2009). These tissue fragments gave rise to highly proliferative epithelium, with supportive myofibroblasts located along the basolateral membrane that could be maintained in culture in vitro for up to 1 year (Ootani et al. 2009). Since the development of the intestinal organoid system, tissues including esophagus, stomach, liver, pancrea, and colon have been used for the development of similar cultures (Barker et al. 2010; DeWard et al. 2014; Jung et al. 2011; Sato et al. 2011a, b; Boj et al. 2015; Engevik et al. 2016; Bertaux-Skeirik et al. 2017; Schumacher et al. 2012, 2015; Bartfeld et al. 2015; Tiriac et al. 2018). A similar protocol was adapted for the generation of human gastric tissue-derived organoids from both normal stomach and cancer epithelium (Bertaux-Skeirik et al. 2017; Engevik et al. 2016; Steele et al. 2019). Collectively, these achievements laid the groundwork for studies describing differentiation and growth of human iPSC and embryonic stem cell-derived organoids (McCracken et al. 2014; Spence et al. 2011; Forster et al. 2014; Hannan et al. 2013; McCracken et al. 2011).

Organoids generated from embryonic stem cells (ESCs) or induced pluripotent stem cells (iPSCs) are cultured through directed differentiation using tissue-specific growth factors and media that replicate the developmental process. In the recent era, human PSC-derived organoids have been developed which have established a new human model culture system with similarities in morphology, growth, cell lineages, and gene expression (reviewed in (Dedhia et al. 2016; McCauley and Wells 2017). In 1998, Thomason et al. reported the first generation of embryonic stem cells (ESCs) that were derived from human blastocysts and capable of rendering three germ layers (Thomson et al. 1998). To this day, these cultures remain in laboratory use by many investigators. In 2007, Takahashi et al. described the induction of pluripotent stem cells (PSCs), also referred to as induced PSCs (iPSCs) (Takahashi et al. 2007). Induced PSCs were derived from adult human fibroblasts via the transduction of four defined transcription factors: OCT3/4 (POU5F1), SOX2, KLF4, and c-MYC (Takahashi et al. 2007). The generation of iPSCs has allowed researchers to not only establish controversy-free PSC lines, but advanced personalized medicine. Patient-specific iPSCs can be engineered to express specific and unique mutations using CRISPR/Cas9 editing and thus model diseases in vitro (Hockemeyer and Jaenisch 2016).

Similar to primary human fetal tissues, human PSC-derived cultures are well suited for developmental studies due to their fetal resemblance. For example, based on the expression of regulators of liver development, hepatocyte organoids generated from PSCs closely resemble early fetal liver rather than adult liver (Takebe et al. 2013; Si-Tayeb et al. 2010; Shan et al. 2013). Similarly, the early iterations of PSC-derived insulin-producing cells represented fetal pancreatic beta cells based on their transcriptional profile and lack of glucose response (Hrvatin et al. 2014). The transcriptome of PSC-derived lung and small intestinal organoids closely resembles that of fetal compared with adult tissues (Dye et al. 2015; Finkbeiner et al. 2015; Watson et al. 2014). Interestingly, the transplantation of PSC-derived intestinal organoids into immunocompromised mice resulted in the maturation of fetal-like tissues into tissues resembling those of adults, as characterized by prominent crypt-villus structures, increased expression of *OLFM4*, brush border enzymes, and markers of mature small intestine including dipeptidyl peptidase 4, glucose transporter type 2, and villin (Finkbeiner et al. 2015; Watson et al. 2014). In contrast, iPSC derived of gastric organoids responds to *H. pylori* infection and physiological stimuli such as histamine and gastric similar to what is seen in the mature cell lineages of the adult stomach (McCracken et al. 2014, 2017; Holokai et al. 2019).

2 Organoids for the Study of Host–Pathogen Interactions

Some microbes invariably result in GI disease, whereas others only cause disease in susceptible individuals. Studies using GI organoids might increase our understanding of susceptibility to and pathogenesis of microbial infections and identify possible therapies. Gastric organoids have been instrumental in these studies because they recapitulate in vivo physiology, respond to infections similar to in vivo tissue, and recapitulate mechanisms of malignant transformation secondary to infection. Organotypic models provide scientific opportunities to not only investigate diseases in novel ways but to investigate diseases in a more significant and complex way. The use of organoids to study host–pathogen interactions has contributed to the field of viral and bacterial pathogenesis significantly. Although cell culture has been useful in evaluating pathogen responses, factors that contribute to viral or bacterial pathogenesis in vivo may be lost in the two-dimensional culture systems.

2.1 Rotavirus, Norovirus, and Zika Virus

One limitation in studying viral pathogenesis is using models that will physiologically replicate the infection process. The standard culture model for studying human rotavirus is an immortalized cell line infected with only one of the well-characterized rotavirus strains. Rotavirus causes diarrheal disease in children throughout the world. Studying a virus with over 110.000 strains has been a scientific challenge. However, the use of organotypic culture models has provided insight into the complex rotavirus host–pathogen interactions. Enteroids are organoids derived from intestinal tissue. The use of enteroids enabled researchers to develop ex vivo structural and functional studies of the human intestine (Ettayebi et al. 2016; Saxena et al. 2016; Sato et al. 2019). Rotavirus-infected enteroids have been used to study the mechanism of rotavirus-induced diarrhea by evaluating organoid water influxes (Saxena et al. 2016) and have shown varying sensitivities to interferon therapies. Such discoveries have introduced the possibility of personalized medicinal treatments in rotaviral research.

Another virus that has been better understood using the organoid culture system is norovirus. The human norovirus (HuNoV) is an intestinal virus that causes severe gastrointestinal disturbances and disease in humans worldwide and is the most common cause of gastroenteritis. Unlike rotavirus, the greatest limitation of studying HuNoV has been the lack of a successful culture model system. With the novel development of cultured HuNoV in iPSC-derived intestinal cells, research is now conducted with HuNoV cultures in enteroids (Sato et al. 2019; Ettayebi et al. 2016).

Aside from developing culture systems to accurately replicate the host response to virus, another challenge of studying viral pathogenesis is predicting which viruses or viral strains will have high reproductive rates and potentially cause an epidemic. The Zika virus, also known as ZIKV, is a mosquito flavivirus that can be transmitted to humans. As the incidence of infection spread in 2016, ZIKV was declared a global health emergency by the World Health Organization. Infection by ZIKV causes mild symptoms in adults but causes fetal brain defects resulting in microencephaly in newborn babies. As epidemics such as the Zika virus occurs, organoid models can be used to design treatments for these diseases. Unlike other cell culture models, brain organoids form fetal brain architecture in vitro (Qian et al. 2017). Human embryonic stem cell-derived organoids demonstrating cerebral cortex morphology were used to investigate fetal brain development impairment during ZIKV infection (Dang et al. 2016, 2019). The impact of ZIKV on stem cell growth and neurogenesis was also investigated in human iPSC-derived brain organoid (Garcez et al. 2016). Adult stem cell-derived and primary tissue-derived organoids can be used to reconstitute complex organ systems such as the brain, study the interactions between the brain and the pathogen, and serve as a basis for regenerative therapies to alleviate and treat the symptoms observed in affected patients.

2.2 *Helicobacter Pylori*

H. pylori infects nearly 50% of the world's population and is the number one risk factor for gastric cancer (Marshall and Warren 1984). Although *H. pylori* infection that is treated with antibiotics is cleared, once a patient has progressed to a metaplastic phenotype, elimination of the bacteria does not reduce the risk of developing gastric cancer (Chen et al. 2016). *H. pylori*-induced pathogenesis is largely mediated by virulence factor cytotoxic-associated gene A (CagA) (Odenbreit et al. 2000). The tropic specificity of *H. pylori* to gastric tissue makes studying the pathophysiology of *H. pylori*-induced diseases both difficult and interesting. Recent advances in organoid technology have demonstrated the successful effort to establish a patient-derived organoid model of *H. pylori* pathogenesis and gastric cancer in context with the patient's immune cells (Holokai et al. 2019; Chakrabarti et al. 2018a, b). In the stomach, histamine stimulates acid secretion from the parietal cells. Using gastric epithelial-derived organoids co-cultured with immortalized stomach mesenchymal cells, Schumacher et al. demonstrated acid secretion in response to histamine; that was reversed by the proton pump inhibitor omeprazole (Schumacher et al. 2015). Given that *H. pylori* causes most cases of peptic ulcer disease by inducing hypersecretion of acid and has been associated with gastric cancer, such an organoid culture system becomes physiologically relevant when studying gastric disease in response to bacterial infection. Similarly, *H. pylori* induces proliferation in PSC-derived gastric organoids but not with mutants expressing a deletion of its virulence factor CagA (Holokai et al. 2019; McCracken et al. 2014).

Using organoid cultures, our laboratory has demonstrated that CagA stimulates Sonic Hedgehog (Shh) signaling within parietal cells, a response that is mediated by NFκB signaling (Schumacher et al. 2012, 2015). We have also shown that the gastric morphogen, Shh, is fundamental to the initiation of gastritis in response to *H. pylori* infection (Schumacher et al. 2012). Upon infection, *H. pylori* induces the secretion of Shh from the acid-secreting parietal cells (Schumacher et al. 2012). Using iPSC- and tissue-derived human gastric organoid cultures, our group has reported that *H. pylori*-induced immune checkpoint molecule programmed cell death 1 ligand (PD-L1) and is mediated by Shh signaling (Holokai et al. 2019). In these studies, we localized PD-L1 specifically to spasmolytic polypeptide (TFF2)-expressing metaplastic (SPEM) cells as a mechanism by which these cells may survive chronic inflammation, for the persistence of infection and progression of the disease to cancer.

Following a sustained increase in Shh secretion and signaling, macrophages and polymorphonuclear (or granulocytic)-myeloid-derived suppressor cells (PMN-MDSCs) are recruited to the infection site (Schumacher et al. 2012; Ding et al. 2016, 2020). Recent characterization of the infiltrating PMN-MDSCs in response to *Helicobacter* infection by the Merchant Laboratory showed that these immunosuppressive cells express the Schlafen (SLFN) family of proteins, a transcriptional target of GLI1, and the endogenous small, non-coding RNA, MiR130b

(Ding et al. 2016, 2020). SLFN4$^+$ myeloid cells express VEGF, IL-1β, and TNF-α that are known factors associated with MDSC regulation, an immunosuppressive cell of relevance to immunotherapy-resistant gastric cancer (Ding et al. 2016, 2020). Multiple suppressive immune cell types including macrophages, myeloid-derived suppressor cells (MDSCs), and regulatory T cells (Tregs) accumulate in gastric cancer (Wang et al. 2013). In humans and mice, two distinct subsets of MDSCs exist, monocytic-MDSCs (M-MDSCs) and polymorphonuclear (or granulocytic)-MDSCs (PMN-MDSCs) (Schouppe et al. 2013; Youn et al. 2008). Of relevance to this review, PMN-MDSCs are known to block CD8 + T cell anti-tumor activity through L-arginine and L-cysteine sequestration as well as production of reactive oxygen species (ROS) (Zhang et al. 2017; Schouppe et al. 2013; Youn et al. 2008). Collectively, these studies have shown that *H. pylori* infection can be modeled in gastric organoids to study mechanisms of gastric disease in response to infection. These studies also emphasize the importance of the study of viral and bacterial infections in the context of the host's immune response. Thus, investigators strive to further develop and refine the in vitro organoid culture system to more closely recapitulate the native in vivo GI environment.

3 Organoid–Immune Cell Co-cultures

Lack of immune cells is a compounding factor in the 3D organoid culture in studies of the tumor microenvironment (TME). With the advancement of immunotherapy to treat different cancer patients, it is essential to investigate the interaction of immune cells with cancer cells in vitro. Organoid-based patient profiling efforts have been developed, including co-culture models. For example, pancreatic cancer organoids have been co-cultured with immune cells outside of the Matrigel Dome (Tsai et al. 2018). A limitation of such a model is that a non-autologous system with immune cells placed outside of the Matrigel Dome does not allow the immune cells to come in contact with the target cancer organoids. Immune/tumor cell adhesion would be important in a system whereby we are studying the interaction between PD-L1 and PD-1. Often anti-CD28-coated plates are used to activate the T cells in the presence of IL-2 to maintain T cell proliferation. These methods have been used in autologous non-small cell lung and colorectal cancer organoid/peripheral blood lymphocyte co-cultures (Dijkstra et al. 2018). In addition, the air–liquid interface (ALI) method has been used to propagate patient-derived organoids (Neal et al. 2018). Examples of organoid–immune cell co-cultures for the study of disease are detailed as follows.

3.1 *Organoid/Macrophage Co-cultures*

Human intestinal-derived organoid immune cell co-cultures were established to study the infection with pathogens. Human enteroid monolayers were derived from duodenum or jejunum tissue from endoscopic biopsies or intestinal surgeries and differentiated by the administration of Wnt3A (Noel et al. 2017). The differentiated and non-differentiated monolayers expressed characteristics of "villi" or "crypts," respectively, and also expressed apical/basolateral polarization. Non-autologous monocyte-derived macrophages were differentiated from peripheral blood mononuclear cells (PBMCs), extracted from patient whole blood. When macrophages were added to the basolateral side of the monolayer plated on transwells, they were able to detect the presence of *E. coli,* added to the apical layer of the co-culture (Noel et al. 2017). Dendrites from the macrophages projected through the monolayer lead to decreased viable bacteria 16 h after culture (Noel et al. 2017). The human-derived enteroid monolayer-macrophage co-culture demonstrates the importance of immune cell–intestinal epithelial cell interaction by secreting important intestinal cytokines, intestinal barrier integrity, and recognition of bacteria (Noel et al. 2017). Thus, refinement of the organoid–immune cell co-cultures could potentially be used to investigate bacterial pathogenesis and subsequent treatment modalities.

3.2 *Organoid Co-culture with Antigen-Presenting Dendritic Cells (DCs)*

Loss of antigen presentation is a major immune escape in cancer; hence, it induces loss of effector T cell function. DCs are professional antigen-presenting cells that play an important role in the generation and sustainability of anti-tumor immune responses. The current immunotherapeutic approaches seek to improve the antigen-presenting ability of DCs and to effectively activate T lymphocytes (Sehgal et al. 2014). Whole tumor cell lysates (TCLs) have been used as the source of tumor antigens for the development of DC vaccines. The first DC vaccine was used for the treatment of prostate cancer (Thara et al. 2011), which was a milestone in the immunotherapeutic application, though the clinical outcomes have been unsatisfactory (Cohn and Delamarre 2014). To improve the efficacy of TCL-based DC vaccines, viruses have been introduced to enhance the immunity of TCLs, to further boost antigen delivery to increase the antigen-presenting ability of DCs (Guillerme et al. 2013). CD4 + and CD8 + T cells become activated and secrete major cytokines, like IFN-γ and IL-2 when they recognize activated DCs, expressing a high level of HLA-DR, and a moderate level of CD1a, CD80, and CD86 expression (Millán et al. 2013; Zhao et al. 2018). Restoring antigen presentation ability of DCs in the tumor microenvironment induces generation of anti-tumor T cells, which may sensitize patients to T cell-based immunotherapies, including checkpoint inhibitors (de Charette et al. 2016). Chakrabarti et al. established a unique autologous murine gastric organoid–immune cell co-culture in a Matrigel Dome, where they showed that tumor-associated antigens efficiently activated antigen-presenting dendritic cells, which then were able to activate naïve T lymphocytes (Chakrabarti et al. 2018b). Under ideal circumstances, immunogenic peptides derived from tumor-associated antigens presented on the surface of tumor cells with MHC-I activate and expand cytotoxic T cells and subsequently result in the lysis of tumor cells (Matsueda and Graham 2014). The development of patient-derived co-culture models may help to establish a promising tool for studies of tumor–immune interaction in varied cancers and could improve understanding of this complex mechanism.

3.3 Organoids Co-cultured with Lymphocytes

Human mammary ductal epithelial organoids derived from non-cancerous breast tissue contained significantly more T lymphocytes than the peripheral blood. The majority of them are CD8 + effector memory T cells (Zumwalde et al. 2016). The human mammary ductal epithelial organoid immune cell co-culture system has the potential to screen T cell reactivity to different types of drugs. T cells were eventually extracted from the organoids and shown to secrete interferon gamma in response to the classical triple-negative breast cancer cell line MDA-MB-468 (Zumwalde et al. 2016). Future studies may prove beneficial to breast cancer immunotherapies.

Mouse studies further investigated organoid–immune cell interactions by deriving intestinal organoids from mouse small intestines as well as extracting intestinal intraepithelial lymphocytes (IEL). Mouse IELs that were extracted from mouse intestines and co-cultured with mouse intestinal organoids survived longer in culture than IELs that were natively present in mouse intestinal organoids. Further, the addition of T cell survival cytokines such as IL-2, IL-7, and IL-15 aided in long-term IEL survival. IELs embedded in Matrigel together with intestinal organoids were shown to migrate toward the organoids and physically interact in culture (Nozaki et al. 2016). This concept of organoid–immune cell co-cultures can be used to study T cell activation, homing, and inhibition.

Mouse organoid–immune cell co-cultures also used immune cells extracted from the spleen to look at T cell activation, maturation, and gut homing (Rogoz et al. 2015). Rogoz et al. used transgenic mice to investigate enteroid-induced T cell activation. CD8 + T cells extracted from mice with a transgenic T cell receptor specific for ovalbumin (Rogoz et al. 2015). These T cells were either cultured with enteroids derived from wildtype C57BL/6 mice or iFABP-tOVA mice that express a truncated, cytosolic OVA under the intestinal fatty acid-binding protein promoter. Interestingly, naïve CD8 + T cells co-cultured with intestinal organoids showed increased expression of gut homing integrins and receptors. Further, these T cells were able to incorporate into the enteroids and move between cells (Rogoz et al.

2015). A mouse autologous organoid–immune cell co-culture was developed to investigate possible therapeutic approaches for gastric cancer using immune checkpoint inhibitors. Immune checkpoint inhibition resulted in organoid death as well as T cell proliferation. This mouse-derived organoid–immune cell co-culture may be a promising preclinical model beneficial to testing immunotherapy for the individual patient (Chakrabarti et al. 2018b).

Thus, the establishment of a tumor-specific microenvironment in vitro that incorporates the host's immune cells may be beneficial for the development of immunotherapeutic approaches for the treatment of disease.

4 The Microbiome and Cancer Development

4.1 Microbial Contribution to Disease Development

It has been shown that the bacteria that colonize the gut directly may influence the system's immune response to different types of cancers. For example, melanoma, non-small cell lung cancer, renal cell carcinoma, and urothelial carcinoma patients that do not respond to immune checkpoint therapies have been shown to be colonized with distinctly different genera of bacteria compared to responsive patients (Routy et al. 2018). Specifically, it was discovered that microbiome diversity, as well as specific genera of bacteria, is crucial to response to anti-PD-1 therapy (Routy et al. 2018). Specific pathogen-free mice bearing either MCA-205 sarcoma or RET melanoma tumors that were treated with ampicillin, colistin, and streptomycin (ATB) did not respond well to anti-PD-1 or anti-CTLA4 therapies (Routy et al. 2018). Patients with non-small cell lung cancer (NSCLC), renal cell carcinoma (RCC), or urothelial carcinoma were treated with anti-PD-1 therapy, and a subset of those patients was also treated ATB (Routy et al. 2018). Patients that received ATB treatment had significantly shorter overall or progression-free survival, in fact, ATB was a predictor of anti-PD-1 therapy resistance in NSCLC and RCC patients (Routy et al. 2018). The bacterial genera that were enriched in the NSCLC and RCC responders were *A. muciniphila,* excluding patients treated with ATB (Routy et al. 2018). Here *A. muciniphila* was associated with a Th1 response in NSCLC and RCC, whereas colonization of NSCLC patients with *E.hirae* colonization had an increased IFN-γ, IL-10, and IL-22 indicating memory T cell response (Routy et al. 2018). These results were confirmed when specific pathogen-free mice bearing RET melanoma tumors or Lewis Llzung carcinoma was treated with ATB and the microbiome was recolonized with *A.muciniphila* and *E. hirae* (Routy et al. 2018). Colonization with *A.muciniphila* and *E.hirae* leads to sensitivity to anti-PD-1 therapy, which was previously inhibited by ATB treatment (Routy et al. 2018). *A.muciniphila* and *E.hirae* recolonization following fecal transplantation from non-responders reinstated anti-PD-1 therapy efficacy (Routy et al. 2018). *Alistipes indistinctus* was able to reinstate anti-PD-1 efficacy in mice transplanted with feces from NSCLC patients that did not respond to anti-PD-1 therapy (Routy et al. 2018). These studies illustrate the importance of the microbiome in cancer progression and response to immune checkpoint inhibitors.

Another foundational study of melanoma revealed that upon fecal transplantation of the feces from Jackson Laboratories mice into Taconic farms mice, tumor volume decreased and CD8 + T cell infiltration increased in mice from Taconic Farms that were subcutaneously injected with B16.SIY melanoma (Sivan et al. 2015). After analyzing the bacterial components of both Jackson Laboratories and Taconic Farm's mouse feces, it was determined that *Bifidobacterium* was significantly associated with a higher CD8 + T cell infiltration into the tumor microenvironment (Sivan et al. 2015). Oral gavage of mice with different bacteria from the *Bifidobacterium* genera leads to improved tumor control and CD8 + T cell infiltration (Sivan et al. 2015). This was due to an increased dendritic cell activation of CD8 + T cells due to an upregulation of MHC class II expression (Sivan et al. 2015). Due to the improvement of CD8 + T cell infiltration, treatment with anti-PD-L1 therapy was more successful in mice with *Bifidobacterium* colonization (Sivan et al. 2015).

Microbiome colonization has been shown to directly impact CD8 + T cell or Th1 cell tumor infiltration, which has a direct effect on the immune response to cancer progression. The inconstancies in bacterial genera between different types of cancers require further study. It may be that different genera of bacteria are associated with different T cell responses. For instance, *Bifidobacteriaceae, Clostridiales,* and *E. hirae* are associated with as CD8 + T cell response; however, *A. muciniphila* is associated with a Th1 response (Sivan et al. 2015; Matson et al. 2018; Routy et al. 2018). Future studies are necessary to determine the mechanisms by which the gut influences cancers in distant organs.

The microbiome composition of human melanoma patients was analyzed (Matson et al. 2018; Gopalakrishnan et al. 2018). The *Bifidobacteriaceae* family of bacteria was enriched in the microbiome of melanoma patients that responded to anti-PD-1 therapy (Matson et al. 2018). This was consistent with the findings of the above mouse studies. Melanoma growth was slower in germ-free mice transplanted with the feces from melanoma patients that responded to anti-PD-1 therapy (Matson et al. 2018). There was an increased frequency of CD8 + T cells in the tumor microenvironment of germ-free mice transplanted with feces from patients that responded to anti-PD-1 therapy (Matson et al. 2018).

Another study of melanoma patient had responded to anti-PD-1 and anti-CTLA4 therapies were colonized with the *Clostridiales* genera of bacteria, whereas the non-responders were colonized with the *Bacteroidales* genera (Gopalakrishnan et al. 2018). In this study, it was also discovered that responders had a higher infiltration of CD8 + T cells as compared to non-responders, but non-responders colonized with *Bacteroidales* had a higher infiltration of T regulatory cells and myeloid-derived suppressor cells (Gopalakrishnan et al. 2018). These results confirm that in melanoma patients, responders to anti-PD-1 therapy are colonized with bacteria that promote CD8 + T cell infiltration of the tumor microenvironment. Non-responders were shown to be colonized with bacteria that are associated with immunosuppressive cells infiltrating the tumor microenvironment (Gopalakrishnan et al. 2018). Thus, the influence of the microbiome on cancer progression is clear, but it is far from a reliable predictor of resistant versus sensitive patients.

4.2 Using Organoids to Gain Insight into the Development of Cancer Therapeutics

The impact of the microbiota on cellular dysfunction and cancer development is evidenced by studies of intestinal organoids. The intestinal microbiota is made up of bacteria, viruses, archaea, and fungal species that are essential for maintaining gut homeostasis and protecting the host against pathogens. Dysregulation of the intestinal microbiota can contribute to the development of colorectal cancer. Organoids cultured in the presence of the microbiome allow for studies of microbe–host interactions and their contribution to disease development. For example, Farin et al. used intestinal organoids treated with microbial antigens and bacteria to identify the dysregulation of Paneth cell function in the progression of disease (Farin et al. 2014). Microbial antigens were introduced to either the organoid culture media allowing for the basolateral exposure, or to sheered organoids luminal exposure. Regardless of the basolateral or luminal exposure, Paneth cell dysregulation was not induced by direct stimulation with microbial antigen or bacterial exposure that demonstrated a key function of IFNγ in Paneth cell dysregulation (Farin et al. 2014). Zhang et al. investigated the effects of *Salmonella* colonization on murine intestinal organoid function and reported that the introduction of the bacteria disrupted tight junctions, reduced the expression of Lgr5, and activated NFκB signaling (Zhang et al. 2014). Another study used microinjection of *Salmonella* to expose the organoid lumen of human iPSC-derived organoids with the bacteria and observed microbial epithelial invasion (Forbester et al. 2015). Organoid cultures have also demonstrated that *Clostridium difficile*, an obligate anaerobe, survives up to 12 h within the lumen, resulting in the disruption of intestinal barrier function (Leslie et al. 2015). Therefore, organoids are an ideal culture for studies that characterize the relationships between microbes and the gastric and intestinal epithelium, and the impact of such relationships has in disease progression and responsiveness to therapies.

5 Conclusions

Organoids can be used to advance cancer therapeutics as a first-pass screen for drug sensitivity in the presence of the host's microbiome. Unlike cancer cell lines that consist of genetic mutations and drift from the native tumor tissue, organoids are developed directly from the patient and are able to most closely replicate the

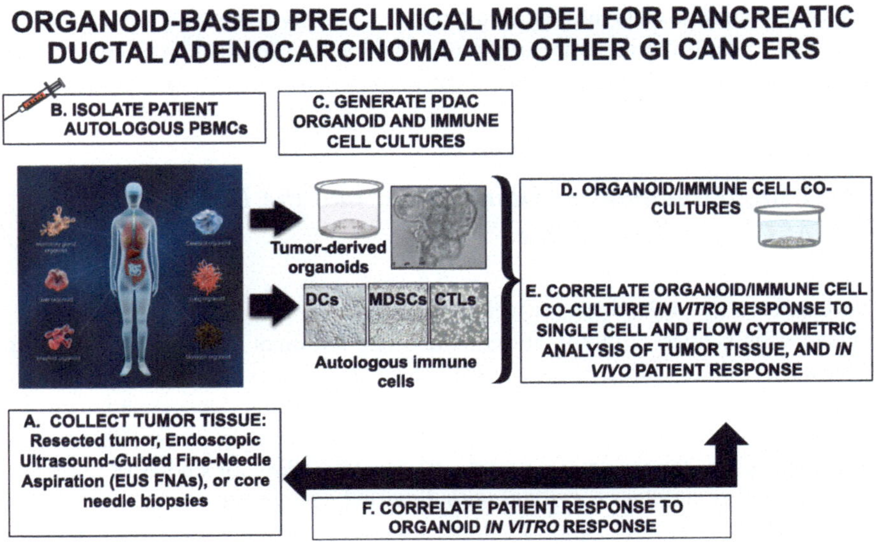

Fig. 1 Establishment of organoid-based preclinical model for pancreatic ductal adenocarcinoma and other GI cancers. **a** Collection of tumors tissues from either, resected tumor, endoscopic ultrasound-guided fine-needle aspiration (EUS FNAs), or core needle biopsies. **b** Isolation of autologous PBMCs from the blood of respective patients. **c** Generation of pancreatic ductal adenocarcinoma (PDAC) organoids from the resected PDAC tumor tissues and development and differentiation of different immune cells, dendritic cells (DCs), myeloid-derived suppressor cells (MDSCs), and cytotoxic T lymphocytes (CTLS) from the autologous PBMCs. **d** Establishment of tumor-derived organoids and autologous immune cell co-cultures within a Matrigel Dome. **e** Comparison of in vitro organoid–immune cell co-culture response to single cell and flow cytometric analysis of the tumor tissue and in vivo patient response. **f** Correlation of in vivo patient response to in vitro organoid response for the development of personalized therapeutic approaches

pathophysiology and TME. Many investigators propose that, before a cancer therapeutic is given to a patient, it can first be tested in organoids derived from the patient's tumor tissue. Several groups have created organoid biobanks from human intestinal and gastric tumor biopsies, for example, to test the efficacy of cancer therapeutics before administering the treatment to the patient (van de Wetering et al. 2015; Chakrabarti et al. 2018b; Steele et al. 2019). Such an approach allows for personalization of cancer treatments in such a way that both the host's genetics and microbiome may be considered in the same culture (Fig. 1). Literature published in recent years shows great promise in allowing us to understand the active role of microbes in disease manifestation and cancer therapy. Organotypic cultures have widespread applications to provide alternate models that not only complement existing experimental systems to study different human diseases but also serve as a prevailing tool for the development of novel therapeutics. These high-throughput model systems provide a better platform beyond other advanced in vitro models.

Indeed, 3D cell-based models have been efficiently used to elucidate novel signaling pathways during normal cellular differentiation and tumor progression (Schmeichel and Bissell 2003; McCracken et al. 2014, 2017; Steele et al. 2019; Chakrabarti et al. 2018b; Michels et al. 2019). Despite different limitations, this promising organoid technology embraces great potential in precisely modeling organ development and human diseases, which can serve as a unique platform for predicting therapy responses and developing personalized medicine (Xu et al. 2018; Steele et al. 2019; Chakrabarti et al. 2018b; Holokai et al. 2019; Tiriac et al. 2018). The recent advancement in stem cell research has helped to establish organoids as "infection models" which will hold major potential for novel therapeutic development for the diagnosis, and treatment of infectious diseases, understanding antimicrobial drug screening, virulence mechanisms, and host–pathogen interactions (Holokai et al. 2019; Bartfeld et al. 2015; Bertaux-Skeirik et al. 2015). The expansion of the organoid model to organoid co-cultures adds an additional layer of innovation and relevance to the organotypic model.

Acknowledgements We are sincerely grateful for the collaboration and support by Dr. James Wells (Division of Developmental Biology, Center for Stem Cell & Organoid Medicine, Division of Endocrinology, Cincinnati Children's Hospital Medical Center) and Dr. Michael Helmrath (Department of Pediatric Surgery, Cincinnati Children's Hospital Medical Center). This work was supported by NIH (NIAID) 5U19AI11649105 (PIs: Weiss and Wells, Project Leader 1: Zavros) and NIH R01DK083402-10 (PI: Zavros).

References

Barker N, Huch M, Kujala P, van de Wetering M, Snippert HJ, van Es JH, Sato T, Stange DE, Begthel H, van den Born M, Danenberg E, van den Brink S, Korving J, Abo A, Peters PJ, Wright N, Poulsom R, Clevers H (2010) Lgr5(+ve) stem cells drive self-renewal in the stomach and build long-lived gastric units in vitro. Cell Stem Cell 6(1):25–36. https://doi.org/10.1016/j.stem.2009.11.013

Bartfeld S, Bayram T, van de Wetering M, Huch M, Begthel H, Kujala P, Vries R, Peters PJ, Clevers H (2015) In vitro expansion of human gastric epithelial stem cells and their responses to bacterial infection. Gastroenterology 148(1):126–136.e126. https://doi.org/10.1053/j.gastro.2014.09.042

Bertaux-Skeirik N, Feng R, Schumacher MA, Li J, Mahe MM, Engevik AC, Javier JE, Peek RM Jr, Ottemann K, Orian-Rousseau V, Boivin GP, Helmrath MA, Zavros Y (2015) CD44 plays a functional role in Helicobacter pylori-induced epithelial cell proliferation. PLoS Pathog 11(2):e1004663. https://doi.org/10.1371/journal.ppat.1004663

Bertaux-Skeirik N, Wunderlich M, Teal E, Chakrabarti J, Biesiada J, Mahe M, Sundaram N, Gabre J, Hawkins J, Jian G, Engevik AC, Yang L, Wang J, Goldenring JR, Qualls JE, Medvedovic M, Helmrath MA, Diwan T, Mulloy JC, Zavros Y (2017) CD44 variant isoform 9 emerges in response to injury and contributes to the regeneration of the gastric epithelium. J Pathol 242(4):463–475. https://doi.org/10.1002/path.4918

Boj SF, Hwang CI, Baker LA, Chio II, Engle DD, Corbo V, Jager M, Ponz-Sarvise M, Tiriac H, Spector MS, Gracanin A, Oni T, Yu KH, van Boxtel R, Huch M, Rivera KD, Wilson JP, Feigin ME, Öhlund D, Handly-Santana A, Ardito-Abraham CM, Ludwig M, Elyada E, Alagesan B, Biffi G, Yordanov GN, Delcuze B, Creighton B, Wright K, Park Y, Morsink FH,

Molenaar IQ, Borel Rinkes IH, Cuppen E, Hao Y, Jin Y, Nijman IJ, Iacobuzio-Donahue C, Leach SD, Pappin DJ, Hammell M, Klimstra DS, Basturk O, Hruban RH, Offerhaus GJ, Vries RG, Clevers H, Tuveson DA (2015) Organoid models of human and mouse ductal pancreatic cancer. Cell 160(1–2):324–338. https://doi.org/10.1016/j.cell.2014.12.021

Chakrabarti J, Holokai L, Syu L, Steele N, Chang J, Dlugosz A, Zavros Y (2018a) Mouse-derived gastric organoid and immune cell co-culture for the study of the tumor microenvironment. Methods Mol Biol 1817:157–168. https://doi.org/10.1007/978-1-4939-8600-2_16

Chakrabarti J, Holokai L, Syu L, Steele NG, Chang J, Wang J, Ahmed S, Dlugosz A, Zavros Y (2018b) Hedgehog signaling induces PD-L1 expression and tumor cell proliferation in gastric cancer. Oncotarget 9(100):37439–37457. https://doi.org/10.18632/oncotarget.26473

Chen HN, Wang Z, Li X, Zhou ZG (2016) Helicobacter pylori eradication cannot reduce the risk of gastric cancer in patients with intestinal metaplasia and dysplasia: evidence from a meta-analysis. Gastric Cancer 19(1):166–175. https://doi.org/10.1007/s10120-015-0462-7

Cohn L, Delamarre L (2014) Dendritic cell-targeted vaccines. Front Immunol 5:255. https://doi.org/10.3389/fimmu.2014.00255

Dang J, Tiwari SK, Lichinchi G, Qin Y, Patil VS, Eroshkin AM, Rana TM (2016) Zika virus depletes neural progenitors in human cerebral organoids through activation of the innate immune receptor TLR3. Cell Stem Cell 19(2):258–265. https://doi.org/10.1016/j.stem.2016.04.014

Dang JW, Tiwari SK, Qin Y, Rana TM (2019) Genome-wide Integrative Analysis of Zika-Virus-Infected Neuronal Stem Cells Reveals Roles for MicroRNAs in Cell Cycle and Stemness. Cell Rep 27 (12):3618–3628 e3615. https://doi.org/10.1016/j.celrep.2019.05.059

de Charette M, Marabelle A, Houot R (2016) Turning tumour cells into antigen presenting cells: the next step to improve cancer immunotherapy? Eur J Cancer 68:134–147. https://doi.org/10.1016/j.ejca.2016.09.010

Dedhia PH, Bertaux-Skeirik N, Zavros Y, Spence JR (2016) Organoid models of human gastrointestinal development and disease. Gastroenterology 150(5):1098–1112. https://doi.org/10.1053/j.gastro.2015.12.042

DeWard AD, Cramer J, Lagasse E (2014) Cellular heterogeneity in the mouse esophagus implicates the presence of a nonquiescent epithelial stem cell population. Cell Rep 9(2):701–711. https://doi.org/10.1016/j.celrep.2014.09.027

Dijkstra KK, Cattaneo CM, Weeber F, Chalabi M, van de Haar J, Fanchi LF, Slagter M, van der Velden DL, Kaing S, Kelderman S, van Rooij N, van Leerdam ME, Depla A, Smit EF, Hartemink KJ, de Groot R, Wolkers MC, Sachs N, Snaebjornsson P, Monkhorst K, Haanen J, Clevers H, Schumacher TN, Voest EE (2018) Generation of tumor-reactive T cells by co-culture of peripheral blood lymphocytes and tumor organoids. Cell 174(6):1586–1598.e1512. https://doi.org/10.1016/j.cell.2018.07.009

Ding L, Hayes MM, Photenhauer A, Eaton KA, Li Q, Ocadiz-Ruiz R, Merchant JL (2016) Schlafen 4-expressing myeloid-derived suppressor cells are induced during murine gastric metaplasia. J Clin Invest 126(8):2867–2880. https://doi.org/10.1172/jci82529

Ding L, Li Q, Chakrabarti J, Munoz A, Faure-Kumar E, Ocadiz-Ruiz R, Razumilava N, Zhang G, Hayes MH, Sontz RA, Mendoza ZE, Mahurkar S, Greenson JK, Perez-Perez G, Hanh NTH, Zavros Y, Samuelson LC, Iliopoulos D, Merchant JL (2020) MiR130b from Schlafen4(+) MDSCs stimulates epithelial proliferation and correlates with preneoplastic changes prior to gastric cancer. Gut. https://doi.org/10.1136/gutjnl-2019-318817

Dye BR, Hill DR, Ferguson MA, Tsai YH, Nagy MS, Dyal R, Wells JM, Mayhew CN, Nattiv R, Klein OD, White ES, Deutsch GH, Spence JR (2015) In vitro generation of human pluripotent stem cell derived lung organoids. Elife 4. https://doi.org/10.7554/elife.05098

Engevik AC, Feng R, Choi E, White S, Bertaux-Skeirik N, Li J, Mahe MM, Aihara E, Yang L, DiPasquale B, Oh S, Engevik KA, Giraud AS, Montrose MH, Medvedovic M, Helmrath MA, Goldenring JR, Zavros Y (2016) The development of spasmolytic polypeptide/TFF2-expressing metaplasia (SPEM) during gastric repair is absent in the aged stomach. Cell Mol Gastroenterol Hepatol 2(5):605–624. https://doi.org/10.1016/j.jcmgh.2016.05.004

Ettayebi K, Crawford SE, Murakami K, Broughman JR, Karandikar U, Tenge VR, Neill FH, Blutt SE, Zeng XL, Qu L, Kou B, Opekun AR, Burrin D, Graham DY, Ramani S, Atmar RL, Estes MK (2016) Replication of human noroviruses in stem cell-derived human enteroids. Science 353(6306):1387–1393. https://doi.org/10.1126/science.aaf5211

Farin HF, Karthaus WR, Kujala P, Rakhshandehroo M, Schwank G, Vries RG, Kalkhoven E, Nieuwenhuis EE, Clevers H (2014) Paneth cell extrusion and release of antimicrobial products is directly controlled by immune cell-derived IFN-γ. J Exp Med 211(7):1393–1405. https://doi.org/10.1084/jem.20130753

Finkbeiner SR, Hill DR, Altheim CH, Dedhia PH, Taylor MJ, Tsai YH, Chin AM, Mahe MM, Watson CL, Freeman JJ, Nattiv R, Thomson M, Klein OD, Shroyer NF, Helmrath MA, Teitelbaum DH, Dempsey PJ, Spence JR (2015) Transcriptome-wide analysis reveals hallmarks of human intestine development and maturation in vitro and in vivo. Stem Cell Rep 4(6):1140–1155. https://doi.org/10.1016/j.stemcr.2015.04.010

Forbester JL, Goulding D, Vallier L, Hannan N, Hale C, Pickard D, Mukhopadhyay S, Dougan G (2015) Interaction of *Salmonella enterica* serovar typhimurium with intestinal organoids derived from human induced pluripotent stem cells. Infect Immun 83(7):2926–2934. https://doi.org/10.1128/iai.00161-15

Forster R, Chiba K, Schaeffer L, Regalado SG, Lai CS, Gao Q, Kiani S, Farin HF, Clevers H, Cost GJ, Chan A, Rebar EJ, Urnov FD, Gregory PD, Pachter L, Jaenisch R, Hockemeyer D (2014) Human intestinal tissue with adult stem cell properties derived from pluripotent stem cells. Stem Cell Rep 2(6):838–852. https://doi.org/10.1016/j.stemcr.2014.05.001

Garcez PP, Loiola EC, Madeiro da Costa R, Higa LM, Trindade P, Delvecchio R, Nascimento JM, Brindeiro R, Tanuri A, Rehen SK (2016) Zika virus impairs growth in human neurospheres and brain organoids. Science 352(6287):816–818. https://doi.org/10.1126/science.aaf6116

Gopalakrishnan V, Spencer CN, Nezi L, Reuben A, Andrews MC, Karpinets TV, Prieto PA, Vicente D, Hoffman K, Wei SC, Cogdill AP, Zhao L, Hudgens CW, Hutchinson DS, Manzo T, Petaccia de Macedo M, Cotechini T, Kumar T, Chen WS, Reddy SM, Szczepaniak Sloane R, Galloway-Pena J, Jiang H, Chen PL, Shpall EJ, Rezvani K, Alousi AM, Chemaly RF, Shelburne S, Vence LM, Okhuysen PC, Jensen VB, Swennes AG, McAllister F, Marcelo Riquelme Sanchez E, Zhang Y, Le Chatelier E, Zitvogel L, Pons N, Austin-Breneman JL, Haydu LE, Burton EM, Gardner JM, Sirmans E, Hu J, Lazar AJ, Tsujikawa T, Diab A, Tawbi H, Glitza IC, Hwu WJ, Patel SP, Woodman SE, Amaria RN, Davies MA, Gershenwald JE, Hwu P, Lee JE, Zhang J, Coussens LM, Cooper ZA, Futreal PA, Daniel CR, Ajami NJ, Petrosino JF, Tetzlaff MT, Sharma P, Allison JP, Jenq RR, Wargo JA (2018) Gut microbiome modulates response to anti-PD-1 immunotherapy in melanoma patients. Science 359(6371):97–103. https://doi.org/10.1126/science.aan4236

Greggio C, De Franceschi F, Figueiredo-Larsen M, Gobaa S, Ranga A, Semb H, Lutolf M, Grapin-Botton A (2013) Artificial three-dimensional niches deconstruct pancreas development in vitro. Development 140(21):4452–4462. https://doi.org/10.1242/dev.096628

Guillerme JB, Boisgerault N, Roulois D, Ménager J, Combredet C, Tangy F, Fonteneau JF, Gregoire M (2013) Measles virus vaccine-infected tumor cells induce tumor antigen cross-presentation by human plasmacytoid dendritic cells. Clin Cancer Res 19(5):1147–1158. https://doi.org/10.1158/1078-0432.Ccr-12-2733

Hannan NR, Fordham RP, Syed YA, Moignard V, Berry A, Bautista R, Hanley NA, Jensen KB, Vallier L (2013) Generation of multipotent foregut stem cells from human pluripotent stem cells. Stem Cell Rep 1(4):293–306. https://doi.org/10.1016/j.stemcr.2013.09.003

Hockemeyer D, Jaenisch R (2016) Induced pluripotent stem cells meet genome editing. Cell Stem Cell 18(5):573–586. https://doi.org/10.1016/j.stem.2016.04.013

Holokai L, Chakrabarti J, Broda T, Chang J, Hawkins JA, Sundaram N, Wroblewski LE, Peek RM Jr, Wang J, Helmrath M, Wells JM, Zavros Y (2019) Increased programmed death-ligand 1 is an early epithelial cell response to helicobacter pylori infection. PLoS Pathog 15(1):e1007468. https://doi.org/10.1371/journal.ppat.1007468

Hrvatin S, O'Donnell CW, Deng F, Millman JR, Pagliuca FW, DiIorio P, Rezania A, Gifford DK, Melton DA (2014) Differentiated human stem cells resemble fetal, not adult, β cells. Proc Natl Acad Sci U S A 111(8):3038–3043. https://doi.org/10.1073/pnas.1400709111

Huch M, Dorrell C, Boj SF, van Es JH, Li VS, van de Wetering M, Sato T, Hamer K, Sasaki N, Finegold MJ, Haft A, Vries RG, Grompe M, Clevers H (2013) In vitro expansion of single Lgr5 + liver stem cells induced by Wnt-driven regeneration. Nature 494(7436):247–250. https://doi.org/10.1038/nature11826

Huch M, Koo BK (2015) Modeling mouse and human development using organoid cultures. Development 142(18):3113–3125. https://doi.org/10.1242/dev.118570

Jung P, Sato T, Merlos-Suárez A, Barriga FM, Iglesias M, Rossell D, Auer H, Gallardo M, Blasco MA, Sancho E, Clevers H, Batlle E (2011) Isolation and in vitro expansion of human colonic stem cells. Nat Med 17(10):1225–1227. https://doi.org/10.1038/nm.2470

Kalabis J, Wong GS, Vega ME, Natsuizaka M, Robertson ES, Herlyn M, Nakagawa H, Rustgi AK (2012) Isolation and characterization of mouse and human esophageal epithelial cells in 3D organotypic culture. Nat Protoc 7(2):235–246. https://doi.org/10.1038/nprot.2011.437

Leslie JL, Huang S, Opp JS, Nagy MS, Kobayashi M, Young VB, Spence JR (2015) Persistence and toxin production by Clostridium difficile within human intestinal organoids result in disruption of epithelial paracellular barrier function. Infect Immun 83(1):138–145. https://doi.org/10.1128/iai.02561-14

Marshall BJ, Warren JR (1984) Unidentified curved bacilli in the stomach of patients with gastritis and peptic ulceration. Lancet 1(8390):1311–1315. https://doi.org/10.1016/s0140-6736(84)91816-6

Matson V, Fessler J, Bao R, Chongsuwat T, Zha Y, Alegre ML, Luke JJ, Gajewski TF (2018) The commensal microbiome is associated with anti-PD-1 efficacy in metastatic melanoma patients. Science 359(6371):104–108. https://doi.org/10.1126/science.aao3290

Matsueda S, Graham DY (2014) Immunotherapy in gastric cancer. World J Gastroenterol 20(7):1657–1666. https://doi.org/10.3748/wjg.v20.i7.1657

McCauley HA, Wells JM (2017) Pluripotent stem cell-derived organoids: using principles of developmental biology to grow human tissues in a dish. Development 144(6):958–962. https://doi.org/10.1242/dev.140731

McCracken KW, Aihara E, Martin B, Crawford CM, Broda T, Treguier J, Zhang X, Shannon JM, Montrose MH, Wells JM (2017) Wnt/β-catenin promotes gastric fundus specification in mice and humans. Nature 541(7636):182–187. https://doi.org/10.1038/nature21021

McCracken KW, Catá EM, Crawford CM, Sinagoga KL, Schumacher M, Rockich BE, Tsai YH, Mayhew CN, Spence JR, Zavros Y, Wells JM (2014) Modelling human development and disease in pluripotent stem-cell-derived gastric organoids. Nature 516(7531):400–404. https://doi.org/10.1038/nature13863

McCracken KW, Howell JC, Wells JM, Spence JR (2011) Generating human intestinal tissue from pluripotent stem cells in vitro. Nat Protoc 6(12):1920–1928. https://doi.org/10.1038/nprot.2011.410

Michels BE, Mosa MH, Grebbin BM, Yepes D, Darvishi T, Hausmann J, Urlaub H, Zeuzem S, Kvasnicka HM, Oellerich T, Farin HF (2019) Human colon organoids reveal distinct physiologic and oncogenic Wnt responses. J Exp Med 216(3):704–720. https://doi.org/10.1084/jem.20180823

Millán O, Rafael-Valdivia L, Torrademé E, López A, Fortuna V, Sánchez-Cabus S, López-Púa Y, Rimola A, Brunet M (2013) Intracellular IFN-γ and IL-2 expression monitoring as surrogate markers of the risk of acute rejection and personal drug response in de novo liver transplant recipients. Cytokine 61(2):556–564. https://doi.org/10.1016/j.cyto.2012.10.026

Neal JT, Li X, Zhu J, Giangarra V, Grzeskowiak CL, Ju J, Liu IH, Chiou SH, Salahudeen AA, Smith AR, Deutsch BC, Liao L, Zemek AJ, Zhao F, Karlsson K, Schultz LM, Metzner TJ, Nadauld LD, Tseng YY, Alkhairy S, Oh C, Keskula P, Mendoza-Villanueva D, De La Vega FM, Kunz PL, Liao JC, Leppert JT, Sunwoo JB, Sabatti C, Boehm JS, Hahn WC, Zheng GXY, Davis MM, Kuo CJ (2018) Organoid modeling of the tumor immune microenvironment. Cell 175(7):1972–1988.e1916. https://doi.org/10.1016/j.cell.2018.11.021

Noel G, Baetz NW, Staab JF, Donowitz M, Kovbasnjuk O, Pasetti MF, Zachos NC (2017) A primary human macrophage-enteroid co-culture model to investigate mucosal gut physiology and host-pathogen interactions. Sci Rep 7:45270. https://doi.org/10.1038/srep45270

Nozaki K, Mochizuki W, Matsumoto Y, Matsumoto T, Fukuda M, Mizutani T, Watanabe M, Nakamura T (2016) Co-culture with intestinal epithelial organoids allows efficient expansion and motility analysis of intraepithelial lymphocytes. J Gastroenterol 51(3):206–213. https://doi.org/10.1007/s00535-016-1170-8

Odenbreit S, Püls J, Sedlmaier B, Gerland E, Fischer W, Haas R (2000) Translocation of helicobacter pylori CagA into gastric epithelial cells by type IV secretion. Science 287 (5457):1497–1500. https://doi.org/10.1126/science.287.5457.1497

Ootani A, Li X, Sangiorgi E, Ho QT, Ueno H, Toda S, Sugihara H, Fujimoto K, Weissman IL, Capecchi MR, Kuo CJ (2009) Sustained in vitro intestinal epithelial culture within a Wnt-dependent stem cell niche. Nat Med 15(6):701–706. https://doi.org/10.1038/nm.1951

Qian X, Nguyen HN, Jacob F, Song H, Ming GL (2017) Using brain organoids to understand Zika virus-induced microcephaly. Development 144(6):952–957. https://doi.org/10.1242/dev.140707

Randall KJ, Turton J, Foster JR (2011) Explant culture of gastrointestinal tissue: a review of methods and applications. Cell Biol Toxicol 27(4):267–284. https://doi.org/10.1007/s10565-011-9187-5

Rogoz A, Reis BS, Karssemeijer RA, Mucida D (2015) A 3-D enteroid-based model to study T-cell and epithelial cell interaction. J Immunol Methods 421:89–95. https://doi.org/10.1016/j.jim.2015.03.014

Routy B, Gopalakrishnan V, Daillère R, Zitvogel L, Wargo JA, Kroemer G (2018) The gut microbiota influences anticancer immunosurveillance and general health. Nat Rev Clin Oncol 15(6):382–396. https://doi.org/10.1038/s41571-018-0006-2

Sato S, Hisaie K, Kurokawa S, Suzuki A, Sakon N, Uchida Y, Yuki Y, Kiyono H (2019) Human norovirus propagation in human induced pluripotent stem cell-derived intestinal epithelial cells. Cell Mol Gastroenterol Hepatol 7(3):686–688.e685. https://doi.org/10.1016/j.jcmgh.2018.11.001

Sato T, Stange DE, Ferrante M, Vries RG, Van Es JH, Van den Brink S, Van Houdt WJ, Pronk A, Van Gorp J, Siersema PD, Clevers H (2011a) Long-term expansion of epithelial organoids from human colon, adenoma, adenocarcinoma, and Barrett's epithelium. Gastroenterology 141 (5):1762–1772. https://doi.org/10.1053/j.gastro.2011.07.050

Sato T, van Es JH, Snippert HJ, Stange DE, Vries RG, van den Born M, Barker N, Shroyer NF, van de Wetering M, Clevers H (2011b) Paneth cells constitute the niche for Lgr5 stem cells in intestinal crypts. Nature 469(7330):415–418. https://doi.org/10.1038/nature09637

Sato T, Vries RG, Snippert HJ, van de Wetering M, Barker N, Stange DE, van Es JH, Abo A, Kujala P, Peters PJ, Clevers H (2009) Single Lgr5 stem cells build crypt-villus structures in vitro without a mesenchymal niche. Nature 459(7244):262–265. https://doi.org/10.1038/nature07935

Saxena K, Blutt SE, Ettayebi K, Zeng XL, Broughman JR, Crawford SE, Karandikar UC, Sastri NP, Conner ME, Opekun AR, Graham DY, Qureshi W, Sherman V, Foulke-Abel J, In J, Kovbasnjuk O, Zachos NC, Donowitz M, Estes MK (2016) Human intestinal enteroids: a new model to study human rotavirus infection, host restriction, and pathophysiology. J Virol 90 (1):43–56. https://doi.org/10.1128/jvi.01930-15

Schmeichel KL, Bissell MJ (2003) Modeling tissue-specific signaling and organ function in three dimensions. J Cell Sci 116(Pt 12):2377–2388. https://doi.org/10.1242/jcs.00503

Schouppe E, Van Overmeire E, Laoui D, Keirsse J, Van Ginderachter JA (2013) Modulation of CD8(+) T-cell activation events by monocytic and granulocytic myeloid-derived suppressor cells. Immunobiology 218(11):1385–1391. https://doi.org/10.1016/j.imbio.2013.07.003

Schumacher MA, Aihara E, Feng R, Engevik A, Shroyer NF, Ottemann KM, Worrell RT, Montrose MH, Shivdasani RA, Zavros Y (2015) The use of murine-derived fundic organoids in studies of gastric physiology. J Physiol 593(8):1809–1827. https://doi.org/10.1113/jphysiol.2014.283028

Schumacher MA, Donnelly JM, Engevik AC, Xiao C, Yang L, Kenny S, Varro A, Hollande F, Samuelson LC, Zavros Y (2012) Gastric Sonic Hedgehog acts as a macrophage chemoattractant during the immune response to Helicobacter pylori. Gastroenterology 142(5):1150–1159. e1156. https://doi.org/10.1053/j.gastro.2012.01.029

Sehgal K, Dhodapkar KM, Dhodapkar MV (2014) Targeting human dendritic cells in situ to improve vaccines. Immunol Lett 162 (1 Pt A):59–67. https://doi.org/10.1016/j.imlet.2014.07.004

Shan J, Schwartz RE, Ross NT, Logan DJ, Thomas D, Duncan SA, North TE, Goessling W, Carpenter AE, Bhatia SN (2013) Identification of small molecules for human hepatocyte expansion and iPS differentiation. Nat Chem Biol 9(8):514–520. https://doi.org/10.1038/nchembio.1270

Si-Tayeb K, Noto FK, Nagaoka M, Li J, Battle MA, Duris C, North PE, Dalton S, Duncan SA (2010) Highly efficient generation of human hepatocyte-like cells from induced pluripotent stem cells. Hepatology 51(1):297–305. https://doi.org/10.1002/hep.23354

Sivan A, Corrales L, Hubert N, Williams JB, Aquino-Michaels K, Earley ZM, Benyamin FW, Lei YM, Jabri B, Alegre ML, Chang EB, Gajewski TF (2015) Commensal Bifidobacterium promotes antitumor immunity and facilitates anti-PD-L1 efficacy. Science 350(6264):1084–1089. https://doi.org/10.1126/science.aac4255

Spence JR, Mayhew CN, Rankin SA, Kuhar MF, Vallance JE, Tolle K, Hoskins EE, Kalinichenko VV, Wells SI, Zorn AM, Shroyer NF, Wells JM (2011) Directed differentiation of human pluripotent stem cells into intestinal tissue in vitro. Nature 470(7332):105–109. https://doi.org/10.1038/nature09691

Stange DE, Koo BK, Huch M, Sibbel G, Basak O, Lyubimova A, Kujala P, Bartfeld S, Koster J, Geahlen JH, Peters PJ, van Es JH, van de Wetering M, Mills JC, Clevers H (2013) Differentiated Troy + chief cells act as reserve stem cells to generate all lineages of the stomach epithelium. Cell 155(2):357–368. https://doi.org/10.1016/j.cell.2013.09.008

Steele NG, Chakrabarti J, Wang J, Biesiada J, Holokai L, Chang J, Nowacki LM, Hawkins J, Mahe M, Sundaram N, Shroyer N, Medvedovic M, Helmrath M, Ahmad S, Zavros Y (2019) An organoid-based preclinical model of human gastric cancer. Cell Mol Gastroenterol Hepatol 7(1):161–184. https://doi.org/10.1016/j.jcmgh.2018.09.008

Strangeways TS, Canti RG (1926) Dark-ground illumination of tissue cells cultivated “in vitro”. Br Med J 2 (3420):155–157. https://doi.org/10.1136/bmj.2.3420.155

Takahashi K, Tanabe K, Ohnuki M, Narita M, Ichisaka T, Tomoda K, Yamanaka S (2007) Induction of pluripotent stem cells from adult human fibroblasts by defined factors. Cell 131 (5):861–872. https://doi.org/10.1016/j.cell.2007.11.019

Takebe T, Sekine K, Enomura M, Koike H, Kimura M, Ogaeri T, Zhang RR, Ueno Y, Zheng YW, Koike N, Aoyama S, Adachi Y, Taniguchi H (2013) Vascularized and functional human liver from an iPSC-derived organ bud transplant. Nature 499(7459):481–484. https://doi.org/10.1038/nature12271

Thara E, Dorff TB, Pinski JK, Quinn DI (2011) Vaccine therapy with sipuleucel-T (Provenge) for prostate cancer. Maturitas 69(4):296–303. https://doi.org/10.1016/j.maturitas.2011.04.012

Thomson JA, Itskovitz-Eldor J, Shapiro SS, Waknitz MA, Swiergiel JJ, Marshall VS, Jones JM (1998) Embryonic stem cell lines derived from human blastocysts. Science 282(5391):1145–1147. https://doi.org/10.1126/science.282.5391.1145

Tiriac H, Belleau P, Engle DD, Plenker D, Deschênes A, Somerville TDD, Froeling FEM, Burkhart RA, Denroche RE, Jang GH, Miyabayashi K, Young CM, Patel H, Ma M, LaComb JF, Palmaira RLD, Javed AA, Huynh JC, Johnson M, Arora K, Robine N, Shah M, Sanghvi R, Goetz AB, Lowder CY, Martello L, Driehuis E, LeComte N, Askan G, Iacobuzio-Donahue CA, Clevers H, Wood LD, Hruban RH, Thompson E, Aguirre AJ, Wolpin BM, Sasson A, Kim J, Wu M, Bucobo JC, Allen P, Sejpal DV, Nealon W, Sullivan JD, Winter JM, Gimotty PA, Grem JL, DiMaio DJ, Buscaglia JM, Grandgenett PM, Brody JR, Hollingsworth MA, O'Kane GM, Notta F, Kim E, Crawford JM, Devoe C, Ocean A, Wolfgang CL, Yu KH, Li E, Vakoc CR, Hubert B, Fischer SE, Wilson JM, Moffitt R, Knox J, Krasnitz A, Gallinger S, Tuveson DA (2018) Organoid profiling identifies common responders to chemotherapy in pancreatic cancer. Cancer Discov 8(9):1112–1129. https://doi.org/10.1158/2159-8290.Cd-18-0349

Tsai S, McOlash L, Palen K, Johnson B, Duris C, Yang Q, Dwinell MB, Hunt B, Evans DB, Gershan J, James MA (2018) Development of primary human pancreatic cancer organoids, matched stromal and immune cells and 3D tumor microenvironment models. BMC Cancer 18 (1):335. https://doi.org/10.1186/s12885-018-4238-4

van de Wetering M, Francies HE, Francis JM, Bounova G, Iorio F, Pronk A, van Houdt W, van Gorp J, Taylor-Weiner A, Kester L, McLaren-Douglas A, Blokker J, Jaksani S, Bartfeld S, Volckman R, van Sluis P, Li VS, Seepo S, Sekhar Pedamallu C, Cibulskis K, Carter SL, McKenna A, Lawrence MS, Lichtenstein L, Stewart C, Koster J, Versteeg R, van Oudenaarden A, Saez-Rodriguez J, Vries RG, Getz G, Wessels L, Stratton MR, McDermott U, Meyerson M, Garnett MJ, Clevers H (2015) Prospective derivation of a living organoid biobank of colorectal cancer patients. Cell 161(4):933–945. https://doi.org/10.1016/j.cell.2015.03.053

Wang L, Chang EW, Wong SC, Ong SM, Chong DQ, Ling KL (2013) Increased myeloid-derived suppressor cells in gastric cancer correlate with cancer stage and plasma S100A8/A9 proinflammatory proteins. J Immunol 190(2):794–804. https://doi.org/10.4049/jimmunol.1202088

Watson CL, Mahe MM, Múnera J, Howell JC, Sundaram N, Poling HM, Schweitzer JI, Vallance JE, Mayhew CN, Sun Y, Grabowski G, Finkbeiner SR, Spence JR, Shroyer NF, Wells JM, Helmrath MA (2014) An in vivo model of human small intestine using pluripotent stem cells. Nat Med 20(11):1310–1314. https://doi.org/10.1038/nm.3737

Xu H, Jiao Y, Qin S, Zhao W, Chu Q, Wu K (2018) Organoid technology in disease modelling, drug development, personalized treatment and regeneration medicine. Exp Hematol Oncol 7:30. https://doi.org/10.1186/s40164-018-0122-9

Youn JI, Nagaraj S, Collazo M, Gabrilovich DI (2008) Subsets of myeloid-derived suppressor cells in tumor-bearing mice. J Immunol 181(8):5791–5802. https://doi.org/10.4049/jimmunol.181.8.5791

Zhang Y, Velez-Delgado A, Mathew E, Li D, Mendez FM, Flannagan K, Rhim AD, Simeone DM, Beatty GL, Pasca di Magliano M (2017) Myeloid cells are required for PD-1/PD-L1 checkpoint activation and the establishment of an immunosuppressive environment in pancreatic cancer. Gut 66(1):124–136. https://doi.org/10.1136/gutjnl-2016-312078

Zhang ZH, Yu LJ, Hui XC, Wu ZZ, Yin KL, Yang H, Xu Y (2014) Hydroxy-safflor yellow A attenuates $A\beta_{1-42}$-induced inflammation by modulating the JAK2/STAT3/NF-κB pathway. Brain Res 1563:72–80. https://doi.org/10.1016/j.brainres.2014.03.036

Zhao L, Niu C, Shi X, Xu D, Li M, Cui J, Li W, Xu J, Jin H (2018) Dendritic cells loaded with the lysate of tumor cells infected with Newcastle Disease Virus trigger potent anti-tumor immunity by promoting the secretion of IFN-γ and IL-2 from T cells. Oncol Lett 16(1):1180–1188. https://doi.org/10.3892/ol.2018.8785

Zumwalde NA, Haag JD, Sharma D, Mirrielees JA, Wilke LG, Gould MN, Gumperz JE (2016) Analysis of immune cells from human mammary ductal epithelial organoids reveals Vδ2 + T cells that efficiently target breast carcinoma cells in the presence of bisphosphonate. Cancer Prev Res (Phila) 9(4):305–316. https://doi.org/10.1158/1940-6207.Capr-15-0370-t

Human Organotypic Models for Anti-infective Research

Astrid Hendriks, Ana Rita Cruz, Elisabetta Soldaini, Andrea Guido Oreste Manetti and Fabio Bagnoli

Contents

Abstract The use of human organotypic models for biomedical research is experiencing a significant increase due to their biological relevance, the possibility to perform high-throughput analyses, and their cost efficiency. In the field of anti-infective research, comprising the search for novel antipathogenic treatments including vaccines, efforts have been made to reduce the use of animal models. That is due to two main reasons: unreliability of data obtained with animal models and

---

Astrid Hendriks, Ana Rita Cruz, These authors contributed equally.

The original version of this chapter was revised: OpenAccess text has been included. The erratum to this chapter is available at https://doi.org/10.1007/82_2019_174.

A. Hendriks · A. R. Cruz · E. Soldaini · A. G. O. Manetti · F. Bagnoli (✉)
GSK, Siena, Italy
e-mail: fabio.x.bagnoli@gsk.com

A. Hendriks · A. R. Cruz
University Medical Center Utrecht, Utrecht, The Netherlands

Current Topics in Microbiology and Immunology (2021) 430: 77–99
https://doi.org/10.1007/82_2018_130

Published Online: 20 September 2018

the increasing willingness to reduce the use of animals in research for ethical reasons. Human three-dimensional (3-D) models may substitute and/or complement in vivo studies, to increase the translational value of preclinical data. Here, we provide an overview of recent studies utilizing human organotypic models, resembling features of the cervix, intestine, lungs, brain, and skin in the context of anti-infective research. Furthermore, we focus on the future applications of human skin models and present methodological protocols to culture human skin equivalents and human skin explants.

1 Introduction

Overall mortality caused by infectious diseases has dramatically decreased in the past decades due to the positive impact of antibiotics, antivirals, antifungals, vaccines, and more recently monoclonal antibodies. The use of in vitro and/or in vivo models in preclinical research aims to predict the level of safety, immunogenicity, and protection of novel anti-infective treatments. Unfortunately, animal models often fail in this respect, as shown by the high number of unsuccessful clinical trials. There is an abundant amount of pathogens with exclusive human-specific properties, thus underlining the need for human in vitro or ex vivo models to study these microbes and to obtain reliable results in preclinical studies. Furthermore, great efforts are being made to reduce, refine, and replace the use of animal models in various fields of biomedical research. Human organotypic models are becoming more popular in both the academic sector as well as industry, producing valuable information to help find novel therapeutic targets. Here, we will present an overview of human organ-specific models used in anti-infective research, focusing on cervical, intestinal, respiratory, brain, and skin models. Moreover, we will dedicate a section of this review to human skin models commenting on their future applications and presenting a methodological protocol for the culture of human skin explants and skin equivalents that can be applied to anti-infective research.

2 Part 1—State of the Art of Organotypic Models for Anti-infective Research

2.1 Cervical Models

Human 3-D organotypic cultures, also known as raft cultures, have been developed in order to mimic human cervical epithelium. These models are frequently composed of primary cervical keratinocytes or cervical carcinoma cell lines seeded on top of a collagen matrix filled with fibroblasts (Delvenne et al. 2001). Organotypic raft cultures are then grown at air–liquid interface in order to promote differentiation and consequent stratification of the epithelium.

Epithelial differentiation is essential for human papillomavirus (HPV) replication (Meyers et al. 1992); thus, cell monolayers are limited models to study the life cycle of HPV and, consequently, to develop antiviral therapies against its replication. For this reason, organotypic epithelial raft cultures have been used to evaluate the efficacy of antivirals in blocking replication of different HPV types. Two commercial preparations of interferon-α (IFN-α) (Sen et al. 2005) and an antisense oligonucleotide targeting the mRNA of an essential protein (ORI-1001) for HPV6 and HPV11 replication (Alam et al. 2005) were tested in this model. These antivirals were able to impair the replication of specific HPV types, and the optimal concentration of the two IFN-α preparations was shown to be different for each of the HPV types. Additionally, the efficacy of bi-specific monoclonal antibodies anti-CD3/anti-EGFR, designed to target epidermal growth factor receptor (EGFR)-positive neoplastic keratinocytes by activating T cells (Ferrini et al. 1993), was also evaluated in a human 3-D model of the cervical epithelium (Renard et al. 2002). After the demonstration that lymphocytes can penetrate in 3-D cultures composed by a cervical carcinoma cell line (Jacobs et al. 1998), the authors proved that T cells, previously incubated with the bi-specific antibodies, induce apoptosis of HPV16-transformed primary keratinocytes, but not of healthy keratinocytes (Renard et al. 2002). These studies highlighted the feasibility and preclinical relevance of organotypic cultures of keratinocytes to test different classes of anti-HPV agents.

Human cervical explants have also been extensively used to screen the efficacy and toxicity of topical antiviral candidates, mainly for human immunodeficiency virus (HIV). The compounds tested include polyanionic molecules, such as the synthetic naphthalenesulfonate polymer PRO2000 and dextrin sulfate (DxS); cellulose acetate 1,2-benzenedicarboxylate (CAP); the carrageenan-based product PC-515; the lysine dendrimer SPL7013; the antimicrobial peptide D2A21; cyanovirin-N; palmitic acid; nonoxynol-9; gramicidin; the pyrimidinedione IQP-0528; and the non-nucleoside reverse transcriptase inhibitor UC781 (Greenhead et al. 2000; Zussman et al. 2003; Fletcher et al. 2005, 2006; Lu et al. 2006; Cummins et al. 2007; Buffa et al. 2009; Lin et al. 2011; Mahalingam et al. 2011). The combinatorial effect of these compounds in preventing HIV-1 infection has also been evaluated in polarized ectocervical explants (Zhang et al. 2015).

Infections caused by HIV are frequently accompanied by other sexually transmitted infections, such as herpes simplex virus (HSV) (Wald and Link 2002; Buve 2010). Highly specific drugs against each of these viruses have been designed, including acyclovir (HSV-specific) and tenofovir (HIV-specific). Surprisingly, clinical trials revealed that acyclovir could reduce HIV-1 levels in HSV-2/HIV-1 coinfected patients (Nagot et al. 2007; Zuckerman et al. 2007; Baeten et al. 2008) and that tenofovir reduced the risk of HSV-2 infection (Cates 2010). The ability of acyclovir to suppress HIV-1 infection was confirmed in human tonsils, lymph nodes, cervicovaginal and colorectal tissues inoculated with HIV-1 and HSV-2 (Lisco et al. 2008), and the effectiveness of tenofovir to impair HSV replication was evaluated in several in vitro and ex vivo models such as organotypic epithelial 3-D rafts, human lymphoid and cervical tissues (Andrei et al. 2011). Therefore, cervical and other types of tissues showed to be suitable models to assess the antiviral effect of microbicides against two different viruses. In a more recent study, the compound PMEO-DAPym (6-phosphonylmethoxyethoxy-2,4-diaminopyrimidine) was also identified to target both HSV and HIV infections (Balzarini et al. 2013). Its potency was also tested in multiple models, including organotypic epithelial raft cultures of human primary keratinocytes and human lymphoid and cervical tissues, showing an overall significant advantage over the existing compounds tenofovir and adefovir, which have been described to impair both HIV and HSV.

In summary, human cervical raft cultures and explants have been exploited as models to test the efficacy of various treatments, to optimize the concentration of microbicides and to assess the benefit of combining antivirals to tackle different viruses. Importantly, organotypic models have been used to overcome the limitations of cell monolayers in reproducing in vivo conditions essential for viral replication.

2.2 Intestinal Models

In the last years, several studies started to use colorectal explant tissues to evaluate the antiretroviral efficacy of topical microbicides marketed or under evaluation in clinical trials for vaginal use. The safety and effectiveness of many antivirals such as CAP, PRO2000, SPL7013, Vena Gel, and the class of reverse transcriptase inhibitors (RTI), which includes the nucleotide-RTI tenofovir, the nucleoside-RTI emtricitabine (FTC) and the non-nucleoside-RTI UC-781, dapivirine (TMC120), and IQP-0528 was assessed in colorectal tissues (Abner et al. 2005; Fletcher et al. 2006; Herrera et al. 2009; Rohan et al. 2010; Herrera et al. 2011; Dezzutti et al. 2012, 2014). When tested in combination, RTI were shown to induce a greater inhibitory potential compared with the compounds tested alone (Herrera et al. 2009, 2011). The effectiveness of other types of drugs, such as the polyanion candidates PRO2000 and DxS, has been also tested in parallel with the RTI activity (Fletcher et al. 2006). More recently, a study combining the non-nucleoside-RTI TMC120 with the HIV entry inhibitor maraviroc proved a synergistic antiretroviral effect in both ectocervical and colonic tissue (Dezzutti et al. 2015). Also, in this case, the use of explant tissue models allows the preclinical assessment of both toxicity and efficacy of topical microbicides, and it facilitates the screening of multiple candidates, alone or in combination, against different viral subtypes, thereby being less laborious than in vivo models.

Another type of human intestinal organotypic model is the 3-D cultured primary intestinal organoid, which is an organ-like tissue generated from pluripotent stem cells isolated from intestinal tissue (Lancaster and Knoblich 2014). Organoids contain various types of cells and recapitulate most aspects of in vivo tissue architecture, like spatial organization and overall microenvironment. Intestinal organoids have been used to study host-pathogen interactions; nevertheless, the number of studies that uses this model to evaluate the efficacy of antimicrobial agents is still limited. Yin et al. showed that primary human intestinal organoids were permissive to rotavirus infection, both to laboratory and patient-derived rotavirus strains. Importantly, clinical isolates demonstrated different sensitivities to the antivirals IFN-α and ribavirin, suggesting that this organotypic model could be valuable to predict the individual responsiveness to antiviral therapy, thus supporting personalized medicine (Yin et al. 2015). Another study implemented a functional assay to determine the efficacy of cholera toxin inhibitors by measuring the swelling of primary human intestinal organoids (Zomer-van Ommen, Pukin et al. 2016).

2.3 Respiratory Models

The respiratory system has a complex architecture and consists of different distinct structures such as the nasal mucosa, the upper and lower bronchial tracts and alveoli. Due to the direct contact with the air, natural exposure to various pathogens is very common. Subsequently, pneumonia remains one of the leading causes of death worldwide, thereby making the respiratory system a key target model for anti-infective research (Global Burden of Disease Study 2015). Certain aspects of the respiratory system, such as the structural complexity and the different cell types involved, make it difficult to fully simulate all features in vitro. Different 3-D models, which mimic specific components or processes of the respiratory system, have been extensively used to test various novel anti-infective agents. For example, the protective role of nasal sprays in maintaining the barrier function of nasal epithelial cells has been shown in vitro using a commercially available 3-D nasal mucosal model (De Servi et al. 2017). The ability of a novel antimicrobial peptide, derived from frog skin, to affect the permeability of lung epithelium has been shown using primary bronchial epithelial cells differentiated in vitro, where the barrier integrity was measured with an electrode determining the transepithelial electrical resistance (TEER) (Chen et al. 2017). A 3-D co-culture system consisting of the human airway epithelial cell line BEAS-2B, immature dendritic cells (DCs) derived from human circulating $CD14^+$ monocytes and human lung fibroblast cell line MRC-5, was recently used to discriminate skin sensitizers from respiratory sensitizers. In this model, OX40L upregulation in DCs was preferentially induced by respiratory sensitizers and was found to be associated with the initiation of the T helper type 2 (Th2) cell response leading to respiratory allergy (Mizoguchi et al. 2017). Similarly, a 3-D lung model, including an epithelial cell line in combination with a human lung fibroblast cell line, was used to test the effect of pooled immunoglobulin (IVIG) in preventing *S. aureus* toxins mediated tissue damage. In addition, live imaging analysis of GFP-expressing epithelial cells within these tissue cultures confirmed the protective effect of IVIG (Mairpady Shambat et al. 2015).

Ex vivo studies of lung-related samples are proving useful in anti-infective research as well. For example, the benefit of the vaccine adjuvant CpG to promote T follicular helper cells, which are important in eliciting a robust anti-influenza antibody response, has been shown in nasopharynx-associated lymphoid tissues

(NALT) derived from healthy children and adults (Aljurayyan et al. 2016). Furthermore, analysis of *Pseudomonas aeruginosa* biofilm aggregates in sputum from cystic fibrosis patients, treated with low concentrations of nitric oxide (NO), showed an enhanced efficacy of antibiotics in treating chronic infection (Howlin et al. 2017). Although beyond the scope of this review, it should be mentioned that numerous in vitro and ex vivo 3-D models are widely being used to study infection of bacteria as well as viruses and fungi and could be important models to test novel anti-infective treatments in the future. In brief, ex vivo models such as human nasal respiratory mucosa have been used in the context of viral infection for studying herpes simplex virus type 1 invasion (Glorieux et al. 2011; Wang et al. 2012). Besides this, infections with *Mycobacterium tuberculosis*, *Aspergillus fumigatus*, and *Burkholderia cenocepacia* have also been extensively studied (Chandorkar et al. 2017; Fonseca et al. 2017; Van den Driessche et al. 2017). A future potential application of these models can be to test the efficacy of nasal nanovaccines, which represent a promising novel anti-infective (Bernocchi et al. 2017). In conclusion, 3-D organotypic human lung cell culture models can provide important information on the efficacy of novel antimicrobials.

2.4 Brain Models

Due to the large anatomical differences between the human brain and other mammals, there is a great need to establish experimental human brain models. However, since the brain contains a dense network of various cell types such as neurons, glial cells, and astrocytes, with complex interaction patterns, it is very challenging to recapitulate this system in vitro. Over the past years, a lot of effort has been made to this cause, e.g., in trying to develop neurospheroids and cerebral organoids (Jorfi et al. 2017). Many studies utilized 3-D human brain models to study the pathogenesis and identify potential therapeutic targets of infectious agents causing damage to the brain. A major advantage of using these models is that they allow for the high-throughput screening of potential anti-infectives, a very important feature in the response to the fast emergence of Zika virus. A pool of existing drugs has been tested for efficacy against Zika and Japanese encephalitis virus infection in in vitro models including fetal-like organoids, human fetal-derived neuronal stem cells, human cortical organoids, and human pluripotent stem cell-derived cortical neural progenitor cells (hNPCs) (Xu et al. 2016; Bullard-Feibelman et al. 2017; Li et al. 2017; Wang et al. 2017; Zhou et al. 2017). Here, it should be noted that the in vitro efficacy of antivirals has been confirmed by additional in vivo experiments in these studies. However, in the case of human-specific pathogens such as varicella zoster virus (VZV), the use of 3-D human brain models is even more critical. Human neuronal systems, derived from embryonic and neural stem cells, have been established to study the various stages of VZV infection and might function as a tool to investigate novel antiviral targets (Markus et al. 2015; Kurapati et al. 2017). Another study aimed to mimic the human blood–brain barrier (BBB) by culturing a human brain microvascular endothelial cell line (HBMEC) in a rotating wall vessel (RWV) bioreactor. This model can be used to study the invasion of pathogens into the central nervous system (Bramley et al. 2017). This system may enable the screening of novel or existing antivirals for their protective effect. Collectively, human brain models are extensively being used as a high-throughput screening tool for anti-infective research, in particular for antivirals. However, data from currently used 3-D models still require additional studies in animal models, due to the very complex characteristics of the human brain.

3 Part 2—in-Depth Overview of Skin Models for Anti-infective Research

The skin is the largest human organ and functions as a barrier against physical injury, infections, and environmental cues (e.g., UV light) and contributes to body temperature regulation (Proksch et al. 2008; Olaniyi et al. 2017). The skin is composed of two major layers: the epidermis and the dermis. Keratinocytes are the main cellular component of the epidermis and its stage of differentiation subdivide the epidermis into four *strata*: *stratum corneum, stratum granulosum, stratum spinosum,* and *stratum basale*. The outermost layer, *stratum corneum*, is populated by terminally differentiated keratinocytes, while *stratum basale* is the inner layer of the epidermis, which is composed of undifferentiated keratinocytes. Although less abundant, Langerhans cells, melanocytes, and cytotoxic T cells can also be found in the epidermis. The dermis is mainly composed of collagen and elastin fibers and is rich in immune cells such as dendritic cells, T cells, macrophages, and mast cells. Fibroblasts, nerves, as well as blood and lymphatic vessels are also present in this layer. In addition, epidermal appendages, such as sweat and sebaceous glands as well as hair follicles, are embedded in the dermal layer. Hair follicles transverse the two compartments of the skin, representing a major route of penetration for everything coming into contact with the skin, including microbes.

Keratinocytes are the first players to act when the skin barrier is challenged. Therefore, 2-D monolayers of keratinocytes have been used to study the primary immune response induced by keratinocytes in multiple fields, ranging from cosmetics to injury and pathogen infection. Nevertheless, monolayers of keratinocytes do not resemble the stratification characteristic of the epidermis. In order to overcome this limitation, reconstructed human epidermis (RHE) models can be built by culturing keratinocytes at the air–liquid interface (Netzlaff et al. 2005).

However, a complete model of the skin is needed to study host–pathogen interactions and anti-infective interventions, including their toxicity, efficacy, and induced immune response. Since animal experimentation is time-consuming, expensive, and its predictive value is limited when applied to humans, the ideal model should reproduce both histology and architecture of the human skin.

Full-thickness skin models, such as human skin equivalents and human skin explants, fulfill these requirements.

3.1 Skin Equivalent Versus Skin Explants

Many studies describe the development of skin equivalent models where the different layers of the skin are rebuilt using primary fibroblasts within a matrix of collagen followed by seeding with keratinocytes. Further differentiation of keratinocytes at air–liquid interface results in stratified layers, closely resembling the human skin. De-vitalized de-epidermalized dermis (DED), obtained by discarding both epidermis and adipose tissue from human skin, can also be used as a more physiological support to grow keratinocytes (Lamb and Ambler 2014; Ventress et al. 2016). In addition, either primary or immortalized keratinocytes can be used to construct the epidermal layer, although the widely used HaCaT cell line does not reach the terminal differentiation stage (Boelsma et al. 1999; Schoop et al. 1999). Human skin explants, commonly collected from cosmetic surgery procedures, represent the most complete skin model.

When selecting a model to evaluate the efficacy or toxicity of an anti-infective compound, it should be taken into account that each of the full-thickness human skin models, which include skin equivalents and skin explants, presents advantages and disadvantages (Table 1).

In order to compensate for the lack of immune cells in the skin equivalent models, Langerhans cells (Dezutter-Dambuyant et al. 2006; Laubach et al. 2011; Ouwehand et al. 2011), dendritic cells (Bechetoille et al. 2007) and macrophages (Bechetoille et al. 2011) have been successfully incorporated into skin equivalent models. Nevertheless, the fact that skin equivalent is built using cells from multiple

Table 1 Advantages and limitations of the use of full-thickness skin models for anti-infective research (Grivel and Margolis 2009; Popov et al. 2014)

Full thickness skin model	Advantages	Limitations
Human Skin Equivalent	Reproducibility Control over the skin components Possibility of genetic manipulation	Lacks most dermal cellular components Genetic heterogeneity of the different skin components
Human Skin Explants	Preserved histology and architecture of the tissue Potential to recapitulate local immune responses	Interdonor variability Limited time of tissue culture Limited availability of the tissue

donors makes the model genetically heterogeneous. On the other hand, the possibility of modulating the cellular components of the skin equivalents allows for studying the role of each cell type. Another advantage of skin equivalent models is the possibility of employing genetic engineering to unveil the function of a specific factor in the context of an infection and immune response, by inducing or depleting its expression in a particular cell type before its introduction into the skin matrix.

Skin explants preserve the histology and the complexity of the skin, being composed of several different resident cells (including immune cells) and appendages. This feature enables the study of the cutaneous local immune response raised against a pathogen or triggered by an anti-infective immune intervention at this site. However, skin explants have a short-term viability and limited availability. Additionally, the explant is not connected to circulation and consequently lacks the ability to recruit immune cells from the blood to the explant, such as neutrophils that are important for microbial clearance. Further, the inter-donor variability of the human skin explants makes interpretation of results more complex, thus requiring a greater number of samples from different donors to be analyzed in order to guarantee validity of the results. On the other hand, this feature is a key to better understand the host-inherent variability leading to differential individual responses, opening the door to personalized medicine.

3.2 *Use of Human Skin Models in Anti-infective Research*

Similar to the respiratory system, the skin has direct contact with the outside environment and therefore acts as a physical and chemical barrier for external threats. The skin is a common initial site of infection, where the first encounter of the immune system with a pathogen often occurs. Consequently, skin models are important for studying infections and anti-infective agents.

Although human skin models have been used to study microbial infection and pathogenesis by a number of authors, there are only a few studies focusing on anti-infective research. Since mouse skin does not resemble human skin in either histology or immunology (Pasparakis et al. 2014), efforts have been made to use human skin equivalent and human skin explant models to test safety and efficacy of antibacterial and antiviral compounds.

Poxviruses have a tropism for skin and mucosal tissues. Thereby, an organotypic culture system, built with human primary keratinocytes derived from neonatal foreskin, was used to prove the inhibitory effect of some acyclic nucleoside phosphonates on the viral cytopathic effect (Snoeck et al. 2002). The same model was also used to evaluate the antiviral activity of the compounds acyclovir, penciclovir, brivudin, foscarnet, and cidofovir against wild-type and mutant alphaherpesviruses HSV type 1 and type 2, and VZV (Andrei et al. 2005). In both studies, the antivirals were added to the medium to mimic systemic delivery of the compounds.

Due to the fact that *Staphylococcus aureus* is the major cause of bacterial skin and soft tissue infections in humans (McCaig et al. 2006) and the widespread emergence of antibiotic-resistant strains, recent studies have tested strategies to block this pathogen at the initial site of the infection. A study performed in a human full-thickness skin equivalent, constructed by keratinocytes and fibroblasts seeded on top of a DED scaffold, showed that a tetraspanin-derived peptide can halve the number of *S. aureus* adherent to keratinocytes as compared to a scrambled control peptide, without impairing keratinocyte viability (Ventress et al. 2016). Although it is still not clear if *S. aureus* infection is the cause or the consequence of atopic dermatitis (AD), the low expression of cutaneous antimicrobial peptides (AMPs) in AD patients is thought to contribute to their increased susceptibility to infections (Ong et al. 2002). Therefore, different strategies have been employed to increase AMP levels to prevent *S. aureus* infection. A study published this year used heat-killed *Lactobacillus johnsonii* NCC 533, a non-replicating probiotic, as a topical treatment to increase the expression of AMP. This treatment reduced the adhesion of *S. aureus* to the RHE model by up to 74% (Rosignoli et al. 2018). Synthetic antimicrobial peptides derived from the natural AMP LL-37, namely P60.4, peptide 10, and SAAP-148, have been tested for their efficiency against multi-drug resistance strains on RHE models (Haisma et al. 2016), human skin equivalents (Haisma et al. 2014), and split-thickness human skin explants (de Breij et al. 2012), respectively. All peptides proved to have antimicrobial and antibiofilm properties against methicillin-resistant *S. aureus* (MRSA), and SAAP-148 was also effective against multi-drug resistant (MDR) *Acinetobacter baumannii* (de Breij et al. 2012). LL-37 formulated in a nano-sized liposome has also been proved to be effective against herpes simplex virus 1 (HSV-1) infection in RHE models (Ron-Doitch et al. 2016). Interestingly, the cytotoxicity induced by the formulation was greater in the monolayer of HaCaT cells than in the RHE model, which could be due to the presence of *stratum corneum*, acting as a protective layer.

Due to the increasing interest in this field of research, we focused on possible future directions with the skin model based on current trends in anti-infective research. The skin, in particular the dermis, can serve as an efficient delivery system of vaccines due to the easy access and presence of many immune cells which can elicit a strong response to vaccine antigens. Intradermal vaccination has already shown to alter Langerhans cell distribution within human skin explants (Pearton et al. 2010), as well as to induce DC maturation (de Gruijl et al. 2006). Similarly, vaccination by microneedles can affect dendritic cell migration (Pearton et al. 2013). Since more evidence is emerging that the gut microbiome can influence immune responses, and hence vaccine efficacy (Littman 2017; Lynn and Pulendran 2017), one could argue that this applies to the skin microbiome as well. This hypothesis could be evaluated using the human skin model, whereby investigating the effect of known skin commensals on the cutaneous immune system. The application of topical compounds, e.g., in treating psoriasis (Kusuba et al. 2017) or other inflammatory diseases, and its effect on susceptibility to infections can be studied in depth with skin models, especially skin explants since the skin surface closely resembles the in vivo situation. In addition, novel anti-infective approaches, such as nano-sponges that absorb secreted bacterial toxins (Hu et al. 2013; Escajadillo et al. 2017) or other innovative antimicrobial delivery systems (Hasan et al. 2017), can be tested for their efficacy with the skin model. Similarly, our group recently showed the protective effect of antibodies against *S. aureus* toxins within the context of the human skin explant model (Olaniyi et al. 2018).

Despite their many advantages, most of the currently used skin models have some limitations, as mentioned before. One of the most prominent limitations is the lack of a systemic flow, thereby leaving out essential immune processes which are key in anti-infective research, i.e., neutrophil recruitment, complement activation, and T cell migration (Pozzi et al. 2015). One way to overcome this deficit is the introduction of the organ on a chip technology, where microfluidics are incorporated in a culture model (Ahadian et al. 2017). Many skin-on-chip models have been established by using different cell types, as reviewed in (van den Broek et al. 2017), and may provide novel insights on therapeutic approaches in the future (Wufuer et al. 2016). Another technique to mimic skin perfusion is an isolated perfused human skin flap model, which has been developed to test drug delivery to the skin. In brief, a human skin explant was linked to an automated perfusion system via one vessel, thereby inducing a controlled, continuous flow. An infrared (IR) camera was used to constantly monitor the skin surface temperature, and thus the degree of skin perfusion (Ternullo et al. 2017). In conclusion, skin models provide a new platform to study the effectiveness of various novel anti-infective compounds, as summarized in Fig. 1.

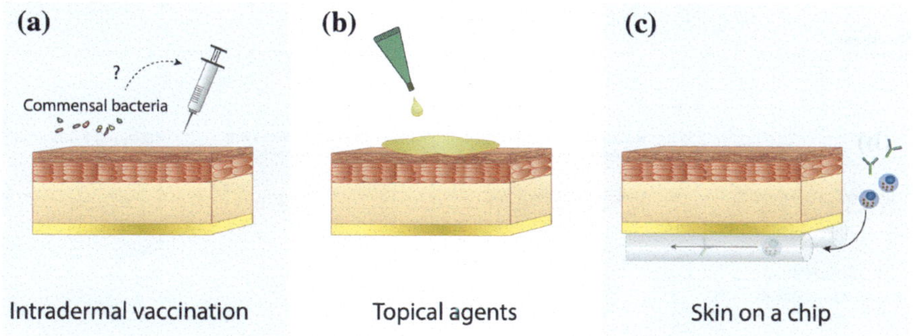

Fig. 1 Proposed future applications of human skin models in anti-infective research. **a** Human skin models can be used to study mechanism of action of intradermal vaccination and its efficacy against dermal pathogens and its interaction with commensal bacteria. **b** Novel topical agents may be tested for efficacy and potential side effects in a human skin model, either skin equivalent or skin explant. **c** Incorporation of microfluidics in the human skin model would mimic the systemic blood flow in the culture model enabling critical immune processes, such as neutrophil recruitment or tissue distribution of antibodies

3.3 Conclusion

Taken together, over the last decade, human 3-D organotypic models have acquired increasing importance and great efforts have been made to improve the culture, handling, and reproducibility of these in vitro models. In contrast, a trend may be observed in which the use of animal models is becoming obsolete. One explanation could be the overall consensus that the use of animal models has to be reduced from an ethical point of view. Secondly, the occurrence of failed studies on novel anti-infectives in the clinical phase underlines the complexity of predicting human response using animal models. Despite the remaining limitations, the use of human organotypic models can be a valuable tool to facilitate the transition from pre-clinical to clinical studies.

3.4 Protocol: Culture of Human Skin Equivalents and Skin Explants for Use in Anti-infective Research

Figure 2 summarizes the procedures to culture full-thickness human skin models for use in anti-infective research.

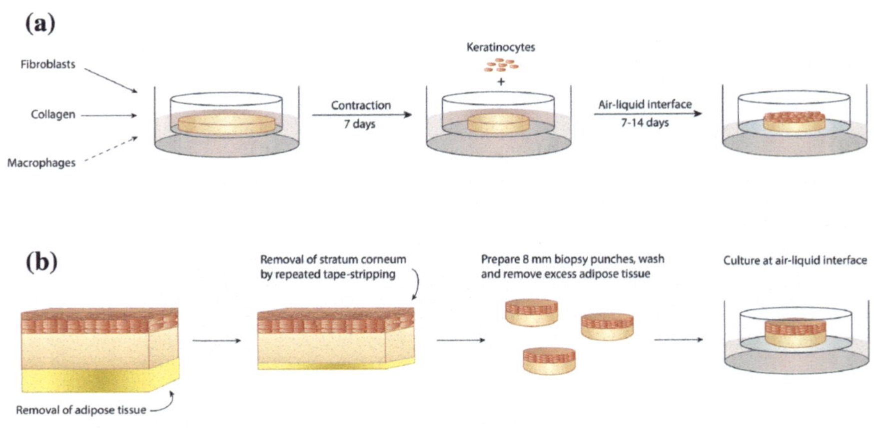

Fig. 2 Schematic overview of the protocol for the culture of human skin equivalents and human skin explants. **a** Human skin equivalent can be cultured in two steps, starting with the dermal component consisting of collagen and fibroblasts. Other immune cells, such as macrophages, can be included at this step. To allow for the fibroblasts to contract, keratinocytes can be seeded after 7 days. Thereafter, differentiation of keratinocytes at air–liquid interface will require 7–14 days, depending on the origin of the keratinocytes. **b** When handling fresh human skin explants four different critical steps are involved. The first is the removal of the adipose tissue. To create a superficial wound for infection protocols, the *stratum corneum* can be removed by repeated tape-stripping followed by the preparation of punches with disposable surgical biopsy punches. The biopsy punches can be washed and cultured at air–liquid interface

3.4.1 Procedure for the Preparation of Human Skin Equivalents

1. **Materials**

   1.1 **Cells**

   - Primary human dermal fibroblasts
   - Human keratinocyte cell line/primary human keratinocytes

   1.2 **Reagents**

   - Collagen type I from bovine hides (3 mg/ml)
   - Sterile phosphate buffered saline (PBS)
   - Trypsin-EDTA (0.05%)
   - 2N NaOH
   - HBSS 10x with phenol-red
   - Fibroblast culture medium:
     - DMEM High glucose
     - 100 U/mL of penicillin and 100 μg/mL of streptomycin
     - 10% fetal bovine serum (FBS) heat inactivated at 56 °C for 30 min
   - Keratinocyte culture medium:
     - DMEM high glucose
     - 100 U/mL of penicillin and 100 μg/mL of streptomycin
   - FAD medium:
     - DMEM/F12 (1:1)
     - 0.4 ug/mL hydrocortisone
     - 50 ug/mL L-ascorbic acid
     - 1×10^{-10} M Cholera toxin
     - 1% antibiotic-antimycotic
     - 10% FBS heat inactivated
   - FAD airlift interface medium (ALI)
     - FAD medium
     - 0.5 mM Na-pyruvate
     - 1.6 mM CaCl2
     - 20 ng/mL TGF-alpha
     - 100 ng/mL GM-CSF

   All media must be filtered and stored at 4 °C, *no longer than* 2 weeks.

   1.3 **Equipment**

   - Cell culture incubator (37 °C, 5% CO_2)
   - Cell culture incubator (37 °C, 0% CO_2)
   - Laminar flow class II cabinet
   - Hemocytometer

- Pipet-aid
- Sterile disposable pipettes
- Single pipette with sterile tips
- Polypropylene tubes, 15 mL and 50 mL
- Glass Pasteur pipettes
- 12 transwell inserts preloaded in 12-well plates; pore size: 0.4 μm, sterile (Transwell®)
- Sterile filter bottles (0.2 μm)
- Ice buckets
- Water bath
- Centrifuge

2. Methods

2.1 Dermal component

Before starting the protocol:

- *Place* 10xHBSS, *collagen and* 2N NaOH *on ice and disposable pipettes in the fridge;*
- *Set an incubator at* 37 °C, 0% CO_2;
- *Calculate the amount of collagen mix needed to fill each insert with* 500 μL. *Considering that collagen mix is composed of collagen:* 10xHBSS*: fibroblast solution in the ratio* 8:1:1, *calculate the required amount of each of the components.*

The following procedure must be performed under a laminar flow cabinet:

1. Harvest fibroblasts and count the number of cells using a hemocytometer
2. Suspend fibroblasts in FBS to a concentration of 1×10^6 cells/mL
3. Keep it at room temperature (RT)

From this step forward work always on ice:

4. Slowly add one part of chilled 10xHBSS to eight parts of chilled collagen in a 50 mL tube, with gentle swirling
5. Adjust the pH of the solution to pH 7.4 using 2N NaOH, dropwise until the color changes to orange/red
   It usually requires 50 μL of 2N NaOH per 10 mL *of collagen mix*
6. Add one part of the fibroblasts solution to the collagen mix and homogenize the gel solution carefully, using a disposable pipette.
7. Add 500 μL of the complete collagen mix onto the filter inserts
   Prevent bubbles by reverse pipetting
8. To allow the gel to solidify, place the 12-well plates for 1 h at 37 °C, 0% CO_2
9. Add 2 mL FAD medium (1.5 mL outside, 0.5 mL inside the insert) and incubate for 7 days at 37 °C, 5% CO_2. Replace the medium every 2-3 days.

2.2 **Epidermal component**

10. Harvest keratinocytes and count the number of cells using a hemocytometer
11. Prepare keratinocytes at 5×10^5 cell/ml in keratinocyte culture medium
12. Remove the medium inside the inserts
13. Add 20 μL of keratinocyte solution to the dermal equivalent
14. Incubate for 2 h at 37 °C, 5% CO_2
15. Add 500 μL of FAD medium inside the inserts
16. Incubate overnight
17. Remove all medium from the wells
18. Add 500 μL FAD ALI medium outside the insert, to allow culture at air–liquid interface
19. Incubate up to 7–14 days (dependent on differentiation status) and change FAD ALI medium every 2–3 days.

3.4.2 Procedure for the Preparation of Human Skin Explants (Olaniyi et al. 2018)

The outermost *stratum* of the epidermis, *stratum corneum*, is the major responsibility for maintaining the physical barrier function of the skin (Baroni et al. 2012). Therefore, this barrier has to be breached so that a pathogen can invade the tissue. Here, we described the use of the tape-stripping strategy as we consider that this method guarantees reproducibility between experiments.

1. **Materials**

1.1 **Reagents**

- Fresh human skin explants
- Sterile phosphate buffered saline (PBS)
- Culture medium:
  - Advanced DMEM supplemented with 4 mM L-glutamine

All media should be filtered and stored at 4 °C, *no longer than* 2 weeks.

1.2 **Equipment**

- Cell culture incubator (37 °C, 5% CO_2)
- Laminar flow class II cabinet
- Pipet-aid
- Sterile disposable pipettes
- Single pipette with sterile tips
- Glass Pasteur pipettes
- 12 transwell inserts preloaded in 12-well plates; pore size: 0.4 μm, sterile (Transwell®)
- Disposable surgical biopsy punches

- Disposable scalpels
- Sterile forceps
- Sterile scissors
- Sterile filter bottles (0.2 μm)
- Water bath
- Adhesive tape
- Dissection board
- Disinfected pins

2. Methods

2.1 Skin processing

Process the tissue immediately after its delivery.
The following procedure must be performed under a laminar flow cabinet:

1. Remove the excess of adipose tissue, using disposable scalpels or sterile scissors
2. Pin the skin in the dissection board
3. Strip the *stratum corneum* 30 times with an adhesive tape
   Make sure that the surface of the skin is dry before starting
4. Take 8 mm (in diameter) punch biopsies using disposable surgical biopsy punches
5. Place the punches inside a six-well plate, previously filled in with culture medium to prevent them to dry
6. Remove the remaining adipose tissue from the punches using a scalpel
7. Add 1 mL of culture medium in the 12-well plate
8. Place the skin punches in the inserts using sterile forceps
9. Culture the explants at air–liquid interface, at 37 °C, 5% CO_2.

Acknowledgements We would like to acknowledge Reuben Olaniyi and Proinnsias Fox at GSK for critical reading of the manuscript.

Funding

This project is funded by the European Union's Horizon 2020 research and innovation program under the Marie Skłodowska-Curie grant agreement No 675106.

Authorship

Fabio Bagnoli had the initial idea of the topic, content, and structure of the review article. Astrid Hendriks and Ana Rita Cruz defined the content and wrote the manuscript. Fabio Bagnoli, Andrea Manetti, and Elisabetta Soldaini reviewed and wrote the manuscript. All authors approved the manuscript before it was submitted.

Declaration of interest

Astrid Hendriks and Ana Rita Cruz are Ph.D. fellows and are enrolled in the Infection and Immunity Ph.D. program, part of the Graduate school of Life Sciences at the University of Utrecht. Fabio Bagnoli, Andrea Manetti, and Elisabetta Soldaini are employees of GSK group of companies. Fabio Bagnoli owns patents on *S. aureus* vaccine candidates and GSK stocks. The authors have no other relevant affiliations or financial involvement with any organization or entity with a financial interest in or financial conflict with the subject matter or materials discussed in the manuscript apart from those disclosed.

References

Abner SR, Guenthner PC, Guarner J, Hancock KA, Cummins JE Jr, Fink A, Gilmore GT, Staley C, Ward A, Ali O, Binderow S, Cohen S, Grohskopf LA, Paxton L, Hart CE, Dezzutti CS (2005) A human colorectal explant culture to evaluate topical microbicides for the prevention of HIV infection. J Infect Dis 192(9):1545–1556

Ahadian S, Civitarese R, Bannerman D, Mohammadi MH, Lu R, Wang E, Davenport-Huyer L, Lai B, Zhang B, Zhao Y, Mandla S, Korolj A, Radisic M (2017) Organ-on-a-chip platforms: a convergence of advanced materials, cells, and microscale technologies. Adv Health Mater

Alam S, Bromberg-White J, McLaughlin-Drubin M, Sen E, Bodily JM, Meyers C (2005) Activity and therapeutic potential of ORI-1001 antisense oligonucleotide on human papillomavirus replication utilizing a model of dysplastic human epithelium. Anticancer Res 25(2A):765–777

Aljurayyan AN, Sharma R, Upile N, Beer H, Vaughan C, Xie C, Achar P, Ahmed MS, McNamara PS, Gordon SB, Zhang Q (2016) A critical role of T follicular helper cells in human mucosal anti-influenza response that can be enhanced by immunological adjuvant CpG-DNA. Antiviral Res 132:122–130

Andrei G, Lisco A, Vanpouille C, Introini A, Balestra E, van den Oord J, Cihlar T, Perno CF, Snoeck R, Margolis L, Balzarini J (2011) Topical tenofovir, a microbicide effective against HIV, inhibits herpes simplex virus-2 replication. Cell Host Microbe 10(4):379–389

Andrei G, van den Oord J, Fiten P, Opdenakker G, De Wolf-Peeters C, De Clercq E, Snoeck R (2005) Organotypic epithelial raft cultures as a model for evaluating compounds against alphaherpesviruses. Antimicrob Agents Chemother 49(11):4671–4680

Baeten JM, Strick LB, Lucchetti A, Whittington WL, Sanchez J, Coombs RW, Magaret A, Wald A, Corey L, Celum C (2008) Herpes simplex virus (HSV)-suppressive therapy decreases plasma and genital HIV-1 levels in HSV-2/HIV-1 coinfected women: a randomized, placebo-controlled, cross-over trial. J Infect Dis 198(12):1804–1808

Balzarini J, Andrei G, Balestra E, Huskens D, Vanpouille C, Introini A, Zicari S, Liekens S, Snoeck R, Holy A, Perno CF, Margolis L, Schols D (2013) A multi-targeted drug candidate with dual anti-HIV and anti-HSV activity. PLoS Pathog 9(7):e1003456

Baroni A, Buommino E, De Gregorio V, Ruocco E, Ruocco V, Wolf R (2012) Structure and function of the epidermis related to barrier properties. Clin Dermatol 30(3):257–262

Bechetoille N, Dezutter-Dambuyant C, Damour O, Andre V, Orly I, Perrier E (2007) Effects of solar ultraviolet radiation on engineered human skin equivalent containing both Langerhans cells and dermal dendritic cells. Tissue Eng 13(11):2667–2679

Bechetoille N, Vachon H, Gaydon A, Boher A, Fontaine T, Schaeffer E, Decossas M, Andre-Frei V, Mueller CG (2011) A new organotypic model containing dermal-type macrophages. Exp Dermatol 20(12):1035–1037

Bernocchi B, Carpentier R, Betbeder D (2017) Nasal nanovaccines. Int J Pharm

Boelsma E, Verhoeven MC, Ponec M (1999) Reconstruction of a human skin equivalent using a spontaneously transformed keratinocyte cell line (HaCaT). J Invest Dermatol 112(4):489–498

Bramley JC, Drummond CG, Lennemann NJ, Good CA, Kim KS, Coyne CB (2017) A three-dimensional cell culture system to model RNA virus infections at the blood-brain barrier. mSphere 2(3)

Buffa V, Stieh D, Mamhood N, Hu Q, Fletcher P, Shattock RJ (2009) Cyanovirin-N potently inhibits human immunodeficiency virus type 1 infection in cellular and cervical explant models. J Gen Virol 90(Pt 1):234–243

Bullard-Feibelman KM, Govero J, Zhu Z, Salazar V, Veselinovic M, Diamond MS, Geiss BJ (2017) The FDA-approved drug sofosbuvir inhibits Zika virus infection. Antiviral Res 137:134–140

Buve A (2010) Can we reduce the spread of HIV infection by suppressing herpes simplex virus type 2 infection? F1000 Med Rep 2

Cates W Jr (2010) After CAPRISA 004: time to re-evaluate the HIV lexicon. Lancet 376(9740): 495–496

Chandorkar P, Posch W, Zaderer V, Blatzer M, Steger M, Ammann CG, Binder U, Hermann M, Hortnagl P, Lass-Florl C, Wilflingseder D (2017) Fast-track development of an in vitro 3D lung/immune cell model to study Aspergillus infections. Sci Rep 7(1):11644

Chen C, Mangoni ML, Di YP (2017) In vivo therapeutic efficacy of frog skin-derived peptides against Pseudomonas aeruginosa-induced pulmonary infection. Sci Rep 7(1):8548

Cummins JE Jr, Guarner J, Flowers L, Guenthner PC, Bartlett J, Morken T, Grohskopf LA, Paxton L, Dezzutti CS (2007) Preclinical testing of candidate topical microbicides for anti-human immunodeficiency virus type 1 activity and tissue toxicity in a human cervical explant culture. Antimicrob Agents Chemother 51(5):1770–1779

de Breij A, Haisma EM, Rietveld M, El Ghalbzouri A, van den Broek PJ, Dijkshoorn L, Nibbering PH (2012) Three-dimensional human skin equivalent as a tool to study Acinetobacter baumannii colonization. Antimicrob Agents Chemother 56(5):2459–2464

de Gruijl TD, Ophorst OJ, Goudsmit J, Verhaagh S, Lougheed SM, Radosevic K, Havenga MJ, Scheper RJ (2006) Intradermal delivery of adenoviral type-35 vectors leads to high efficiency transduction of mature, CD8 + T cell-stimulating skin-emigrated dendritic cells. J Immunol 177(4):2208–2215

De Servi B, Ranzini F, Pique N (2017) Protective barrier properties of Rhinosectan(R) spray (containing xyloglucan) on an organotypic 3D airway tissue model (MucilAir): results of an in vitro study. Allergy Asthma Clin Immunol 13:37

Delvenne P, Hubert P, Jacobs N, Giannini SL, Havard L, Renard I, Saboulard D, Boniver J (2001) The organotypic culture of HPV-transformed keratinocytes: an effective in vitro model for the development of new immunotherapeutic approaches for mucosal (pre)neoplastic lesions. Vaccine 19(17–19):2557–2564

Dezutter-Dambuyant C, Black A, Bechetoille N, Bouez C, Marechal S, Auxenfans C, Cenizo V, Pascal P, Perrier E, Damour O (2006) Evolutive skin reconstructions: from the dermal collagen-glycosaminoglycan-chitosane substrate to an immunocompetent reconstructed skin. Biomed Mater Eng 16(4 Suppl):S85–S94

Dezzutti CS, Russo J, Wang L, Abebe KZ, Li J, Friend DR, McGowan IM, Rohan LC (2014) Development of HIV-1 rectal-specific microbicides and colonic tissue evaluation. PLoS ONE 9 (7):e102585

Dezzutti CS, Shetler C, Mahalingam A, Ugaonkar SR, Gwozdz G, Buckheit KW, Buckheit RW Jr (2012) Safety and efficacy of tenofovir/IQP-0528 combination gels—a dual compartment microbicide for HIV-1 prevention. Antiviral Res 96(2):221–225

Dezzutti CS, Yandura S, Wang L, Moncla B, Teeple EA, Devlin B, Nuttall J, Brown ER, Rohan LC (2015) Pharmacodynamic activity of Dapivirine and Maraviroc single entity and combination topical gels for HIV-1 prevention. Pharm Res 32(11):3768–3781

Escajadillo T, Olson J, Luk BT, Zhang L, Nizet V (2017) A red blood cell membrane-camouflaged nanoparticle counteracts streptolysin O-Mediated virulence phenotypes of invasive group a streptococcus. Front Pharmacol 8:477

Ferrini S, Cambiaggi A, Sforzini S, Marciano S, Canevari S, Mezzanzanica D, Colnaghi MI, Grossi CE, Moretta L (1993) Targeting of T lymphocytes against EGF-receptor + tumor cells by bispecific monoclonal antibodies: requirement of CD3 molecule cross-linking for T-cell activation. Int J Cancer 55(6):931–937

Fletcher P, Kiselyeva Y, Wallace G, Romano J, Griffin G, Margolis L, Shattock R (2005) The nonnucleoside reverse transcriptase inhibitor UC-781 inhibits human immunodeficiency virus type 1 infection of human cervical tissue and dissemination by migratory cells. J Virol 79(17): 11179–11186

Fletcher PS, Elliott J, Grivel JC, Margolis L, Anton P, McGowan I, Shattock RJ (2006a) Ex vivo culture of human colorectal tissue for the evaluation of candidate microbicides. AIDS 20(9): 1237–1245

Fletcher PS, Wallace GS, Mesquita PM, Shattock RJ (2006b) Candidate polyanion microbicides inhibit HIV-1 infection and dissemination pathways in human cervical explants. Retrovirology 3:46

Fonseca KL, Rodrigues PNS, Olsson IAS, Saraiva M (2017) Experimental study of tuberculosis: from animal models to complex cell systems and organoids. PLoS Pathog 13(8):e1006421
Global Burden of Disease Study C (2015) Global, regional, and national incidence, prevalence, and years lived with disability for 301 acute and chronic diseases and injuries in 188 countries, 1990–2013: a systematic analysis for the Global Burden of Disease Study 2013. Lancet 386 (9995):743–800
Glorieux S, Bachert C, Favoreel HW, Vandekerckhove AP, Steukers L, Rekecki A, Van den Broeck W, Goossens J, Croubels S, Clayton RF, Nauwynck HJ (2011) Herpes simplex virus type 1 penetrates the basement membrane in human nasal respiratory mucosa. PLoS ONE 6(7): e22160
Greenhead P, Hayes P, Watts PS, Laing KG, Griffin GE, Shattock RJ (2000) Parameters of human immunodeficiency virus infection of human cervical tissue and inhibition by vaginal virucides. J Virol 74(12):5577–5586
Grivel JC, Margolis L (2009) Use of human tissue explants to study human infectious agents. Nat Protoc 4(2):256–269
Grulich AE, Zablotska I (2010) Commentary: probability of HIV transmission through anal intercourse. Int J Epidemiol 39(4):1064–1065
Haisma EM, de Breij A, Chan H, van Dissel JT, Drijfhout JW, Hiemstra PS, El Ghalbzouri A, Nibbering PH (2014) LL-37-derived peptides eradicate multidrug-resistant Staphylococcus aureus from thermally wounded human skin equivalents. Antimicrob Agents Chemother 58 (8):4411–4419
Haisma EM, Goblyos A, Ravensbergen B, Adriaans AE, Cordfunke RA, Schrumpf J, Limpens RW, Schimmel KJ, den Hartigh J, Hiemstra PS, Drijfhout JW, El Ghalbzouri A, Nibbering PH (2016) Antimicrobial Peptide P60.4Ac-Containing Creams and Gel for Eradication of Methicillin-Resistant Staphylococcus aureus from Cultured Skin and Airway Epithelial Surfaces. Antimicrob Agents Chemother 60(7):4063–4072
Hasan S, Thomas N, Thierry B, Prestidge CA (2017) Controlled and Localized Nitric Oxide Precursor Delivery from Chitosan Gels to Staphylococcus aureus Biofilms. J Pharm Sci
Herrera C, Cranage M, McGowan I, Anton P, Shattock RJ (2009) Reverse transcriptase inhibitors as potential colorectal microbicides. Antimicrob Agents Chemother 53(5):1797–1807
Herrera C, Cranage M, McGowan I, Anton P, Shattock RJ (2011) Colorectal microbicide design: triple combinations of reverse transcriptase inhibitors are optimal against HIV-1 in tissue explants. AIDS 25(16):1971–1979
Howlin RP, Cathie K, Hall-Stoodley L, Cornelius V, Duignan C, Allan RN, Fernandez BO, Barraud N, Bruce KD, Jefferies J, Kelso M, Kjelleberg S, Rice SA, Rogers GB, Pink S, Smith C, Sukhtankar PS, Salib R, Legg J, Carroll M, Daniels T, Feelisch M, Stoodley P, Clarke SC, Connett G, Faust SN, Webb JS (2017) Low-Dose Nitric Oxide as targeted anti-biofilm adjunctive therapy to treat Chronic Pseudomonas *aeruginosa* Infection in Cystic Fibrosis. Mol Ther
Hu CM, Fang RH, Copp J, Luk BT, Zhang L (2013) A biomimetic nanosponge that absorbs pore-forming toxins. Nat Nanotechnol 8(5):336–340
Jacobs N, Moutschen MP, Franzen-Detrooz E, Boniver V, Boniver J, Delvenne P (1998) Organotypic culture of HPV-transformed keratinocytes: a model for testing lymphocyte infiltration of (pre)neoplastic lesions of the uterine cervix. Virchows Arch 432(4):323–330
Jorfi M, D'Avanzo C, Kim DY, Irimia D (2017) Three-dimensional models of the human brain development and diseases. Adv Health Mater
Kurapati S, Sadaoka T, Rajbhandari L, Jagdish B, Shukla P, Ali MA, Kim YJ, Lee G, Cohen JI, Venkatesan A (2017) Role of the JNK pathway in varicella-zoster virus lytic infection and reactivation. J Virol 91(17)
Kusuba N, Kitoh A, Dainichi T, Honda T, Otsuka A, Egawa G, Nakajima S, Miyachi Y, Kabashima K (2017) Inhibition of IL-17-committed T cells in a murine psoriasis model by a vitamin D analogue. J Allergy Clin Immunol
Lamb R, Ambler CA (2014) Modified methods for growing 3-D skin equivalents: an update. Methods Mol Biol 1195:171–182

Lancaster MA, Knoblich JA (2014) Organogenesis in a dish: modeling development and disease using organoid technologies. Science 345(6194):1247125

Laubach V, Zoller N, Rossberg M, Gorg K, Kippenberger S, Bereiter-Hahn J, Kaufmann R, Bernd A (2011) Integration of Langerhans-like cells into a human skin equivalent. Arch Dermatol Res 303(2):135–139

Li C, Deng YQ, Wang S, Ma F, Aliyari R, Huang XY, Zhang NN, Watanabe M, Dong HL, Liu P, Li XF, Ye Q, Tian M, Hong S, Fan J, Zhao H, Li L, Vishlaghi N, Buth JE, Au C, Liu Y, Lu N, Du P, Qin FX, Zhang B, Gong D, Dai X, Sun R, Novitch BG, Xu Z, Qin CF, Cheng G (2017) 25-Hydroxycholesterol protects host against Zika virus infection and its associated microcephaly in a mouse model. Immunity 46(3):446–456

Lin X, Paskaleva EE, Chang W, Shekhtman A, Canki M (2011) Inhibition of HIV-1 infection in ex vivo cervical tissue model of human vagina by palmitic acid; implications for a microbicide development. PLoS ONE 6(9):e24803

Lisco A, Vanpouille C, Tchesnokov EP, Grivel JC, Biancotto A, Brichacek B, Elliott J, Fromentin E, Shattock R, Anton P, Gorelick R, Balzarini J, McGuigan C, Derudas M, Gotte M, Schinazi RF, Margolis L (2008) Acyclovir is activated into a HIV-1 reverse transcriptase inhibitor in herpesvirus-infected human tissues. Cell Host Microbe 4(3):260–270

Littman DR (2017) Do the microbiota influence vaccines and protective immunity to pathogens? If so, is there potential for efficacious microbiota-based vaccines?" Cold Spring Harb Perspect Biol

Lu H, Zhao Q, Wallace G, Liu S, He Y, Shattock R, Neurath AR, Jiang BS (2006) Cellulose acetate 1,2-benzenedicarboxylate inhibits infection by cell-free and cell-associated primary HIV-1 isolates. AIDS Res Hum Retroviruses 22(5):411–418

Lynn DJ, Pulendran B (2017) The potential of the microbiota to influence vaccine responses. J Leukoc Biol

Mahalingam A, Simmons AP, Ugaonkar SR, Watson KM, Dezzutti CS, Rohan LC, Buckheit RW Jr, Kiser PF (2011) Vaginal microbicide gel for delivery of IQP-0528, a pyrimidinedione analog with a dual mechanism of action against HIV-1. Antimicrob Agents Chemother 55(4): 1650–1660

Mairpady Shambat S, Chen P, Nguyen Hoang AT, Bergsten H, Vandenesch F, Siemens N, Lina G, Monk IR, Foster TJ, Arakere G, Svensson M, Norrby-Teglund A (2015) Modelling staphylococcal pneumonia in a human 3D lung tissue model system delineates toxin-mediated pathology. Dis Model Mech 8(11):1413–1425

Markus A, Lebenthal-Loinger I, Yang IH, Kinchington PR, Goldstein RS (2015) An in vitro model of latency and reactivation of varicella zoster virus in human stem cell-derived neurons. PLoS Pathog 11(6):e1004885

McCaig LF, McDonald LC, Mandal S, Jernigan DB (2006) Staphylococcus aureus-associated skin and soft tissue infections in ambulatory care. Emerg Infect Dis 12(11):1715–1723

Meyers C, Frattini MG, Hudson JB, Laimins LA (1992) Biosynthesis of human papillomavirus from a continuous cell line upon epithelial differentiation. Science 257(5072):971–973

Mizoguchi I, Ohashi M, Chiba Y, Hasegawa H, Xu M, Owaki T, Yoshimoto T (2017) Prediction of chemical respiratory and contact sensitizers by OX40L expression in dendritic cells using a novel 3D coculture system. Front Immunol 8:929

Nagot N, Ouedraogo A, Foulongne V, Konate I, Weiss HA, Vergne L, Defer MC, Djagbare D, Sanon A, Andonaba JB, Becquart P, Segondy M, Vallo R, Sawadogo A, Van de Perre P, Mayaud P, Group AS (2007) Reduction of HIV-1 RNA levels with therapy to suppress herpes simplex virus. N Engl J Med 356(8):790–799

Netzlaff F, Lehr CM, Wertz PW, Schaefer UF (2005) The human epidermis models EpiSkin, SkinEthic and EpiDerm: an evaluation of morphology and their suitability for testing phototoxicity, irritancy, corrosivity, and substance transport. Eur J Pharm Biopharm 60(2): 167–178

Olaniyi RO, Pancotto L, Grimaldi L, Bagnoli F (2018) Deciphering the pathological role of *Staphylococcal* alpha-Toxin and Panton-Valentine Leukocidin using a novel ex vivo human skin model. Front Immunol 9:951

Olaniyi R, Pozzi C, Grimaldi L, Bagnoli F (2017) *Staphylococcus aureus*-associated skin and soft tissue infections: anatomical localization, epidemiology, therapy and potential prophylaxis. Curr Top Microbiol Immunol 409:199–227

Ong PY, Ohtake T, Brandt C, Strickland I, Boguniewicz M, Ganz T, Gallo RL, Leung DY (2002) Endogenous antimicrobial peptides and skin infections in atopic dermatitis. N Engl J Med 347(15):1151–1160

Ouwehand K, Spiekstra SW, Waaijman T, Scheper RJ, de Gruijl TD, Gibbs S (2011) Technical advance: langerhans cells derived from a human cell line in a full-thickness skin equivalent undergo allergen-induced maturation and migration. J Leukoc Biol 90(5):1027–1033

Pasparakis M, Haase I, Nestle FO (2014) Mechanisms regulating skin immunity and inflammation. Nat Rev Immunol 14(5):289–301

Pearton M, Kang SM, Song JM, Anstey AV, Ivory M, Compans RW, Birchall JC (2010) Changes in human Langerhans cells following intradermal injection of influenza virus-like particle vaccines. PLoS ONE 5(8):e12410

Pearton M, Pirri D, Kang SM, Compans RW, Birchall JC (2013) Host responses in human skin after conventional intradermal injection or microneedle administration of virus-like-particle influenza vaccine. Adv Healthc Mater 2(10):1401–1410

Popov L, Kovalski J, Grandi G, Bagnoli F, Amieva MR (2014) Three-dimensional human skin models to understand staphylococcus aureus skin colonization and infection. Front Immunol 5:41

Pozzi C, Lofano G, Mancini F, Soldaini E, Speziale P, De Gregorio E, Rappuoli R, Bertholet S, Grandi G, Bagnoli F (2015) Phagocyte subsets and lymphocyte clonal deletion behind ineffective immune response to *Staphylococcus aureus*. FEMS Microbiol Rev 39(5):750–763

Proksch E, Brandner JM, Jensen JM (2008) The skin: an indispensable barrier. Exp Dermatol 17 (12):1063–1072

Renard I, Mezzanzanica D, Canevari S, Ferrini S, Boniver J, Delvenne P, Jacobs N (2002) Anti-CD3/anti-epidermal growth factor receptor-bispecific antibody retargeting of lymphocytes against human neoplastic keratinocytes in an autologous organotypic culture model. Am J Pathol 160(1):113–122

Rohan LC, Moncla BJ, Ayudhya RPKN, Cost M, Huang Y, Gai F, Billitto N, Lynam JD, Pryke K, Graebing P, Hopkins N, Rooney JF, Friend D, Dezzutti CS (2010) In vitro and ex vivo testing of tenofovir shows it is effective as an HIV-1 microbicide. PLoS ONE 5(2):e9310

Ron-Doitch S, Sawodny B, Kuhbacher A, David MM, Samanta A, Phopase J, Burger-Kentischer A, Griffith M, Golomb G, Rupp S (2016) Reduced cytotoxicity and enhanced bioactivity of cationic antimicrobial peptides liposomes in cell cultures and 3D epidermis model against HSV. J Control Release 229:163–171

Rosignoli C, Thibaut de Menonville S, Orfila D, Beal M, Bertino B, Aubert J, Mercenier A, Piwnica D (2018) A topical treatment containing heat-treated Lactobacillus johnsonii NCC 533 reduces Staphylococcus aureus adhesion and induces antimicrobial peptide expression in an in vitro reconstructed human epidermis model. Exp Dermatol

Schoop VM, Mirancea N, Fusenig NE (1999) Epidermal organization and differentiation of HaCaT keratinocytes in organotypic coculture with human dermal fibroblasts. J Invest Dermatol 112(3):343–353

Sen E, McLaughlin-Drubin M, Meyers C (2005) Efficacy of two commercial preparations of interferon-alpha on human papillomavirus replication. Anticancer Res 25(2A):1091–1100

Snoeck R, Holy A, Dewolf-Peeters C, Van Den Oord J, De Clercq E, Andrei G (2002) Antivaccinia activities of acyclic nucleoside phosphonate derivatives in epithelial cells and organotypic cultures. Antimicrob Agents Chemother 46(11):3356–3361

Ternullo S, de Weerd L, Flaten GE, Holsaeter AM, Skalko-Basnet N (2017) The isolated perfused human skin flap model: A missing link in skin penetration studies? Eur J Pharm Sci 96: 334–341

van den Broek LJ, Bergers L, Reijnders CMA, Gibbs S (2017) Progress and future prospectives in skin-on-chip development with emphasis on the use of different cell types and technical challenges. Stem Cell Rev 13(3):418–429

Van den Driessche F, Vanhoutte B, Brackman G, Crabbe A, Rigole P, Vercruysse J, Verstraete G, Cappoen D, Vervaet C, Cos P, Coenye T (2017) Evaluation of combination therapy for *Burkholderia cenocepacia* lung infection in different in vitro and in vivo models. PLoS ONE 12(3):e0172723

Ventress JK, Partridge LJ, Read RC, Cozens D, MacNeil S, Monk PN (2016) Peptides from Tetraspanin CD9 Are Potent Inhibitors of *Staphylococcus Aureus* Adherence to Keratinocytes. PLoS ONE 11(7):e0160387

Wald A, Link K (2002) Risk of human immunodeficiency virus infection in herpes simplex virus type 2-seropositive persons: a meta-analysis. J Infect Dis 185(1):45–52

Wang S, Liu Y, Guo J, Wang P, Zhang L, Xiao G, Wang W (2017) Screening of FDA-Approved Drugs for Inhibitors against Japanese Encephalitis Virus Infection. J Virol

Wang X, Zhang N, Glorieux S, Holtappels G, Vaneechoutte M, Krysko O, Zhang L, Han D, Nauwynck HJ, Bachert C (2012) Herpes simplex virus type 1 infection facilitates invasion of Staphylococcus aureus into the nasal mucosa and nasal polyp tissue. PLoS ONE 7(6):e39875

Wufuer M, Lee G, Hur W, Jeon B, Kim BJ, Choi TH, Lee S (2016) Skin-on-a-chip model simulating inflammation, edema and drug-based treatment. Sci Rep 6:37471

Xu M, Lee EM, Wen Z, Cheng Y, Huang WK, Qian X, Tcw J, Kouznetsova J, Ogden SC, Hammack C, Jacob F, Nguyen HN, Itkin M, Hanna C, Shinn P, Allen C, Michael SG, Simeonov A, Huang W, Christian KM, Goate A, Brennand KJ, Huang R, Xia M, Ming GL, Zheng W, Song H, Tang H (2016) Identification of small-molecule inhibitors of Zika virus infection and induced neural cell death via a drug repurposing screen. Nat Med 22(10): 1101–1107

Yin Y, Bijvelds M, Dang W, Xu L, van der Eijk AA, Knipping K, Tuysuz N, Dekkers JF, Wang Y, de Jonge J, Sprengers D, van der Laan LJ, Beekman JM, Ten Berge D, Metselaar HJ, de Jonge H, Koopmans MP, Peppelenbosch MP, Pan Q (2015) Modeling rotavirus infection and antiviral therapy using primary intestinal organoids. Antiviral Res 123:120–131

Zhang W, Hu M, Shi Y, Gong T, Dezzutti CS, Moncla B, Sarafianos SG, Parniak MA, Rohan LC (2015) Vaginal microbicide film combinations of two reverse transcriptase inhibitors, EFdA and CSIC, for the prevention of HIV-1 sexual transmission. Pharm Res 32(9):2960–2972

Zhou T, Tan L, Cederquist GY, Fan Y, Hartley BJ, Mukherjee S, Tomishima M, Brennand KJ, Zhang Q, Schwartz RE, Evans T, Studer L, Chen S (2017) High-content screening in hPSC-neural progenitors identifies drug candidates that inhibit Zika virus infection in fetal-like organoids and adult brain. Cell Stem Cell

Zomer-van Ommen DD, Pukin AV, Fu O, Quarles van Ufford LH, Janssens HM, Beekman JM, Pieters RJ (2016) Functional characterization of cholera toxin inhibitors using human intestinal organoids. J Med Chem 59(14):6968–6972

Zuckerman RA, Lucchetti A, Whittington WL, Sanchez J, Coombs RW, Zuniga R, Magaret AS, Wald A, Corey L, Celum C (2007) Herpes simplex virus (HSV) suppression with valacyclovir reduces rectal and blood plasma HIV-1 levels in HIV-1/HSV-2-seropositive men: a randomized, double-blind, placebo-controlled crossover trial. J Infect Dis 196(10):1500–1508

Zussman A, Lara L, Lara HH, Bentwich Z, Borkow G (2003) Blocking of cell-free and cell-associated HIV-1 transmission through human cervix organ culture with UC781. AIDS 17(5):653–661

Part II
Three Dimensional Human Organotypic Models for Studying Non-infectious Diseases

Progress on Reconstructed Human Skin Models for Allergy Research and Identifying Contact Sensitizers

Charlotte Rodrigues Neves and Susan Gibbs

Contents

Abstract Contact with the skin is inevitable or desirable for daily life products such as cosmetics, hair dyes, perfumes, drugs, household products, and industrial and agricultural products. Whereas the majority of these products are harmless, a number can become metabolized and/or activate the immunological defense via innate and adaptive mechanisms resulting in sensitization and allergic contact dermatitis upon following exposures to the same substance. Therefore, strict safety (hazard) assessment of actives and ingredients in products and drugs applied to the skin is essential to determine I) whether the chemical is a potential sensitizer and if so II) what is the safe concentration for human exposure to prevent sensitization from occurring. Ex vivo skin is a valuable model for skin penetration studies but due to logistical and viability limitations the development of in vitro alternatives is

C. Rodrigues Neves · S. Gibbs
Department of Dermatology, VU Medical Center, Amsterdam, The Netherlands

S. Gibbs
Department of Oral Cell Biology, Academic Centre for Dentistry Amsterdam (ACTA), University of Amsterdam and Vrije Universiteit Amsterdam, Amsterdam, The Netherlands

S. Gibbs (✉)
Dermatology Lab, 0/2 Building, de Boelelaan 1108, 1081 Hz Amsterdam, The Netherlands
e-mail: s.gibbs@vumc.nl

Current Topics in Microbiology and Immunology (2021) 430: 103–129
https://doi.org/10.1007/82_2018_88

Published Online: 23 June 2018

required. The aim of this review is to give a clear overview of the organotypic in vitro skin models (reconstructed human epidermis, reconstructed human skin, immune competent skin models incorporating Langerhans Cells and T-cells, skin-on-chip) that are currently commercially available or which are being used in a laboratory research setting for hazard assessment of potential sensitizers and for investigating the mechanisms (sensitization key events 1–4) related to allergic contact dermatitis. The limitations of the models, their current applications, and their future potential in replacing animals in allergy-related science are discussed.

Abbreviations

ACD	Allergic Contact Dermatitis
AOP	Adverse Outcome Pathway
DC	Dendritic Cell
DNCB	2,4-Dinitrochlorobenzene
DPRA	Direct Peptide Reactivity Assay
EC3	Estimated Concentration of a substance expected to produce a stimulation index of 3
EC50	Half maximal Effective Concentration
ECM	Extra Cellular Matrix
FAM	Fluorescein Amidite
GSH	Glutathione
HaCaT	Spontaneously transformed aneuploid immortal keratinocyte cell line
h-CLAT	Human Cell Line Activation Test
hOSEC	Human Organotypic Skin Explant Cultures
HUVEC	Human Umbilical Vein Endothelial Cell
ICD	Irritant Contact Dermatitis
ISDN	Isosorbide Dinitrate
LC	Langerhans Cell
LEM	Leiden Epidermal Model
LLNA	Local Lymph Node Assay
MO-LC	Monocyte Derived Langerhans Cell
MTT	(3-(4,5-dimethylthiazol-2-yl)-2,5-diphenyltetrazolium bromide) tetrazolium reduction *assay*
MUSST	Myeloid U936 Skin Sensitization Test
MUTZ-LC	Acute Myelomonocytic Leukemia Cell (Mutz-3)-derived Langerhans Cell
NEM	N/TERT Epidermal Model
NOEL	No Observed Effect Level
OECD	Organization for Economic Co-operation and Development
ORG	Organotypic Skin Culture
OXA	Oxazolone
PDMS	Polydimethylsiloxane
RhE	Reconstructed Human Epidermis
RhS	Bilayered Reconstructed Human Skin

SDS	Sodium Dodecyl Sulfate
SI	Stimulation Index
TEWL	Transepidermal Water Loss
UPS-RhE	Reconstructed Human Epidermis developed by University of São Paulo
VUmc	Vrije Universiteit Medical Center

1 Introduction

Approximately 15–20% of the general population have allergic contact dermatitis (ACD) resulting from environmental chemical exposure, making this skin disease a major health problem. ACD is a delayed-type hypersensitivity immune reaction mediated by T-cells, resulting from repeated exposure of the skin to an allergen (Peiser et al. 2012). Development occurs in two phases: in the first (induction) phase, exposure to a chemical allergen causes immunological priming known as skin sensitization. The second (elicitation) phase is triggered if a sensitized person is again exposed to the same allergen (or a cross-reactive allergen). Therefore, prevention of ACD should be prioritized, warranting extensive efforts into understanding the cellular mechanisms of sensitization in order to identify sensitizers, predict the potency of a sensitizer, and most importantly to predict the safe (no response) concentration of a potential sensitizer. The first part in the process leading to sensitization is the penetration of a chemical hapten through the *Stratum corneum* of the skin into the viable layers of the underlying epidermis (Fig. 1, key event 1). Here, the chemical will activate keratinocytes to secrete cytokines, e.g., IL1a, IL-18 (key event 2) attracting dendritic cells (DCs). After exposure to an allergen or an irritant, Langerhans cells (LCs) residing in the epidermis migrate toward the dermis. During sensitization, but not irritation, haptens penetrating the skin either directly activate LC or form a hapten–carrier protein complex that is taken up by skin-resident LC. These hapten-loaded LCs will change into a mature phenotype by upregulating their surface maturation markers such as CXCR4, CD86, and CD83, start increasing their CXCL 8 secretion, and begin their migration from the epidermis to the dermis. In the dermis, they further continue to mature and now also start upregulating CCR7 which enables them together with CXCR4 to migrate toward the lymph nodes (key event 3).

The central event in immune sensitization is the presentation of antigen by DC to antigen-responsive T-cells in the local lymph node which results in T-cell priming (memory) (key event 4). This process is now thought to be orchestrated by the activation and maturation state of DC and their cytokine and chemokine products, and also by factors released by other cell types, including keratinocytes and fibroblasts.

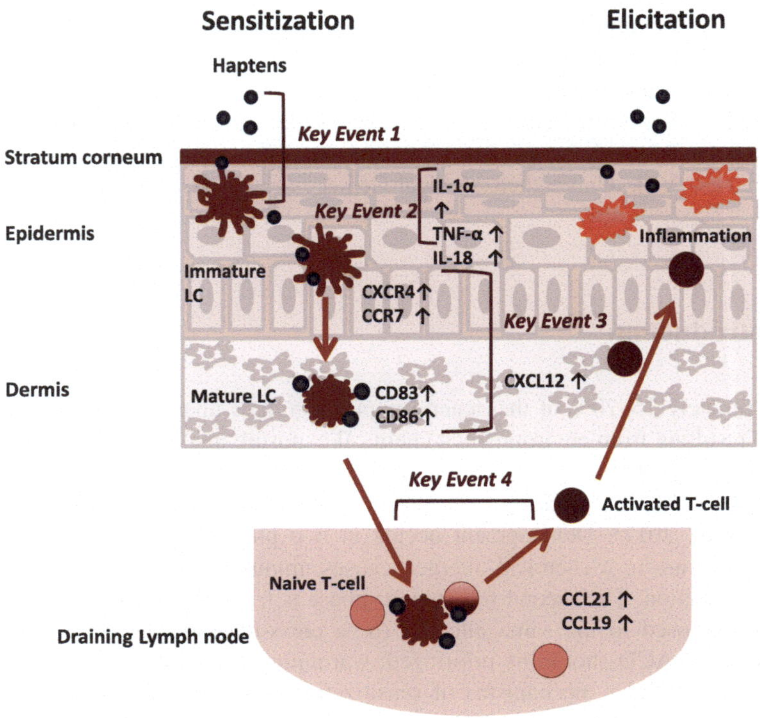

Fig. 1 Diagram showing adverse outcome pathway for sensitization. (Key event 1) The penetration of a chemical hapten through the skin's *S. corneum* into the viable layers of the underlying epidermis; (Key event 2) Activation of keratinocytes resulting in cytokine secretion, e.g., IL1a, IL-18, TNF-a; (Key event 3) Langerhans cell (LC) activation (migration: upregulation of CCR7 (CCL19, CCL21) and CXCR4 (CXCL12) and maturation: upregulation of CD83 and CD86) directly or by a hapten–carrier protein complex); and (Key event 4) Presentation of the antigen by matured LC to antigen-responsive T-lymphocytes in the draining lymph nodes which stimulate proliferation and maturation of T-lymphocytes into primed effector and memory T-lymphocytes are shown

In the past, animals were used to determine whether or not a chemical was a potential sensitizer. The murine local lymph node assay (LLNA) and the guinea pig Buehler test were the most frequently used tests (Api et al. 2015). However, these animal models were subject to ethical considerations and gave a number of false positives and negatives most probably due to differences between human and animal skin physiology and immunity (Seok et al. 2013; Mak et al. 2014). Since the complete European marketing ban in 2013 for risk assessment of cosmetic products which contain ingredients that have been tested on animals, there has been an urgent need to develop alternative methods to identify chemical sensitizers. Because of the high complexity of skin sensitization biology, as documented in the OECD adverse outcome pathway (AOP), only a combination of test methods covering key events 1–4 are expected to lead to a test strategy that will enable the industry to conduct safety assessment of chemicals regarding sensitization potential and potency without the use of animals (Rovida et al. 2015; OECD 2012) (Fig. 1).

Therefore, a major worldwide objective is to develop and provide alternatives to animal testing for the prediction of the skin-sensitizing potential of chemicals within test strategies. The most promising in vitro methods available at this time are the direct peptide reactivity assay (DPRA) (Gerberick et al. 2004), KeratinoSens™ (Delaine et al. 2011; Andreas et al. 2011; Natsch et al. 2010), the Myeloid U937 skin sensitization test (MUSST) (Ade et al. 2005), and the cell line activation test h-CLAT (Ashikaga et al. 2006). However, major limitations have already been identified in current assays which use conventional submerged skin culture. These include solubility problems in particular hydrophobic compounds such as oils and surfactants, and how to test mixtures. These limitations may be overcome by developing skin models which more closely represent human skin physiology. Since a major route of penetration of substances applied to the skin is via the *S. corneum* of a differentiated epidermis, there is increasing interest in developing organotypic skin models for allergenicity testing. This review describes the current state of the art for organotypic skin models which are being developed for hazard assessment of potential sensitizers. An overview of the use of ex vivo skin for allergenicity testing is described, followed by the progress on developing cultured organotypic models of increasing complexity ranging from reconstructed human epidermis to immune competent skin models with integrated Langerhans cells or T-cells. The common feature of these models is that they are all cultured at the air–liquid interface which promotes epidermal differentiation and stratification, thus enabling chemicals to be applied topically to the *S. corneum*, mimicking human exposure, and enabling exposure of hydrophobic compounds.

2 Ex Vivo Skin

Freshly excised skin obtained from standard surgical procedures is a valuable source of intact human skin for allergy research (Rustemeyer et al. 2003; Jacobs et al. 2004, 2006; Lehe et al. 2003; Pistoor et al. 1996; Schmook et al. 2001). Ex vivo skin has a relatively intact barrier function most suitable for penetration studies and contains LC which can become activated and migrate from the epidermis into the dermis upon chemical exposure (Fig. 2). Explant skin cultures have been used to study the characteristics of chemical-induced migration of LC either as full thickness human organotypic skin explant cultures (hOSEC) (Rustemeyer et al. 2003; Lehe et al. 2003; Pistoor et al. 1996; Jacobs et al. 2002) or as a partial thickness epidermal explant culture (Ouwehand et al. 2008). A more accelerated migration of LC out of the epidermis occurs upon treatment with contact allergens compared to non-sensitizers (Lehe et al. 2003). Organotypic skin explant cultures (with methyl-green pyronine labeled KC RNA?) derived from human skin,

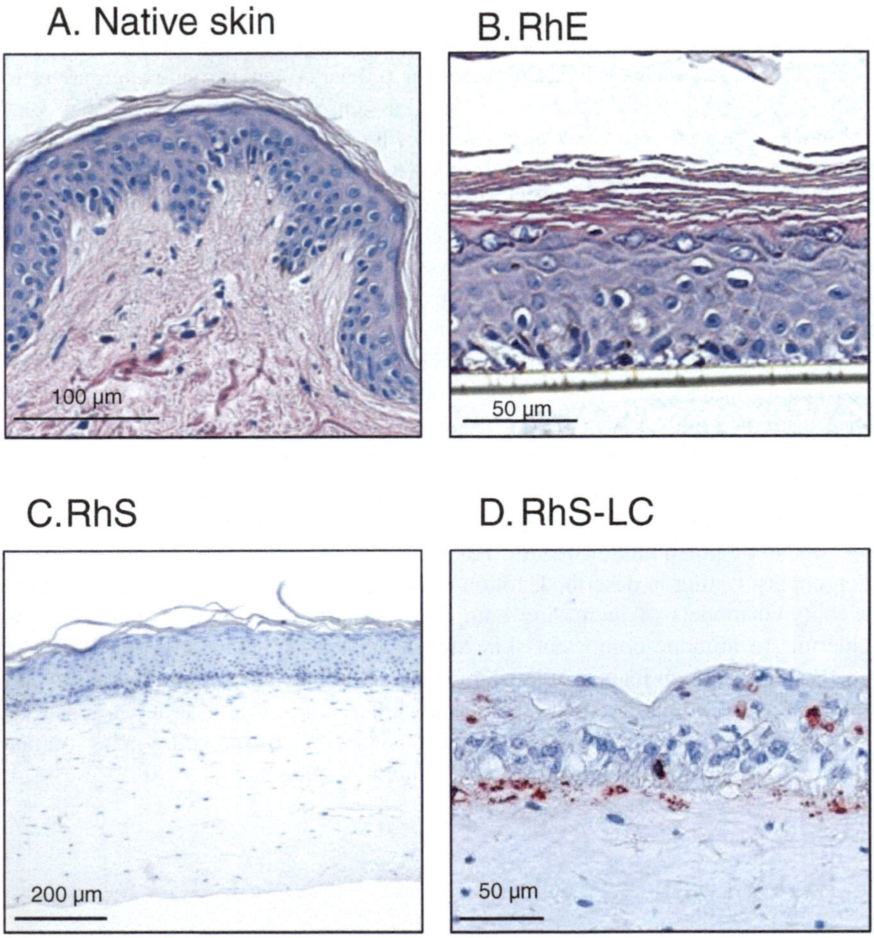

Fig. 2 Histology (hematoxylin–eosin staining) of (**a**) native skin (biopsy), (**b**) reconstructed human epidermis (RhE) on a polycarbonate filter, (**c**) reconstructed epidermis on fibroblast-populated dermis (RhS), and (**d**) An HLA-DR immunohistochemical staining of MUTZ-LC integrated in reconstructed epidermis on fibroblast-populated collagen hydrogel (RhS-LC) (stratum corneum was lost during immunohistochemical staining procedure)

porcine skin, or rabbit skin were able to predict the irritant hazard potential of 22 chemicals with 100, 95, and 93% accuracy, respectively (Jacobs et al. 2002). The major problem with implementing ex vivo skin models outside of the field of basic research is the logistics of supply of fresh skin to the laboratory and the short viability period of the skin (<24 h), particularly if functional LC maturation and migration readouts are required. Furthermore, the skin is subject to donor variability due to the undefined numbers and types of cells present making it very difficult to make into a standardized assay.

3 Reconstructed Human Epidermis (RhE)

RhE is the simplest form of organotypic skin model with an epidermal barrier function. RhE is constructed by seeding keratinocytes onto a porous membrane (transwell). After several days of submerged culture, the RhE is exposed to the air from above with the culture medium below. This method of culture stimulates the basal keratinocytes to differentiate and migrate upward to form a stratified epidermis with *S. corneum* (Fig. 2). The method is easily standardized and this has resulted in a number of commercial RhE now being available (Table 1). The EpiSkin™, EpiDerm™, SKinEthic™RhE, and EpiCs® RhE models have already been validated assays for skin irritation and skin corrosion testing according to the OECD test guidelines 439 and 432, respectively (Cottrez et al. 2015, 2016; Saito et al. 2013; Andres et al. 2017; Gibbs et al. 2013). RhE can be used to investigate key sensitization events 1 and 2 in which penetration of a chemical hapten through the *S. corneum* into the viable layers of the epidermis is the first step of the sensitization process. Keratinocytes respond by secreting cytokines such as IL-1 α, TNF-α, and IL-18 (Fig. 1). The prediction (label) and quantification (potency, characterization) of the sensitization hazard of various sensitizers and non-sensitizers have been tested in a number of studies using different biomarker readouts. In the SENS-IS and EpiSensA assays, changes in gene expression are analyzed upon topical exposure with a chemical (Table 1). The SENS-IS assay has been studied most extensively using the EpiSkin™ model. In a ring study with three participating laboratories, 150 chemicals were tested by analyzing the expression of a carefully selected panel of 65 genes (Cottrez et al. 2015, 2016). This study highlighted the transferability and robustness (repeatability) of the SENS-IS assay to detect sensitizers with a high predictivity and 100% reproducibility between laboratories. The sensitivity, specificity, and accuracy of this assay were all well above 90% compared to the LLNA, guinea pig, and human data based on literature sources obtained by Basketter et al. (2014), suggesting it to be a serious alternative to in vivo sensitization testing. It could be argued that the use of such a broad set of genes will increase expense and complicate implementation in a broad setting. The EpiSensA assay is also based on gene analysis (Saito et al. 2013, 2017). However, only the predictive performance of five genes, all related to cellular stress, is examined. Three out of five genes, namely ATF3, DNAJB4, and GCLM, but not HSPA6 and HSPH1 showed a high accuracy, 100, 93.8, and 87.5, respectively, when tested with 16 chemicals suggesting that they could be useful markers for skin sensitization. Most probably because of the smaller set of gene biomarkers used, the overall sensitivity, specificity, and accuracy were slightly lower compared to the SENS-IS assay. The assay was developed making use of the MaTtek EpiDerm™ model (Saito et al. 2013). Later, the assay was also performed in the LabCyte EPI-MODEL24 SIT. IL-8 was added to the set of genes that were analyzed with the goal of decreasing false negatives (Saito et al. 2017). A total of 72 chemicals (43 hydrophilic chemicals and 29 lipophilic chemicals including 11 pre/pro-haptens) were evaluated this time. The sensitivity, specificity, and accuracy were 93, 100,

Table 1 Reconstructed human epidermis models for skin sensitization testing

Assay	RhE model	Commercially available?	Readout/biomarkers	Test substances	Refs.
SENS-IS	[b]EpiSkin™	Yes (Skinethics)	• **Sensitization** assessment of 65 genes (41 SENS-IS groups, 24 in ARE group) by qRT-PCR. – 100% between-laboratory reproducibility (N = 3) – Sensitivity 95.8%, specificity 96.5%, accuracy 96% • **Potency Classification** minimum concentration inducing a sensitization signal (weak, moderate, strong, extreme) – Accuracy LLNA = 92.6%, accuracy human data = 90.6% – 89.5% between-laboratory reproducibility	*S* = 27, NS = 23	Cottrez et al. (2015, 2016)
EpiSensA	[b]EpiDerm™	Yes (MatTek Corporation)	• **Sensitization** assessment of 3 genes (ATF3, DNAJB4 and GCLM) by qRT-PCR. – accuracy (100%, 93.8% and 87.5% per gene respectively) • No **Potency classification**	S = 8, NS = 4, pre-/pro-haptens = 4	Saito et al. (2013)
	[a]LabCyte EPI-MODEL24 SIT	Yes (J-TEC)	• **Sensitization** assessment of 4 genes (ATF3, DNAJB4, IL-8 and GCLM) by qRT-PCR. – Sensitivity 94%, specificity 78%, accuracy 90% • **Potency classification** tissue viability (MTT, EC) – Accuracy LLNA = 82%	29 lipophilic chemicals, 43 hydrophilic chemicals (incl 11 pre-/pro-haptens)	Saito et al. (2017)
EE-IL-18 potency assay	[b]SkinEthic™ RhE	Yes (Skinethics) (available in different stages of maturity)	• **Sensitization** by measurement of basal IL-18 release (Elisa) • **Potency Classification** tissue viability (MTT, (EC_{50}) and IL-1α (EE-IL-1$\alpha_{10\times}$ value, ELISA)	18 substances	Andres et al. (2017), dos Santos et al. (2011)
	VUmc-EE In-house model	No	• **Sensitization** by measurement of basal IL-18 release (Elisa) – Accuracy 95% (N = 2) • **Potency classification** IL-1α release with Elisa (EE-IL-1$\alpha_{10\times}$ value) and cell viability by MTT (EE-EC_{50})	S = 16	Gibbs et al. (2013, dos Santos et al. (2011)

(continued)

Table 1 (continued)

Assay	RhE model	Commercially available?	Readout/biomarkers	Test substances	Refs.
	[b]EpiCs® (previously (EST1000™)	Yes (CellSystems)	• **Sensitization** by measurement of basal IL-18 release (Elisa) – Accuracy 95% (N = 2) • **Potency classification** tissue viability (MTT, EE-EC_{50}), IL-1α (EE-IL-$1\alpha_{10\times}$ value), and IL-18 release (ELISA)	S = 13, NS = 9	Andres et al. (2017), Gibbs et al. (2013), dos Santos et al. (2011)
	[b]EpiDerm™	Yes (MatTek Corporation)	• **Sensitization** by measurement of basal IL-18 release (Elisa) • **Potency classification** tissue viability (MTT, EE-EC_{50})	S = 9, NS = 3	Gibbs et al. (2013)
	LEM and NEM in-house model	No	• **Sensitization** activation of the Keap1-Nrf2-ARE pathway (Western Blot for Nrf2 expression and GLCM and OSGIN-1 upregulation in NEM (qPCR)	S = DNCB, NS = SDS	Alloul-Ramdhani et al. (2014)
SenCeeTox® (by Cyprotex)	[b]SkinEthic™ RhE	Yes (Skinethics) (available in different stages of maturity)	• **Sensitization** ***(only in EpiDerm™)*** identification primary signaling pathway, qPCR (CCL5, CCL27, and TNF-α) • **Potency** assessment of 8 Nrf2/ARE, 2 Nrf1/MRE and 1 AhR/XRE(pro-haptens) genes (qPCR) per category (extreme, strong, moderate, weak, or non-sensitizer) and tissue viability (MTT), oxidative stress (DCFDA assay), and chemical reactivity (GSH)	S = 9, NS = 2	McKim et al. (2012)
	[b]EpiDerm™	Yes (MatTek Corporation)			

[1]All epidermal models are made of (NHEKs) normal human epidermal keratinocytes (from adult or neonate foreskin origin) except for the NEM model of which the epidermis is reconstructed out of a human N/TERT keratinocyte cell line and the VUmc-EE also contains melanocytes

[2]All epidermal equivalents were cultured on an inert polycarbonate filter except for Episkin™ which is cultured on a collagen type I matrix with a thin layer of type VI human collagen

[3]*S* Sensitizer, *NS* Non-sensitizer

[a]Skin models were also validated for skin irritation testing (OECD Test Guidelines 439)

[b]Skin models were also validated for skin irritation testing (OECD Test Guidelines 439) and skin corrosion testing (OECD Test Guidelines 431)

and 93% for lipophilic chemicals and 96, 75, and 88% for hydrophilic chemicals. These values are higher than values obtained from existing validated in vitro tests (DPRA, h-CLAT, and KeratinoSens) (Saito et al. 2017) which did not distinguish between hydrophilic and lipophilic chemicals. Also, most probably different panels of test chemicals were used making direct comparisons difficult. Five sensitizers (OXA, Benzoyl peroxide, benzyl cinnamate, lilial, and tridecane) were positive only for the induction of IL-8. Three of these five sensitizers were lipophilic, showing a high capability of the EpiSensA assay to detect lipophilic chemicals and thereby overcoming one of the major limitations of the existing validated in vitro tests. Also, all of the 11 tested pre/pro-haptens, including 6 hydrophilic pre/pro-haptens (Isoeugenol, ethylene diamine, diethylene triamine, resorcinol, cinnamic alcohol, and eugenol) that often showed false-negative results, could be detected in the EpiSenzA assay explaining its relatively high predictivity.

The RhE IL-18 assay is based on the ability of contact sensitizers but not respiratory sensitizers or contact irritants to be able to increase intracellular production and release IL-18 (Gibbs et al. 2013). The assay was developed by combining the NCTC 544 keratinocyte IL-18 assay which could label a sensitizer (Corsini et al. 2009, 2013; Galbiati et al. 2011) with the RhE potency assay which assesses sensitizer potency (characterization) based on the irritant property of the chemical (Teunis et al. 2013; dos Santos et al. 2011; Spiekstra et al. 2009). The resulting assay was an RhE assay based on a single biomarker (IL-18) which was easily transferable from the Vrije Universiteit medical center (VUmc) in-house RhE model to commercially available RhE (SkinEthic™ RhE, EpiDerm™, and EpiCS®). Therefore, the RhE IL-18 assay is a relatively simple and robust method to assess a sensitizer label as well as potency (Gibbs et al. 2013). Using the VUmc RhE, chemicals were labeled (YES/NO) as sensitizer if a threshold of more than fivefold IL-18 release was reached. This threshold needs to be set for each different RhE model. The potency of the chemical, also known as the in vitro estimation of expected sensitization induction level, was assessed by interpolating in vitro EC50 (chemical concentration required to reduce viability by 50%) and IL-18 SI2 (chemical concentration required to increase IL-18 release by twofold) with LLNA EC3 and human NOEL values from standard reference curves generated using DNCB (extreme) and benzocaine (weak) (Galbiati et al. 2017; Gibbs 2017). Notably, whereas a good prediction was observed when traditional test panel chemicals were tested, when the assay was challenged with metal salts representative of leachables from medical devices, it was found that these metals which are very difficult to test in vitro and in vivo also fell outside of the applicability domain of this assay (Gibbs et al. 2013; Galbiati et al. 2017; Gibbs 2017). In addition to interpolating in vitro EC50 and IL-18 SI2 with LLNA EC3 and human NOEL values, a simple binary prediction model was developed for assessing sensitizer potency based on the irritant potential of the chemical (EC50 value and/or IL-1alpha release) or IL-18 SI2 (Gibbs 2017; Teunis et al. 2014). A ring study with four laboratories showed 77% accuracy with the binary prediction model for sensitizer potency (EC50 $\geq$ 7 mg/ml = weak to moderate sensitizer and EC50 $<$ 7 mg/ml = strong to extreme sensitizer) (Teunis et al. 2014). In all laboratories, human RhE EC50 data showed better correlation to human data than to mouse LLNA-EC_3 data. A low intra- and inter-experiment variability between laboratories and the different RhE models was observed.

Another promising assay is the SenCeeTox assay (McKim et al. 2012). EpiDerm™ and the SkinEthic™ RhE models were used to categorize chemical sensitizers by combining solubility, chemical reactivity, cytotoxicity, and activation of the Nrf2/ARE pathway. The expressions of eight Nrf2/ARE, one AhR/XRE, and two Nrf1/MRE controlled genes were measured using qRT-PCR. The fold induction at six exposure concentrations of a training set of 11 chemical sensitizers (representing extreme/strong-, moderate-, weak-, and non-sensitizing potency categories) was combined with glutathione (GSH) reactivity and cytotoxicity (MTT assay) data to determine the sensitization potential of the compounds and to establish the sensitivity of the two RhE models. Thereafter, a set of seven low-solubility chemicals and extracts used in the manufacture of medical devices were assessed. The ability of the assay to accurately place the compounds in one of the four potency categories was 71%. In addition to the RhE assays described above, several non-commercialized RhE are being used to develop assays that might be used in the future for sensitization testing of chemicals. For example, the Leiden epidermal model and the N/TERT epidermal model have also been used to test sensitization by measuring the Keap1-Nrf2-ARA activation pathway but only one sensitizer and one non-sensitizer have been tested to date (Alloul-Ramdhani et al. 2014). A larger panel of chemicals needs to be tested in order to determine whether the Leiden epidermal model and N/TERT epidermal model will be a suitable tool for chemical labeling. Also, a new RhE model has been presented as an open-source protocol by the University of São Paulo (Pedrosa et al. 2017). This UPS-RhE in-house model showed 85.7% specificity, 100% sensitivity, and 92.3% accuracy with a high within-laboratory reproducibility (92.3%) when thirteen chemicals were tested for their skin irritation potential according to the EpiSkin™ protocol (OECD TG 439) (Pedrosa et al. 2017).

Taken together, a number of commercially available and in-house RhE are being used to develop very promising assays which are currently entering different phases of validation. RhE is stable during transport making them extremely interesting for laboratories without the in-house know-how to construct and culture RhE. When using different types of RhE within an assay, the prediction model thresholds need to be calibrated beforehand. For example, some differences in the induction level of marker genes were observed in the EpiSenza assay between the LabCyte EPI-MODEL SIT24 and the EpiDerm™ model. The induction levels of ATF3 and DNJAB4 in the LabCyte EPI-MODEL SIT24 were slightly lower than in the EpiDerm™ model. Also, differences were observed between the IL-18 stimulation indexes in the RhE IL-18 assay when comparing the results obtained from different types of RhE (Gibbs et al. 2013). These differences are most probably due to different barrier properties of the *S. corneum* (Kano et al. 2011; Ponec et al. 2002), affecting the penetration of the topically applied chemicals and the metabolic activity within the different RhE due to the use of different culture media. RhE has the advantage over conventional submerged cultures in that they have a *S. corneum* to enable hydrophobic and chemicals of poor solubility to be tested using relevant vehicles which are used in in vivo studies. Furthermore, they release keratinocyte-derived (pro)inflammatory cytokines, are metabolically competent to a certain extent and do have a barrier function although these properties are not yet as well developed as those found in healthy native skin (Netzlaff et al. 2005). However, they still have limitations in that they only incorporate one cell type (keratinocytes) and do not have a dermal compartment which would make the model more representative on human skin. This is important since it is the in vitro cross talk between keratinocytes and fibroblasts which drives the inflammatory cytokine response in skin models lacking immune cells (Spiekstra et al. 2005, 2007).

4 Bilayered Reconstructed Human Skin (RhS)

Bilayered reconstructed skin models are constructed by seeding keratinocytes on a fibroblast-populated dermal matrix, e.g., a collagen hydrogel, collagen–elastin matrix, and donor dermis (Spiekstra et al. 2005, 2007; van den Broek et al. 2012; Gibbs et al. 2006). As with RhE, after several days of submerged culture, RhS is lifted to the air–liquid interface so that the keratinocytes differentiate to form a stratified epithelium with *S. corneum* (Fig. 2). As with RhE, the epidermis of RhS addresses chemical penetration (key event 1) and keratinocyte activation (key event 2) (Fig. 1). In RhS, interaction between fibroblasts and keratinocytes results in the formation of the basal membrane (El Ghalbzouri et al. 2005). In particular, fibroblasts are needed for dermis extracellular matrix secretion and optimal localization of dermal–epidermal junction components such as type VII collagen and laminin V (Marionnet et al. 2006). Importantly, keratinocytes upon contact with a chemical produce IL-1α which triggers dermal fibroblasts to produce a cascade of cytokines and chemokines, thus initiating the inflammatory response which in vivo results in LC migration into the dermis and immune cell invasion into the skin (Spiekstra et al. 2005, 2007; Ouwehand et al. 2011a, b). There are currently three commercially available RhS and a number of in-house models (Table 2). The EpidermFT™ has been used to determine the skin irritation potential of surfactants by assessment of the release of the primary cytokine interleukin IL-1α after exposure to 46 commercial skin cleansers (containing 224 nonionic or anionic surfactant-containing formulations) (Walters et al. 2016). The IL-1α release measured in vitro was compared to clinical TEWL (transepidermal water loss) measurements and showed good correlation ($R^2 = 0{,}66$). Another commercially available RhS is the Phenion® FT model. The barrier property (key event 1) of the Phenion® FT was compared with ex vivo pig skin after topical application of testosterone, caffeine, nicotine, and benzoic acids. The Phenion® FT model turned out to be more permeable than pig skin but its barrier properties were comparable to those of RhE (EpiDerm^{FT}, SkinEthic™ RhE, and EpiSkin™) (Ackermann et al. 2010). Also, the StrataTest® showed consistent IL-1α release between batches after

Table 2 Reconstructed bilayered human skin models (RhS)

RhS model	Commercially available?	Cells	Readout/biomarkers	Test substances	Refs.
EpidermFT™	Yes (MatTek Corporation)	• NHEK (adult or neonatal) and NHFB • Donor-matched • Collagen hydrogel	• **Irritation** – Tissue viability (MTT) – Primary cytokine interleukin-1α release. • good correlation between cytokine release in *vitro* and clinical TEWL ($R^2 = 0.66$)	224 nonionic, amphoteric, and/or anionic surfactant-containing formulations (46 commercial skin cleansers)	Walters et al. (2016)
Phenion® FT	Yes (Henkel)	• NHEK and NHFB • Donor-matched • Collagen hydrogel	• **Barrier property** permeability measurements percutaneous absorption -Liquid scintillation counting -Histological scoring of damage	Testosterone, caffeine, benzoic acid, and nicotine	Ackermann et al. (2010)
Organotypic skin cultures (ORGs)	In house (University of Antioquia)	• NHEK and NHFB • Fibrin hydrogel	• **Corrosion** tissue viability (MTT) • **Irritation** – Tissue viability (MTT) – Pro-inflammatory cytokines (IL-8, IL-1b, IL-6, IL-10, IL12p70, and TNF)	Corrosion: SDS, KOH, eugenol, and $NaHCO_3$ Irritant: 3 surfactants, SDS, Triton- ×100, Tween-20, and a new drug.	Morales et al. (2016)
StrataTest®	Yes (Stratatecht)	• NIKS® keratinocyte progenitors • NHFB • Collagen hydrogel	• **Irritation** – Tissue viability (MTT) – By measuring IL-1α release with Elisa (consistent between batches) • **Barrier function measurement** by electrical impedance (reduced compared to uninjured skin, but superior compared to tape-stripped skin)	SDS, Triton- ×100	Rasmussen et al. (2010)

(continued)

Table 2 (continued)

RhS model	Commercially available?	Cells	Readout/biomarkers	Test substances	Refs.
VUmc skin equivalent	In house (VU medical center)	• NHEK and NHFB (adult or neonatal) • Melanocytes • Collagen hydrogel	• **Epidermal cytotoxicity** – Decrease in KC RNA by pyronine Y staining • **(IL-α and TNF-α dependency of) allergen- and irritant-increased cytokine secretion** – CCL27, CCL20 CXCL8 secretion (Elisa) and IL-α and TNF-α blocking, and key cytokines involved in LC migration (IL-18, CCL2, CCL5, IL-6, and CXCL12)	$NiSO_4$, $K_2Cr_2O_7$, and SDS, cinnamaldehyde, and TNF-α	Spiekstra et al. (2005), Kosten et al. (2015)

exposure to SDS (Rasmussen et al. 2010). The organotypic skin culture (ORG) in-house model of the University of Antioquia has been used for testing corrosive and irritation potency of a small panel of 11 substances by measuring general pro-inflammatory cytokine release and cell viability (MTT assay) (Morales et al. 2016). One out of four substances tested for corrosion, SDS, was incorrectly classified as corrosive and two out of three surfactants, Triton-×100 and Tween 20, were incorrectly classified as irritants. A possible explanation for this could be the exposure time to the chemical which was based on the protocols used for the EpiDerm^{FT} model, resulting in damage to the fibrin matrix. Also, the high levels of basal pro-inflammatory cytokine release in this model could be related to the use of the fibrin matrix (fibrin induces a secretory phenotype in fibroblasts, leading to pro-inflammatory cross talk with keratinocytes (Martinez et al. 2006)). The VUmc in-house RhS was used to investigate the alarm signals after exposure to a chemical by measuring key cytokines that initiate the infiltration of immune cells such as immature dendritic cells, T-cells, B-cells, and neutrophils into the skin (key event 2). It was shown that the exposure to an allergen (nickel sulfate and potassium dichromate) as well as to an irritant (SDS), with or without the presence of IL-1α or TNF-α neutralizing antibodies, results in an IL-1α and TNF-α dependent increase in CCL20 and CXCL8 secretion and IL-1α and TNF-α independent CCL27 secretion. This data suggests that skin-residential keratinocytes and fibroblasts respond to allergen as well as irritant exposure by releasing mediators that initiate immune cell infiltration. Whether this may facilitate an ACD or ICD reaction depends further on the properties of the chemical and how it interacts with immune cells, e.g., LC (Spiekstra et al. 2005; Kosten et al. 2015). Limited sensitization testing has been performed on these RhSs so far, which may be attributed to difficulties with stable transportation hindering commercialization or more complex methodology required for constructing in-house models. Whereas RhE and RhS have found an important niche in allergy research and sensitizer identification, a major limitation in these models is that they lack integrated immune cells which play a pivotal role in all human skin disease including allergy.

5 Immune Competent Skin Models

RhE and RhS models are starting to progress to include immune cells and thereby include in a single model key events 1, 2, and 3 (Fig. 1). Current commercially available RhE and RhS all lack immune cells. However, more complex in-house models with integrated LC and/or T-cells are being developed (Table 3). The first model to be developed was an RhE with integrated CD34^{+} cord blood progenitor-derived LC, which was used to study the reactivity of LC to topically applied allergens (Facy et al. 2005). The LC adopted an activated morphology (higher Langerin staining in the body and shorter dendrites), and the epidermis showed increased IL-1β and CD86 mRNA expression when exposed to sensitizers compared to irritants. This RhE-LC model developed by L'Oréal showed some donor variation with the majority of the donors responding to the sensitizers (Table 3). The major limitations, however, with this model were (I) the dermis compartment was not present and therefore no LC migration occurred and (II) it was very logistically complicated to construct due to the dependence of cultured primary keratinocytes and cord blood-derived LC.

In order to overcome the limitations of the RhE-LC model containing cord blood-derived LC, VUmc developed an RhS with integrated MUTZ-3-derived LC (Fig. 2). MUTZ-3 is an acute myeloid leukemia-derived human cell line with $CD34^+$ proliferating progenitor cells which can be differentiated into LC (MUTZ-3 LC) in a cytokine-dependent fashion. MUTZ-3 LCs closely resemble their native counterparts, both phenotypically and functionally (Kosten et al. 2015; Masterson et al. 2002; Santegoets et al. 2008; dos Santos et al. 2009). By incorporating MUTZ-LC into an RhS (reconstructed epidermis on fibroblast-populated collagen hydrogel), a unique model was developed which (I) enabled the distinct mechanisms of migration of LC into the dermis to be investigated in a standardized manner (key event 3) (Ouwehand et al. 2010, 2011; Kosten et al. 2015) and (II) overcame the complicated logistics of using primary cell-derived LC. Using the RhS-containing functional MUTZ-LC, the only assay until now has been developed which can distinguish sensitizers from irritants based on the different mechanisms of LC migration and phenotypic plasticity. Migration of sensitizer-exposed maturing $CXCR4^+$, $CD86^+$ MUTZ-LC, into the dermis and consequent increase in CCR7 expression is blocked with neutralizing antibodies to CXCL12, whereas migration of irritant-exposed non-maturing MUTZ-LC is blocked by neutralizing antibodies to CCL5. Within the dermis, the irritant-exposed MUTZ-LC undergoes a phenotypic switch to a macrophage-like cells ($CD1A^-/CD14^+/CD68^+$) under the influence of IL-10 (Ouwehand et al. 2011; Kosten et al. 2015; de Gruijl et al. 2006). Indeed, the model was able to correctly label three surfactants (SDS, Tween 80, and Triton ×100) as true irritants, whereas these surfactants tend to score as false positives in current in vitro and animal models. Most importantly, it is the incorporation of fibroblasts into the RhS which permits this LC migration due to their secretion of CXCL12 and CCL5 upon chemical exposure which provides the chemotactic gradient for LC. Furthermore, by reconstructing human oral mucosa in the same way as RhS, different migratory mechanisms of mucosa LC can now be compared with that of skin upon sensitizer exposure (Kosten et al. 2016). In a recent comparative study, monocyte-derived LC (Mo-LC) were compared with MUTZ-3-derived LC incorporated in RhS using cytokine secretion and mRNA as a readout for LC phenotype (Bock et al. 2017). Whereas both types of LC showed phenotypic changes upon chemical exposure, the limitation of using primary Mo-LC in RhS does add a hurdle for further widespread implementation. Also, blood-derived LC are influenced by donor variability which is not the case for cell line-derived LC. However, like $CD34^+$-derived LC, Mo-DC do represent donor-specific biological variances which reflect biological diversity in the in vivo situation. To what extent the functionality of the different types of LC-like cells and donor variation affects their interaction with T-cells (key event 4) still needs to be elucidated.

Table 3 Immune competent skin models

Immune competent skin models	Cells	Readout/biomarkers	Test substances	Refs.
Skin Equivalent-LC (In house, VU medical center)	• NHEK and NHFB • Melanocytes • Mutz-3-derived LC • Collagen hydrogel	• **LC migration (allergen and irritant induced)** – Percentage of CD1a cells in epidermis/dermis (fluorescence density) – Anti-CXCL12/anti-CCL5 blocks allergen/irritant-induced migration • **LC phenotyping** – Flow cytometry CFSE labeled and CD1a positive cells in dermis – Maturation: upregulation CCR7 and IL1α transcripts (qPCR) and upregulation CD83 (immunohistochemistry) – Phenotypic switch blocked by IL-10 • CXCL8 (Elisa) uniformly increased after exposure to sensitizers but not non-sensitizers	Allergens: $NiSO_2$, resorcinol, and cinnamaldehyde Irritants: TritonX-100, SDS, Tween 80, salicylic acid, isopropanol, DMSO, BCl_2, and phenol)	Santegoets et al. (2008), dos Santos et al. (2009), Ouwehand et al. (2010)
Skin equivalent populated with $CD4^+$ T-cells (In house, Radboud University Nijmegen medical center)	• NHEK • Allogeneic-activated $CD4^+$ T-cells • Decellularized dermis	• **Interaction KC and T-cells** – Activation status of T-cells (Flow cytometry) and immunostaining psoriasis-associated proteins hBD2, KRT16, elafin, and differentiation markers filaggrin and involucrin • Upon **direct contact** epidermal inflammation: induction psoriasis-associated genes DEFB4, PI3, LCE3A, KRT16, S1000A7, and	Medium supplemented with Cyclosporin A and ATRA (all-trans retinoic acid)	van den Bogaard et al. (2014)

(continued)

Table 3 (continued)

Immune competent skin models	Cells	Readout/biomarkers	Test substances	Refs.
		pro-inflammatory cytokines *IL6, IL8*, and *IL23* (qPCR) • Upon **indirect contact** upregulation of CCL2 and CXCL10 in KC – Anti-inflammatory drugs counteract the epidermal inflammatory phenotype and downregulate psoriasis-associated genes		
Reconstructed epidermis with CD34 (+)-derived Langerhans cells (in house, L'Oréal Recherche)	• NHEK • $CD34^+$-derived LC • Episkin® collagen dermal support	• **LC phenotyping (allergen and irritant induced)** – Maturation: higher Langerin and CD1a staining in cellular body and shorter dendrites in 87% of donors (immunohistochemistry) – Allergen (not irritant) induced IL-1β and CD86 mRNA overexpression – IL-1β/CD86 mRNA over-expressions in 57% of donors (RT-PCR)	TNF-α, IL-1b, allergens: DNFB, oxazolone, pPD, NiSO4, eugenol, benzocaine, Irritants: SLS, benzalkonium chloride	Facy et al. (2005)
RHS with Mo-LC or MUTZ-LC (In-house model, University of Berlin)	• NHEK and NHFB (neonatal) • Collagen hydrogel • Mutz-3 or monocyte-derived LC	• **LC phenotyping (allergen and irritant induced)** – Increased IL-6, IL-8, and IL-18 secretion (ELISA) – Maturation: Increased ATF3, IL-1β, CD83, CXCR4, and PD-L1 gene expression • **LC migration (allergen and irritant induced)** – CD1a and CD207 positive cells in dermal compartment (flow cytometry)	DNCB, pPD, isoeugenol and SDS	Bock et al. (2017)

Skin models incorporating T-cells are in the very early phase of development. Van den Bogaard et al. developed a model which mimics psoriasis by enabling $CD4^{+}$T-cells, activated with anti-CD3/CD28 monoclonal antibody-coated beads, to migrate into the dermis of reconstructed skin (RhE on acellular dermis in a transwell culture system) (van den Bogaard et al. 2014) (Table 3). The reconstructed skin did not contain fibroblasts, and therefore no direct cell-to-cell contact between the T-cells and the keratinocytes occurred, thus enabling allogenic cells to be cultured together without cytotoxic effects. Before incorporation into the model, the T-cells were reported to be able to produce INF-γ, TNF-α, IL-17, and IL-22. At day two after initiating migration, gene expression analysis still showed relatively abundant INF-γ expression compared to IL-17, IL-22, and TNF-α. Also, pro-inflammatory cytokine and chemokine production by keratinocytes (IL-6, IL-8, IL-23, and CXCL-10) was highest at day 2. It was found that direct contact between keratinocytes and T-cells was not necessary for cross talk since soluble factors produced by T-cells were the main stimulus for the inflammatory phenotype. Since no model has yet incorporated both LC and T-cells, currently no in vitro skin model exists which can be used to study all four key events of the sensitization process.

6 Skin-on-a-Chip

Current commercially available or in-house skin models are based on static culture systems. Novel microphysiological systems are being developed, creating the possibility of culturing human skin models in a systemically controlled microenvironment in which homeostasis can be created and maintained (van den Broek et al. 2017; Watson et al. 2017; Ahadian 2017). The perfusion of the systems is expected to introduce shear stress, clear secreted products, increase the barrier function, create biomolecular gradients, and permit interaction with distant cells and may possibly even enable the influx and outward migration of immune cells from and to the lymph nodes. In addition, the maintenance and testing period of (commercially) available skin equivalents may possibly be prolonged by using bioreactor platforms. Several simplistic models for substance penetration in skin have been developed. For example, a model has been described in which epidermal, dermal, and endothelial human cells were co-cultured in a microfluidic device separated by porous membranes that allowed communication between the different cell type monolayers (Wufuer et al. 2016). Other models describe cultured monolayers of HaCat-KC cells, co-cultured with U937 cells or just human epidermal keratinocytes alone in a microfluidic device to assess the irritation potential and toxicity of chemical compounds (Wufuer et al. 2016; Ramadan and Ting 2016). A number of skin-on-a-chip models have been demonstrated to be promising for future substance testing (Table 4). Percutaneous penetration in RhS has been demonstrated where microfluidic channels (gravity driven) were used to collect the penetrated substances (FAM-tagged oligonucleotides). This model showed that the barrier function remained consistent for over 3 weeks of culturing (Abaci et al. 2015). The model was exposed to a clinically relevant

Table 4 Skin-on-a-chip models

Skin-on-a-chip	Cells	Readout/biomarkers/test substance	Refs.
Multi-organ chip (TissUse, technical university of Berlin)	• NHEK • NHFB • Human dermal endothelial cells (HDMECs) from neonate foreskin	• **Skin with liver, intestine, and kidney viable >28 days** • **Cell viability** measured by live cell viability to demonstrate microvascular system functionality HDMECs elongated and oriented into flow direction • **Improved vitality compared to static conditions** shown by more viable (TUNEL-apoptosis negative) and proliferating (Ki67-positive) cells • **peristaltic on-chip micropump**	Maschmeyer (2015), Maschmeyer et al. (2015), Wagner et al. (2013), Atac et al. (2013)
Skin-on-a-chip with perfusable vascular channels (University of Tokyo)	• NHEK • NHFB • Human umbilical vein endothelial cells (HUVECs) • Collagen hydrogel	• **Barrier function** measured by repelled PBS droplets • **Capacitance** of epidermal and dermal layer (LCR meter) • **Function of vascular channels** by measuring cell distribution (fluorescence DAPI) • **Percutaneous absorption** HPLC of supernatant from beneath or from vascular channels • **Peristaltic pump** • **Caffeine, ISDN**	Mori et al. (2017)
3D multicellular microfluidic chip model (Hongik University, Seoul)	• HaCaT (human KC cell line)/ NHEK (neonate) • NHFB (neonate) donor-matched • Human umbilical vein endothelial cells (HUVECs) • Collagen hydrogel	• **Cell viability** measured by calcein AM • **Diffusion coefficient** diffusion of FITC-dextran (fluorescence intensity), glucose concentration (submerged/air-exposed) and oxygen concentration (flow/no flow)	Lee et al. (2017)

(continued)

Table 4 (continued)

Skin-on-a-chip	Cells	Readout/biomarkers/test substance	Refs.
RhS with gravity-driven microchannels (Cornell University, New York)	• NHEK (neonate) • NHFB (neonate) • Collagen hydrogel	• **Barrier function** transdermal transport of FAM-tagged oligonucleotides measured by fluorescence spectrometry showed consistency for 3 weeks. • **No Integrated micropump** • **Exposure to Doxorubicin** via microfluidics	Abaci et al. (2015)
Rhs with perfusable vasculature using iPSC and 3D technology (Colombia University Medical Center)	• NHEK (neonate) • NHFB (neonate) • iPSC-derived endothelial cells • collagen hydrogel • 3D-printed alginate molds (vasculature)	• **Barrier function** diffusion rate of fluorescein isothiocyanate (FITC)-dextran solution through microchannels (fluorescence intensity/time)	Abaci et al. (2016)
BioVaSc system with vascularized porcine jejunum (Translational Center Würzburg)	• Decellularized porcine jejunum (collagen type I and II • NHEK (neonate) • NHFB(neonate) • Primary endothelial cells	• **Barrier function** electrical impedance measuring 192Ω/cm^2 (nonsignificantly lower than native tissue) • **Vascular perfusion** electrical impedance measuring at 100 mmHg inflow at perfusate of 100 µl/(min cm^2) which correlates with skin capillary blood flow	Groeber et al. (2016)
Multi-organ chip using EpiDerm^{FT} (TissUse, Technical University of Berlin)	• EpiDerm^{FT} • Subcutaneous tissue (neonate) • Ex vivo skin biopsies • Single-hair follicular unit extracts (FUEs)	• **Tissue viability:** Ki67 and TUNEL assay for proliferation and apoptosis	Atac et al. (2013)

concentration of the anticancer drug Doxorubicin via the microfluidics, thus mimicking systemic application. Later, physiological relevance of the model was improved by the addition of perfusable vascularization using induced pluripotent stem cell (iPSC)-derived endothelial cells and 3D printing technology (Abaci et al. 2016). Other skin-on-a-chip models have also introduced vasculature, e.g., Lee et al. introduced fluidic channels using PDMS (polydimethylsiloxane) and HUVEC in RhS with HaCat-KC (Lee et al. 2017). Mori et al. developed a perfusable skin-on-a-chip model with HUVEC-lined nylon wires within the dermal compartment, mimicking the blood vessel–tissue barrier and also preventing ECM contraction (Mori et al. 2017). In this model, the percutaneous absorption of caffeine and ISDN was measured in the medium collected from beneath the skin equivalent or from the vascular channels. Another very promising model has been described in which the scaffold is constructed from decellularized porcine jejunum (BioVaSc matrix) with intact vasculature placed in a bioreactor system (Groeber et al. 2016). This model enables the interaction of cellular and noncellular compartments of the bloodstream with different layers of tissues and might be an excellent tool for immunological research and in particular to study the migration of lymphocytes in allergic contact dermatitis. Taken together, skin-on-a-chip models are expected to have added value for substance testing in allergenicity research. However, until now, only proof of concept has been established and biomarker analysis still needs to be defined. Technology areas which require attention include sensor integration and user-friendliness. Furthermore, biological optimization, such as extracellular matrix modification to reduce gel contraction and medium composition changes for an improved culture period, is still required.

7 Conclusions and Future Perspectives

In addition to single-organ skin models, multi-organ platforms are also under development in order to accurately predict drug toxicity, systemic absorption, and metabolism of drugs in, e.g., the skin, small intestine, liver, and kidneys. A four-organ chip including a preformed skin model and human intestine has been co-cultured with liver lobules and human proximal tubule epithelial cells (kidney) by Maschmeyer et al. (Maschmeyer 2015; Maschmeyer et al. 2013, 2015). Atac et al. (2013) are developing a multi-organ platform in which the commercially available EpiDerm^{FT} model is cultured in combination with subcutaneous tissue in a microfluid device together with single-hair follicular units. Hair shafts are an easy route for chemical compounds to penetrate the skin which makes such a model very interesting for allergenicity testing.

In conclusion, reconstructed human skin models for allergy research and identifying contact sensitizers are rapidly advancing and are already replacing the use of animals in many areas of substance testing and research. RhE has enabled the testing of hydrophobic chemicals and chemicals of poor solubility. They are metabolically competent to a certain extent and have a barrier function. They are easily standardized which has already resulted in a number of commercially available RhE and have already been validated for skin irritation and skin corrosion testing. They are limited because they only incorporate keratinocytes. RhS allows us to study the crosstalk between keratinocytes and fibroblasts, which drives the inflammatory cytokine response. Commercialization of RhS is hindered due to difficulties with stable transportation. Most RhE and RhS lack integrated immune cells. Immune competent models with integrated LC or T-cells are being developed to include key events 1, 2, and 3 (Fig. 1) in a single model. So far, no model has incorporated both LC and T-cells. Skin-on-a-chip has introduced the possibility to culture skin models in perfusable microphysiological culture systems in which homeostasis can be controlled and maintained. The different levels of complexity of the models ranging from RhE to skin-on-chip will enable a suitable model to be selected to fit each different research question.

Acknowledgements The authors have no conflicts of interest to declare. This manuscript was supported in part by EuroStars project (TESHI 8855).

References

Abaci HE et al (2015) Pumpless microfluidic platform for drug testing on human skin equivalents. Lab Chip 15(3):882–888

Abaci HE et al (2016) Human skin constructs with spatially controlled vasculature using primary and iPSC-derived endothelial cells. Adv Healthcare Mat 5(14):1800–1807

Ackermann K et al (2010) The Phenion full-thickness skin model for percutaneous absorption testing. Skin Pharmacol Physiol 23(2):105–112

Ade N et al (2006) Activation of U937 cells by contact sensitizers: CD86 expression is independent of apoptosis. J Immunotoxicol 3(4):189–197

Ahadian S, et al (2017) Organ-on-a-chip platforms: a convergence of advanced materials, cells, and microscale technologies. Adv Healthc Mater

Alloul-Ramdhani M, Tensen CP, El Ghalbzouri A (2014) Performance of the N/TERT epidermal model for skin sensitizer identification via Nrf2-Keap1-ARE pathway activation. Toxicol In Vitro 28(5):982–989

Andreas N et al (2011) The intra- and inter-laboratory reproducibility and predictivity of the KeratinoSens assay to predict skin sensitizers in vitro: results of a ring-study in five laboratories. Toxicol In Vitro 25(3):733–744

Andres E et al (2017) Preliminary performance data of the RHE/IL-18 assay performed on SkinEthic RHE for the identification of contact sensitizers. Int J Cosmet Sci 39(2):121–132

Api AM, Basketter D, Lalko J (2015) Correlation between experimental human and murine skin sensitization induction thresholds. Cutaneous Ocular Toxicol 34(4):298–302

Ashikaga T et al (2006) Development of an in vitro skin sensitization test using human cell lines: the human cell line activation test (h-CLAT) I. Optimization of the h-CLAT protocol. Toxicol In Vitro 20(5):767–773

Atac B et al (2013) Skin and hair on-a-chip: in vitro skin models versus ex vivo tissue maintenance with dynamic perfusion. Lab Chip 13(18):3555–3561

Basketter DA et al (2014) Categorization of chemicals according to their relative human skin sensitizing potency. Dermatitis 25(1):11–21

Bock S et al (2017) Characterization of reconstructed human skin containing Langerhans cells to monitor molecular events in skin sensitization. Toxicol In Vitro 46:77–85

Corsini E et al (2009) Use of IL-18 production in a human keratinocyte cell line to discriminate contact sensitizers from irritants and low molecular weight respiratory allergens. Toxicol In Vitro 23(5):789–796
Corsini E et al (2013) NCTC 2544 and IL-18 production: a tool for the identification of contact allergens. Toxicol In Vitro 27(3):1127–1134
Cottrez F et al (2015) Genes specifically modulated in sensitized skins allow the detection of sensitizers in a reconstructed human skin model. Development of the SENS-IS assay. Toxicol In Vitro 29(4):787–802
Cottrez F et al (2016) SENS-IS, a 3D reconstituted epidermis based model for quantifying chemical sensitization potency: reproducibility and predictivity results from an inter-laboratory study. Toxicol In Vitro 32:248–260
de Gruijl TD et al (2006) A postmigrational switch among skin-derived dendritic cells to a macrophage-like phenotype is predetermined by the intracutaneous cytokine balance. J Immunol 176(12):7232–7242
Delaine T et al (2011) Structure-Activity Relationship between the in vivo skin sensitizing potency of analogues of phenyl glycidyl ether and the induction of Nrf2-dependent luciferase activity in the keratinosens in vitro assay. Chem Res Toxicol 24(8):1312–1318
dos Santos GG et al (2009) Progress on the development of human in vitro dendritic cell based assays for assessment of the sensitizing potential of a compound. Toxicol Appl Pharmacol 236 (3):372–382
dos Santos GG et al (2011) A potential in vitro epidermal equivalent assay to determine sensitizer potency. Toxicol In Vitro 25(1):347–357
El Ghalbzouri A et al (2005) Basement membrane reconstruction in human skin equivalents is regulated by fibroblasts and/or exogenously activated keratinocytes. J Invest Dermatol 124 (1):79–86
Facy V et al (2005) Reactivity of Langerhans cells in human reconstructed epidermis to known allergens and UV radiation. Toxicol In Vitro 19(6):787–795
Galbiati V et al (2011) Further development of the NCTC 2544 IL-18 assay to identify in vitro contact allergens. Toxicol In Vitro 25(3):724–732
Galbiati V et al (2017) Development of an in vitro method to estimate the sensitization induction level of contact allergens. Toxicol Lett 271:1–11
Gerberick GF et al (2004) Development of a peptide reactivity assay for screening contact allergens. Toxicol Sci 81(2):332–343
Gibbs S, et al (2017) Assessment of metal sensitizer potency with the reconstructed human epidermis IL-18 assay. Toxicology
Gibbs S et al (2006) Autologous full-thickness skin substitute for healing chronic wounds. Br J Dermatol 155(2):267–274
Gibbs S et al (2013) An epidermal equivalent assay for identification and ranking potency of contact sensitizers. Toxicol Appl Pharmacol 272(2):529–541
Groeber F et al (2016) A first vascularized skin equivalent for as an alternative to animal experimentation. Altex 33(4):415–422
Jacobs JJ et al (2002) An in vitro model for detecting skin irritants: methyl green-pyronine staining of human skin explant cultures. Toxicol In Vitro 16(5):581–588
Jacobs JJL et al (2004) Assessment of contact allergens by dissociation of irritant and sensitizing properties. Toxicol In Vitro 18(5):681–690
Jacobs JJL et al (2006) Skin irritants and contact sensitizers induce Langerhans cell migration and maturation at irritant concentration. Exp Dermatol 15(6):432–440
Kano S et al (2011) Comparison of several reconstructed cultured human skin models by microscopic observation: their usefulness as an alternative membrane for skin in drug permeation experiments. AATEX 16(2):51–58
Kosten IJ et al (2015a) Gingiva equivalents secrete negligible amounts of key chemokines involved in langerhans cell migration compared to skin equivalents. J Immunol Res 2015:627125

Kosten IJ et al (2015b) MUTZ-3 derived Langerhans cells in human skin equivalents show differential migration and phenotypic plasticity after allergen or irritant exposure. Toxicol Appl Pharmacol 287(1):35–42

Kosten IJ et al (2016) MUTZ-3 Langerhans cell maturation and CXCL12 independent migration in reconstructed human gingiva. Altex 33(4):423–434

Lee S et al (2017) Construction of 3D multicellular microfluidic chip for an in vitro skin model. Biomed Microdevices 19(2):22

Lehe CL et al (2003) A two-centre evaluation of the human organotypic skin explant culture model for screening contact allergens. Altern Lab Anim 31(6):553–561

Mak IW, Evaniew N, Ghert M (2014) Lost in translation: animal models and clinical trials in cancer treatment. Am J Transl Res 6(2):114–118

Marionnet C et al (2006) Interactions between fibroblasts and keratinocytes in morphogenesis of dermal epidermal junction in a model of reconstructed skin. J Invest Dermatol 126(5):971–979

Martinez V et al (2006) Evaluation of eye and skin irritation of arginine-derivative surfactants using different in vitro endpoints as alternatives to the in vivo assays. Toxicol Lett 164(3):259–267

Maschmeyer I, et al (2015) Chip-based human liver-intestine and liver-skin co-cultures—a first step toward systemic repeated dose substance testing in vitro. Eur J Pharm Biopharm 95(Pt A):77–87

Maschmeyer I et al (2015) A four-organ-chip for interconnected long-term co-culture of human intestine, liver, skin and kidney equivalents. Lab Chip 15(12):2688–2699

Masterson AJ et al (2002) MUTZ-3, a human cell line model for the cytokine-induced differentiation of dendritic cells from CD34+ precursors. Blood 100(2):701–703

McKim JM, Keller DJ, Gorski JR (2012) An in vitro method for detecting chemical sensitization using human reconstructed skin models and its applicability to cosmetic, pharmaceutical, and medical device safety testing. Cutan Ocular Toxicol 31(4):292–305

Morales M et al (2016) Evaluation of fibrin-based dermal-epidermal organotypic cultures for in vitro skin corrosion and irritation testing of chemicals according to OECD TG 431 and 439. Toxicol In Vitro 36:89–96

Mori N, Morimoto Y, Takeuchi S (2017) Skin integrated with perfusable vascular channels on a chip. Biomaterials 116:48–56

Natsch A et al (2010) Chemical basis for the extreme skin sensitization potency of (E)-4-(ethoxymethylene)-2-phenyloxazol-5(4H)-one. Chem Res Toxicol 23(12):1913–1920

Netzlaff F et al (2005) The human epidermis models EpiSkin, SkinEthic and EpiDerm: an evaluation of morphology and their suitability for testing phototoxicity, irritancy, corrosivity, and substance transport. Eur J Pharm Biopharm 60(2):167–178

OECD (2012) OECD series on testing and assessment no. 168: the adverse outcome pathway for skin sensitisation initiated by covalent binding to proteins. Part 1: Scientific evidence. Part 2: Use of the AOP to develop chemical categories and integrated assessment and testing approaches. OECD, Paris

Ouwehand K et al (2008) CXCL12 is essential for migration of activated Langerhans cells from epidermis to dermis. Eur J Immunol 38(11):3050–3059

Ouwehand K et al (2010) Comparison of a novel CXCL12/CCL5 dependent migration assay with CXCL8 secretion and CD86 expression for distinguishing sensitizers from non-sensitizers using MUTZ-3 Langerhans cells. Toxicol In Vitro 24(2):578–585

Ouwehand K et al (2011a) Technical advance: langerhans cells derived from a human cell line in a full-thickness skin equivalent undergo allergen-induced maturation and migration. J Leukoc Biol 90(5):1027–1033

Ouwehand K et al (2011b) Irritant-induced migration of Langerhans cells coincides with an IL-10-dependent switch to a macrophage-like phenotype. J Invest Dermatol 131(2):418–425

Pedrosa TDN et al (2017) A new reconstructed human epidermis for in vitro skin irritation testing. Toxicol In Vitro 42:31–37

Peiser M et al (2012) Allergic contact dermatitis: epidemiology, molecular mechanisms, in vitro methods and regulatory aspects. Current knowledge assembled at an international workshop at BfR, Germany. Cell Mol Life Sci 69(5):763–781

Pistoor FHM et al (1996) Novel predictive assay for contact allergens using human skin explant cultures. Am J Pathol 149(1):337–343

Ponec M et al (2002) Characterization of reconstructed skin models. Skin Pharmacol Appl Skin Physiol 15(Suppl 1):4–17

Ramadan Q, Ting FC (2016) In vitro micro-physiological immune-competent model of the human skin. Lab Chip 16(10):1899–1908

Rasmussen C et al (2010) The StrataTest (R) human skin model, a consistent in vitro alternative for toxicological testing. Toxicol In Vitro 24(7):2021–2029

Rovida C et al (2015) Integrated Testing Strategies (ITS) for safety assessment. Altex 32(1):25–40

Rustemeyer T et al (2003) Comparison of two in vitro dendritic cell maturation models for screening contact sensitizers using a panel of methacrylates. Exp Dermatol 12(5):682–691

Saito K et al (2013) Development of a new in vitro skin sensitization assay (Epidermal Sensitization Assay; EpiSensA) using reconstructed human epidermis. Toxicol In Vitro 27 (8):2213–2224

Saito K et al (2017) An in vitro skin sensitization assay termed EpiSensA for broad sets of chemicals including lipophilic chemicals and pre/pro-haptens. Toxicol In Vitro 40:11–25

Santegoets SJ et al (2008) Human dendritic cell line models for DC differentiation and clinical DC vaccination studies. J Leukoc Biol 84(6):1364–1373

Schmook FP, Meingassner JG, Billich A (2001) Comparison of human skin or epidermis models with human and animal skin in in-vitro percutaneous absorption. Int J Pharm 215(1–2):51–56

Seok J et al (2013) Genomic responses in mouse models poorly mimic human inflammatory diseases. Proc Natl Acad Sci U S A 110(9):3507–3512

Spiekstra SW et al (2005) Induction of cytokine (interleukin-1alpha and tumor necrosis factor-alpha) and chemokine (CCL20, CCL27, and CXCL8) alarm signals after allergen and irritant exposure. Exp Dermatol 14(2):109–116

Spiekstra SW et al (2007) Wound-healing factors secreted by epidermal keratinocytes and dermal fibroblasts in skin substitutes. Wound Repair Regen 15(5):708–717

Spiekstra SW et al (2009) Potential method to determine irritant potency in vitro—comparison of two reconstructed epidermal culture models with different barrier competency. Toxicol In Vitro 23(2):349–355

Teunis M et al (2013) Transfer of a two-tiered keratinocyte assay: IL-18 production by NCTC2544 to determine the skin sensitizing capacity and epidermal equivalent assay to determine sensitizer potency. Toxicol In Vitro 27(3):1135–1150

Teunis MAT et al (2014) International ring trial of the epidermal equivalent sensitizer potency assay: reproducibility and predictive capacity. Altex-Altern Animal Experimentation 31 (3):251–268

van den Bogaard EH et al (2014) Crosstalk between Keratinocytes and T Cells in a 3D Microenvironment: a model to study inflammatory skin diseases. J Invest Dermatol 134 (3):719–727

van den Broek LJ et al (2012) Development, validation and testing of a human tissue engineered hypertrophic scar model. Altex 29(4):389–402

van den Broek LJ et al (2017) Progress and future prospectives in skin-on-chip development with emphasis on the use of different cell types and technical challenges. Stem Cell Rev 13(3):418–429

Wagner I et al (2013) A dynamic multi-organ-chip for long-term cultivation and substance testing proven by 3D human liver and skin tissue co-culture. Lab Chip 13(18):3538–3547

Walters RM et al (2016) In vitro assessment of skin irritation potential of surfactant-based formulations by using a 3-D skin reconstructed tissue model and cytokine response. Atla-Altern Lab Animals 44(6):523–532

Watson DE, Hunziker R, Wikswo JP (2017) Fitting tissue chips and microphysiological systems into the grand scheme of medicine, biology, pharmacology, and toxicology. Exp Biol Med 242 (16):1559–1572

Wufuer M et al (2016) Skin-on-a-chip model simulating inflammation, edema and drug-based treatment. Sci Rep 6:37471

Cancer Explant Models

Christian T. Stackhouse, George Yancey Gillespie and Christopher D. Willey

Contents

C. T. Stackhouse
Department of Radiation Oncology, Department of Neurosurgery,
University of Alabama, Birmingham, USA
e-mail: ctstackh@uab.edu

G. Y. Gillespie
Department of Neurosurgery, University of Alabama, Birmingham, USA
e-mail: gygillespie@uabmc.edu

C. D. Willey (✉)
Department of Radiation Oncology, University of Alabama, Birmingham, USA
e-mail: cwilley@uabmc.edu

Current Topics in Microbiology and Immunology (2021) 430: 131–160
https://doi.org/10.1007/82_2019_157

Published Online: 20 March 2019

Abstract Overcoming the challenges of understanding and treating cancer requires reliable patient-derived models of cancer (PDMCs). For decades, cancer research and therapeutic development relied primarily on cancer cell lines because of their prevalence, reproducibility, and simplicity to maintain. However, findings from research conducted in cell lines are rarely recapitulated in vivo and seldom directly translatable to patients. The tumor microenvironment (TME), tumor-stromal interactions, and associations with host immune cells produce profound changes in tumor phenotype and complexity not captured in traditional monolayer cell culture. In this chapter, we present various cancer explant models and discuss their applicability based on specific research aims. We discuss the appropriateness of these models for basic science questions, drug screening/development, and for personalized, precision medicine. We also consider logistical factors such as resource cost, technical difficulty, and accessibility. We finish this chapter with a practical guide intended to help the reader select the cancer explant model system(s) that best address their research aims.

1 Introduction

As we enter a new age of high-throughput molecular sequencing and personalized, precision medicine, it has become increasingly clear that understanding human disease requires both macroscopic and microscopic approaches. Single-cell sequencing (SCS) has revealed vast intratumoral heterogeneity and complex cellular relationships both among cancer cells and between cancer and normal cells (Winterhoff et al. 2017). SCS has enabled researchers to track tumor evolution and gain insights into the molecular processes underlying tumor phenotypic characteristics (Ellsworth et al. 2017). Pan-cancer genome-wide association studies (GWAS) in human and mouse models have identified pleiotropic associations elucidating canonical pathways involved in carcinogenesis, therapy resistance, and metastasis (Fehringer et al. 2016; Hung et al. 2017; Martens-de Kemp et al. 2017). Rapidly advancing technologies in nucleic acid sequencing and imaging are enabling the integration of these data to better understand correlations between in situ tumor phenotypes and tumor-stromal-immune expression networks (Na and Choi 2017). Advances in computer science, biomedical data science, and machine learning have revolutionized the study of cancer through advanced modeling and *in silico* screening (Huang et al. 2017a, b; Vidic et al. 2017). However, so far, we have not witnessed advancements in cancer treatments or improvements in patient outcomes to match the rate of advancement in our technologies (Perry and Lawrence 2017). Successfully translating preclinical therapies to clinical application requires faithful patient-like models of the actual disease (Willey et al. 2015). Preclinical models must suitably recapitulate human disease, including tumor heterogeneity, tumor microenvironment, and interactions with host stromal cells (Daher and de Groot 2017; Decker et al. 2017; Quail and Joyce 2017). Figure 1 represents some of the in situ tumor complexity to be considered when formulating the research question and selecting the appropriate cancer explant model. Selection of the optimal cancer explant model is dependent on the nature of the research question and the expected or intended outcomes of the study. In this chapter, we will evaluate cancer explant model systems according to their (i) applicability to the study design, (ii) biological relevance to the in situ condition and, (iii) practical accessibility/utility in cancer research.

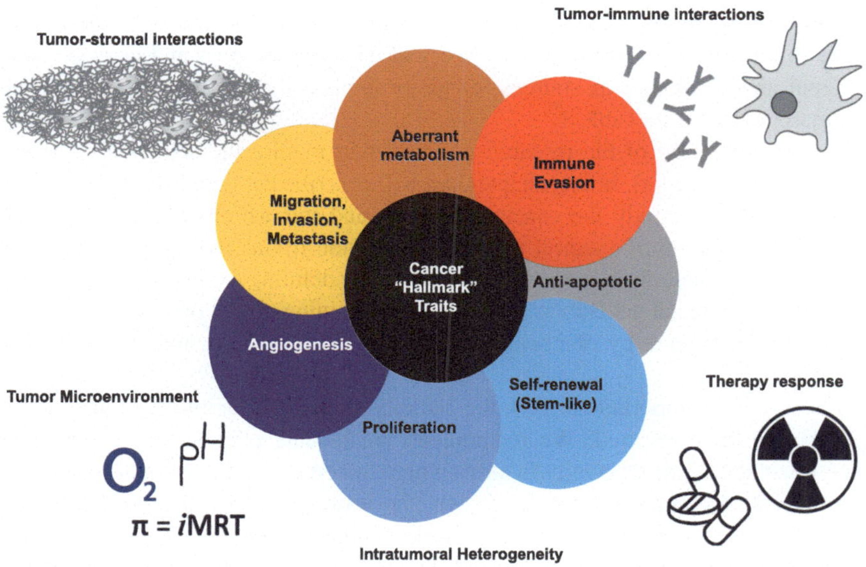

Fig. 1 Introduction to the hallmark phenotypic traits of cancer and the complexity of in situ conditions that should be considered in selecting a cancer explant model. In addition to the hallmark phenotypic traits listed, higher order systems considerations including intratumoral heterogeneity, tumor microenvironment (hypoxia, pH, and osmotic pressure indicated), tumor–stromal interactions, tumor–immune interactions, and therapy response should also be considered

2 Framing the Research Question

The first step in experimental design is to define the problem. The research question may be broad for example: "What are the role(s) of non-coding RNA in facilitating therapeutic resistance in glioblastoma (GBM) tumors?" This question must then be broken down into testable units: The lncRNA MALAT1 confers resistance to temozolomide in GBM by suppressing miR-203 expression (Chen et al. 2017). The next step is to determine your expectations and desired outcomes. If the hope is to discover a novel compound to inhibit MALAT1 activity, a model conducive to high-throughput screening such as cell culture techniques would be appropriate. If the desire is to clinically translate that compound, then a model which provides a more realistic replication of in situ conditions is required to evaluate drug delivery, efficacy, and safety. An animal model or organotypic explant model may be ideal in this case.

Once the question and expectations are defined, the source of the cancer explant material(s) must be determined. Primary tissues can be difficult to obtain due to cost, accessibility, and viability, but they can provide insights into the molecular state and diversity of the patient tumor. Established cell lines are affordable and versatile, but they have lost the complexity and heterogeneity seen in the primary tumor and have magnified the expression of genetic aberrations that help select the fastest growing cells. Animal models can be used to produce murine tumors analogous to the human tumor in an immunologically intact host; alternatively, human tumors can be grafted into various murine models to maintain the human tissue in an orthotopic environment. We will describe the pros and cons of using different animal models as sources for tumor explant models later in the chapter, summarizing them in Table 1.

The interrogation of the research question will require the selection of appropriate explant model(s). In this chapter, we will describe the applicability of various explant models to study the cancer hallmarks and interactions at the appropriate levels. The selection of a cancer tumor model is non-trivial and highly complex as represented in Table 2. We recommend starting by defining exclusion criteria based on your experimental goals then judging each remaining model on applicability and feasibility determined by practical considerations such as cost and other resource constraints.

Finally, it is important to test the validity and reproducibility of the model system you have selected. We recommend small-scale pilot testing of any new experimental design, especially in cases where materials are rare or non-renewable. The individual components of the design, the source material, the explant model, and the molecular assays may all be well validated, but the synthesis of these components (i.e., the novelty of the study) will almost certainly require calibrations. In the last section of this chapter, we describe a generic start-to-finish strategy for developing optimal cancer explant systems summarized in Fig. 2.

Table 1 Descriptions of different sources of tumor material

	In vitro		Animal models					
	Cell lines		Transplantable		Induced		GEMM/transgenic	
	Established	Primary culture	Xenografts	Syngrafts	Chemically virally	Physically	Spontaneous	Inducible
Patient-derived	Yes	Yes	Yes	No	No	No	No	No
Cost	Low	Moderate	High	High	High	High	High	High
Availability	High	Moderate	Moderate	Moderate	Moderate	Low	Moderate	Moderate
Reproducibility	High	Moderate	Moderate	Moderate	Moderate	Moderate	High	High
Heterogeneity	No	Yes	Yes	Variable	Variable	Variable	Variable	Variable
Tumor-Stromal	No	Limited	Limited	Yes	Yes	Yes	Yes	Yes
Immune Competent	No	No	No	Variable	Yes	Yes	Yes	Yes
In situ *Complexity*	Low	High	High	Moderate	High	High	Moderate	Moderate

Table 2 Overview and evaluation criteria for various cancer explant models

	In vitro			Ex vivo		In vivo			
	Cell Culture	Transwell	Spheroids	Microtumors & Organoids	Tumor-on-a-chip	CAM	GEMM	PDX	Organotypic Slice
Cellular level	✓	✓	✓	✓	X	X	X	X	X
Cell cycle: Proliferation and apoptosis	✓	✓	✓	✓	X	X	X	X	X
Cell mobility	✓	✓	X	X	✓	✓	X	X	X
DNA damage repair	✓	✓	✓	✓	X	X	✓	✓	✓
Aberrant metabolism	✓	✓	✓	✓	X	X	✓	✓	✓
Signaling	✓	✓	✓	✓	✓	✓	✓	✓	✓
Gene expression networks	✓	✓	✓	✓	✓	✓	✓	✓	✓
Molecular profiling	✓	✓	✓	✓	✓	✓	X	✓	✓
Tumor/system level	□	□	□	□	✓	✓	✓	✓	✓
Tumor–stromal interactions	□	□	□	□	□	X	✓	□	✓
Tumor microenvironment	□	□	□	□	✓	X	✓	✓	✓
Intratumoral heterogeneity	□	X	□	□	□	X	□	✓	✓
Tumor–immune interactions	□	□	□	□	□	X	✓	X	✓
Cancer stem cells	✓	□	✓	✓	□	X	X	✓	✓
Migration/invasion	✓	✓	X	X	✓	✓	✓	✓	✓
Circulating tumor cells	□	X	X	X	✓	✓	✓	✓	X
Metastasis	X	X	X	X	X	✓	✓	✓	X

(continued)

Table 2 (continued)

	In vitro			Ex vivo		In vivo			
	Cell Culture	Transwell	Spheroids	Microtumors & Organoids	Tumor-on-a-chip	CAM	GEMM	PDX	Organotypic Slice
Angiogenesis	□	□	X	X	□	✓	□	□	□
Therapy response/ resistance	□	□	✓	✓	✓	✓	✓	✓	✓
Translational/ preclinical level	□	□	□	✓	✓	✓	✓	✓	✓
Precision medicine	□	□	□	X	✓	□	✓	✓	✓
High-throughput screening	✓	✓	□	✓	□	□	X	X	X
Drug development	□	□	□	□	✓	✓	✓	✓	✓
Drug delivery	X	X	X	X	✓	X	✓	✓	X
Drug safety	X	X	X	□	✓	□	✓	✓	X
Drug efficacy	□	□	✓	✓	✓	✓	✓	✓	✓

Key
✓ Appropriate
□ Possible
X **Not Recommended**

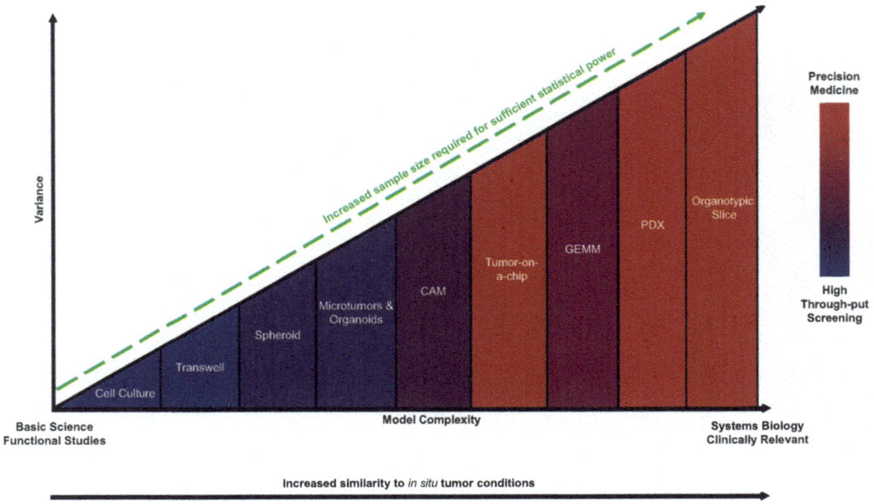

Fig. 2 Cancer explant model selection guide. The x-axis represents increasing model complexity which combines technical complexity and biological complexity. The y-axis represents increasing total variance including biological and technical variances. The solid, black trendline indicates that total variance increases with model complexity. Increasing variance will decrease the ability to distinguish effect size, thus decreasing statistical power. The green, dashed trendline represents the requirement for increased sample size to achieve sufficient power. The models are separated in nine bins under the black, solid trendline ordered by increasing model complexity. Each bin is colored to represent a spectrum of clinical/translational applicability. More heat (red) indicates suitability for precision medicine applications. Less heat (blue) indicates suitability for high-throughput screening applications. The solid, black line below the x-axis indicates that the above cancer explant models increase in similarity to the in situ tumor conditions from left to right

3 Cancer Hallmarks

The primary "hallmarks" of cancer (Fig. 1) are intratumoral heterogeneity, self-renewal (stem-like), aberrant proliferation and metabolism, abnormal angiogenesis, migratory and invasive cell properties, metastasis, immune evasion, and avoidance of apoptosis (Hanahan and Weinberg 2000; Liu and Cheng 2013; Pavlova and Thompson 2016). Further complexity comes from consideration of tumor–immune interactions, tumor–stromal interactions, tumor microenvironment conditions, and response to damage from therapy. Decide which of these hallmarks or interactions are relevant to your research question: A functional study may focus on a single hallmark such as angiogenesis or a study may investigate multiples of these aspects. It is important to keep in mind the complexity of the overall system regardless of the functional level that is the focus of the study in order to provide a broader context.

3.1 Cancer Stem Cells

There remains mystery and contention around the mechanisms of tumorigenesis and tumor organizational structure. The existence of cancer stem cells has been argued semantically, and therefore many researchers have settled on using the term "tumor initiating cells" (TICs) to refer to those cells within the tumor that possess the properties of pluripotency/multipotency and self-renewal (Cermeño and García 2016; Qureshi-Baig et al. 2017). Solid tumors display a high degree of intratumoral heterogeneity with cells ranging in state of differentiation and replicative capacity (Zhang et al. 2017). There are tumor cell organizational models which postulate a hierarchical structure, a stochastic hypothesis, and/or retro-differentiation programing (Melzer et al. 2017).

3.2 Angiogenesis

Therapeutic management and prognostic consideration for many cancers consider the general phenotype of tumor being highly angiogenic and having abnormal vasculature (Viallard and Larrivee 2017). Abnormal tumor vasculature affects perfusion and drug retention, creates hypoxic regions, and can facilitate an immunosuppressive TME (Chen et al. 2017). Normalization of tumor vasculature may enhance the efficacy of contemporary chemotherapies; especially immunotherapy in tumors (Chen et al. 2017; Johansson-Percival et al. 2017). Vascular phenotypes from pathology and functional imaging can be both prognostic and predictive of outcomes in some cancers (Ing et al. 2017).

3.3 Proliferation and Avoidance of Apoptosis

Cancer is characterized by abnormal cell cycle regulation leading to rampant cellular growth/divisions and an ability to avoid the induction of normal apoptotic mechanisms. Dysregulation and dysfunction are the words that best summarize carcinogenesis.

3.4 Migration/Invasion

Migrating tumor cells which invade normal stroma are one of the major obstacles in treatment leading often to tumor recurrence. Cancer cell motility depends on their ability to associate with, degrade, and restructure the surrounding extracellular matrix (ECM). This requires the expression of proteolytic proteins such as matrix metalloproteinases (MMPs), focal adhesion kinases, and integrins (Das, Monteiro, Barai, Kumar, & Sen, 2017).

3.5 *Metastasis*

Metastasis is related to tumor cell migration and invasion, as they are necessary for initiation, but not sufficient for the metastatic phenotype. Some cancers are highly invasive of normal tissue such as neuroblastomas and glioblastomas, but these tumors rarely metastasize. To metastasize, tumor cells must first migrate from the primary site, then enter the lymph or circulatory systems as circulating tumor cells where they must evade the host immune system. They must then exit circulation and establish a new tumor in the secondary site. Cancer-specific tissue specificity such as primary breast to brain metastasis is common (particularly to the posterior fossa); yet this tropism remains poorly understood (Narkhede et al. 2017).

3.6 *Metabolism*

Cancer is often characterized by rapid, dysregulated growth requiring the "reprogramming" of cellular metabolism to exploit the often nutrient-poor tumor microenvironment (Pavlova and Thompson 2016). Aberrant tumor metabolism in situ has been evaluated using FDG PET imaging which was subsequently correlated with expression networks in the TME for head and neck squamous cell carcinoma (HNSCC) (Na and Choi 2017). It is apparent that investigation of metabolic networks within the tumor should consider cancer cell metabolism as well as changes in tumor-associated stromal cells which may play supportive roles for the tumor and TME (Otte et al. 2016).

3.7 *Therapy Resistance and Recurrence*

Therapeutic evasion or resistance is a major hurdle in therapeutic management of recurrent tumors which are intractable and highly aggressive. Recurrent tumors are most often responsible for patient mortality. In some cases, aggressive therapy can lead to poorer patient outcomes by selecting for therapy resistant, aggressive sub-populations of cells, or by initiating growth in response to the release of mitogens, for example, following surgical resection (Lannin and Wang 2017).

3.8 Hypoxia

Hypoxia and large regions of necrosis in solid tumors are generally poor prognostic indicators. Hypoxia-inducible factors often lead to pro-tumor transcriptional activation leading to increased growth and poorer patient outcomes. Hypoxia also correlates with therapy resistance (Yasui et al. 2017).

3.9 Basic Science, Translational, and Clinical Research

The selection of a cancer explant model will depend on the nature of the research question itself. Basic science studies are hypothesis-driven, typically highly specific, and often focus on understanding the underlying mechanisms or process. They can also be aimed at developing new research methods or models. Translational research is best summarized as "bench-to-bedside" or vice versa. It has the specific goal of generating knowledge that can be translated to the clinic as a novel therapeutic agent, treatment strategy, or medical device. Clinical research is very focused on patient outcomes. Clinical research and clinical trials are highly regulated as they frequently involve studying patients directly or collecting biological materials from them. The aims of clinical research are to test the safety and efficacy of drugs and to develop diagnostic, prognostic, or predictive tests to detect and track the progression of disease.

4 Tumor Material Sources

It is important to identify the source of tumor material that will be used in the study. The source could be human tumors derived from patients or can be tumors from other species such as murine tumors that are analogous to the human cancer. Table 1 summarizes some of the different tumor sources.

4.1 Solid Tumors Versus Liquid Tumors

The vast majority of cancers are solid tumors and so there are many cancer explant models that can be used to characterize them. There are often many TME niches in solid tumors, each with distinctly behaving sub-populations of cells. Abnormal vasculature, stromal invasion, and necrosis all lead to distinct sub-regional microenvironments. This adds spatial–temporal complexity, which some tumor explant models are better able to simulate.

A minority of "liquid" tumors, namely leukemia and lymphoma, are more challenging to recapitulate in explant models, but a number of murine models of lymphoma and leukemia have been developed (Kohnken et al. 2017). Established human explant models of these cancers are rare, resulting in a reliance on access to primary tissues. This chapter will focus on cancer explant models of solid tumors which are more prevalent and better characterized.

4.2 Renewable or Consumable

Primary tissue can either be used to establish long-term cancer explant models such as cell lines in vitro or "xenolines" in vivo. Alternatively, they can be utilized for a limited number of assays at the end of which the material is completely consumed. In these cases, it is recommended to cryopreserve aliquots of cells, snap frozen tissue, or cellular extracts. The formation of "biobanks" of primary tissue for direct interrogation and drug screening is an idea that is gaining popularity, but is still hampered by non-renewability, technology, and logistical hurdles (Jackson and Thomas 2017). Some cancers or tumors are inherently more difficult to establish in vitro and/or in vivo which introduces model-specific biases often for rapidly growing cells which can adapt to the given model. Primary tissue cultures will often recapitulate the diversity of the original tumor, but heterogeneity and phenotypic relevance to the primary tumor will decline rapidly the longer the material is out of the in situ environment and declines further with increasing passages.

4.3 Patient Primary Materials

Access to primary patient tumor material is not available to all researchers at all institutions. The use of primary and patient-derived materials is heavily regulated by Institutional Review Boards (IRB) that ensures compliance with institutional, state, and federal regulatory mandates. Red-tape aside, the quality of isolated patient primary tissue can vary considerably based on the conditions of the surgery itself, largely on the handling of the material and the timespan from resection to utilization. For molecular studies, it is best for the material to be snap frozen as quickly as possible following resection to preserve the in situ molecular profile. For culture or cultivation, tissues could be placed into appropriate medium on ice, but the key factor for successful cultivation or characterization is time (Healy 2017).

4.4 *Syngeneic and Allogeneic Murine Tumors*

Murine tumors that are implanted into the same species are classified as either syngeneic or allogeneic based on the strain of the animal (as defined by the haplotype of the Major Histocompatibility Complex of the host and the tumor). Syngeneic tumors arise from the same strain as the host, while allogeneic tumors are from a different strain of the same species of host. As such, both murine tumor types have the advantage of including host stromal–tumor interactions. A syngeneic model has the added advantage of studying host immune–tumor interactions as the tumor or graft is genetically identical to the host, eliminating immune rejection and closely replicating the normal immune state of the host. An allogeneic graft originates from an animal of the same species as the host, but it may result in an abnormal or hyper-immune response (Lentinget al. 2017) and thus should be used in immunocompromised hosts.

4.5 *Carcinogen or Virally-Induced*

Murine tumors can be induced using chemical carcinogens or by high-energy ionizing radiation (Rivina et al. 2014). 20-methylcholanthrene (20-MCA) belongs to a family of polycyclic aromatic hydrocarbons (PAHs) which are considered common environmental mutagens responsible for carcinogenesis in a variety of species (Kumar et al. 2012). The chemical carcinogen N-methyl-N-nitrosourea (MNU) induces DNA damage and increases c-Myc expression in mouse embryonic fibroblast leading to the observation that Fbxw7 haploinsufficiency increases the risk for gastric carcinogenesis (Jiang et al. 2017). A mutagenic screen of C57BL/6 J mice with a related chemical carcinogen, N-ethyl-N-nitrosourea (ENU), produced novel mutant mice (Bcl2WE/WE) which can be used to study cancer immunotherapies (Viant et al. 2017). 20-MCA has been used to generate GL-261 mouse glioblastoma cells and ENU to generate RG-2 rat glioblastomas which are genetically heterogeneous and replicative of the conditions of human gliomagenesis by preserving the TME including the intact blood–brain barrier and immune competence (Lenting et al. 2017). ENU-induced gliomas are highly relevant; however, these models are not easily reproducible due to their spontaneous and diverse nature exhibiting biological variance similar to that seen in human GBM (Lenting et al. 2017). Other induced models can often bear carcinogen molecular signatures that can confound the molecular phenotype and tumorigenic mechanisms of *de novo* tumors. Induced-transplantable models also frequently trigger a large immunogenic response that can confound the study (Lenting et al. 2017). It is, therefore, important to understand the limitations of the model employed.

Oncogenic viruses can also be used to generate spontaneous murine tumors. Mouse mammary tumor virus (MMTV) has been used to generate spontaneous tumors in inbred mouse strains. Murine models have also been created to study human papillomavirus-associated genital cancers (Decrausaz et al. 2011).

4.6 Genetically Engineered Mouse Models/Transgenics

Genetically engineered mouse models (GEMMs) and transgenic mice are versatile and relevant as these models produce analogous murine tumors, allowing for the study of host stromal–tumor interactions and host immunity–cancer interactions recently reviewed (Kersten et al. 2017). Murine tumors can also be a substitute source when patient-derived human tumors are not accessible or are difficult to establish.

The simplest form of GEMMs is produced by selective breeding techniques such as inbreeding which go back centuries. Syngeneic mouse or rat gliomas can also be transplanted into rodents as allografts. However, modern GEMMs typically refer to *de novo* tumors produced through specific germline or non-germline genetic engineering strategies. Original GEMM models used transgenic overexpression of suspected oncogenes, so called oncomice, but recent approaches have utilized more elegant conditional systems and embryonic stem cell manipulation. GEMMs can also be generated with germline editing using tools such as CRISPR-Cas9 to accelerate GEMM development. Nevertheless, these tumors do arise more slowly than transplanted or chemically induced models, but permits the study of early tumor development and often can be followed longitudinally over the full span of tumor development. Some GEMMs result in spontaneous tumor formation by driving oncogenic processes mimicking oncogenesis in humans. Transgenic mice do often require maintenance of breeding colonies and the generation of a novel model can be expensive. Tumor development is often slow, can be multi-focal, and penetrance is often incomplete. These factors lead to increased cost, resource demand, and reduced statistical power. The National Cancer Institute (NCI) has created a resource for investigators to learn about GEMMs and other animal models called the Electronic Models Information, Communication, and Education (eMICE). The NCI Mouse Repository (https://frederick.cancer.gov/science/technology/mouserepository) can provide frozen embryos of various murine strains to the scientific community for academic, non-profit, and commercial use. Other murine models for cancer research can be obtained from the Jackson Laboratory (Bar Harbor, ME) or other such commercial sources.

4.7 *Cancer Cell Lines*

The availability of primary patient materials is one major obstacle for building effective disease research models. Many scientists employ immortalized cell lines which are readily accessible, relatively cheap, and easy to work with. Some groups will establish their own spheroid or 2D monolayer cultures from primary patient tissues, but not all tissues are receptive to or stable in these conditions. This can lead to "dish" biases that reduce cellular diversity and induce phenotypic changes that may confound processes which occur in vivo. Established cell lines are very well characterized owing to sometimes decades of study. They are largely homogenous meaning they generate predictable and reproducible results. Cancer cell lines are available from multiple sources commercially and academically. The NIH Division of Cancer Treatment and Diagnosis Developmental Therapeutics Program (DTP) located at the National Cancer Institute in Frederick, Maryland, maintains a large repository of tumors and tumor cell lines available to the research community. The NCI also maintains a national repository of patient-derived models (PDMs) including PDXs.

5 Cancer Explant Models

New explant models are being developed constantly. The following are representative, but not exhaustive characterizations of popular cancer explant models. Every model has specific strengths and weakness. It is important to consider the trade-offs between complexity and variance which can affect the power of a given study. Every model will also have inherent biases that should be accounted for when planning experiments and interpreting results. Table 2 provides recommendations for the use of certain cancer explant models based on a qualitative assessment of these considerations.

5.1 *Organotypic Slice*

Organotypic slice culture has long been a staple of basic neuroscience research to study cell morphology, electrophysiology, and neurocircuitry (Humpel 2015). Brain slices from freshly killed mice can remain viable for hours in perfusion chambers containing media which simulates cerebral spinal fluid. The pH and partial pressures of gases can be regulated to replicate in situ conditions. Slices can be assayed using calcium imaging, manipulated by the addition of compounds, and/or electrically stimulated with electrophysiology measurements. Slices can later be fixed for histology or in situ hybridization studies.

Organotypic slice or raft cultures are a cross section of the in situ state, allowing for spatial–temporal study of tumor–stromal and tumor–immune cell interactions. Organotypic cerebellar slice culture was used to study pediatric medulloblastoma tumor cell invasion into normal stroma (Neve et al. 2017). The major trade-offs in these models are the technical complexity and resource cost compared to more traditional in vitro methods. Organotypic slice as an ex vivo cancer model is rapidly evolving as its translational potential in drug development and precision medicine has been recognized (Naipal et al. 2016).

5.2 *Xenografts*

Xenografting (xenos—Greek "foreign") or cross-species transplantation of tissues or organs has a long history in science, medicine, and folklore dating back centuries (Cooper 2012). Xenotransfusions and xenotransplantations of blood or derived factors have been applied clinically dating back to the seventeenth century and are still of interest today (Cooper 2012; Gao et al. 2017). The first successful human tumor to murine xenograft was reported in 1969 (Fiebig and Burger 2002). The use of xenografts as cancer explant models has gained popularity over the past decade as the need for more reliable preclinical models has been realized. Patient-derived xenografts (PDX) are gaining popularity as models for drug development and radiation research (Lai et al. 2017; Willey et al. 2015). PDX exist for colorectal cancer (Brown et al. 2016), glioblastoma (William et al. 2017), prostate cancer (McLean et al. 2017), ovarian cancer (Heo et al. 2017), melanoma (Kemper et al. 2016), neuroblastoma (Braekeveldt and Bexell 2017), and many others (Bagby et al. 2016; Davies et al. 2017; Dobrolecki et al. 2016; Lai et al. 2017; Morgan et al. 2017). The PDX model is favored because of its replication of many features of the in situ tumor microenvironment (TME) conditions and the faithful recapitulation of patient tumor genomic and phenotypic characteristics (Scott et al. 2016). Establishing PDX from primary patient tissue is technically straightforward if nuanced; however, access to high-quality, viable tissue, and varying rates of success for the establishment of a stable graft can make this model difficult and expensive compared to other less-sophisticated models (Choi et al. 2014, 2016; Ledur et al. 2017).

Xenografts can be established from readily available cancer cell lines or from patient primary tissues. Human cells are grafted into immune-compromised animals such as athymic nude (*nu/nu*), C.B-1.7 scid, or NOD-scid mice. However, non-patient-derived xenografts from established immortalized cell lines that have been selected over decades for growth in vitro often fail to faithfully recapitulate the diversity or phenotype of patient tumors. Cells transplanted from a norm-oxic, nutrient-rich medium undergo rapid shifts to adapt to the more "stringent" in vivo environment. This selective pressure reduces clonal diversity, further introducing a bias for fast-growing, rapidly adapting subclonal populations. PDX established directly from patient tissue possess higher diversity and complexity than cell lines and retain to a much greater extent the patterns of aberrant gene expression observed in human tumors. These cells are already adapted to the rigors of an in vivo environment that may be characterized by hypoxia, low glucose, and low perfusion. PDX can be serially passaged in immunocompromised hosts as "patient derived xenolines" which maintain increased complexity and leads to molecular phenotypes that closely resemble the in situ profile even after many passages.

Xenografts can either be placed heterotopically, such as a subcutaneous (e.g., flank) graft of glioma cells, or orthotopically: the placement of human glioma cells intracranially. An orthotopic xenograft (PDOX) is superior in replicating the in situ TME conditions; thus, PDOXs more closely resemble the patient tumor phenotype and are more accurate predictors of response to therapy (Hoffman 2015; William et al. 2017). Depending on the disease site and goals of your study, it may be more difficult to use the PDOX model and so the heterotopic xenograft may be more appropriate. It is common to track tumor growth in PDX experiments. A subcutaneous xenograft can be measured reasonably accurately and inexpensively with calipers. The growth of an orthotopic PDX must be measured indirectly with bioluminescence or imaging modalities such as CT or MRI (Jimenez-Gonzalez et al. 2017). Advances in imaging and computational technologies now enable more advanced techniques including SPECT, PET, or MR spectroscopy (Cao et al. 2017). We have found in our glioma xenolines that passaging cells between the flank and intracranial environments do not significantly compromise molecular characteristics from the patient tumor at least for tumors at low-to-moderate passage levels. The PDX model falls somewhat short of GEMMs in the loss of host stromal-tumor interactions and in the lack of immune competence in the hosts (Kersten et al. 2017). Humanized mouse models of PDX are rare, difficult to establish, and expensive but could be valuable for immune-oncology drug discovery (Verma et al. 2017). A major concern will be to ensure the humanized host mouse has the same MHC haplotype as the patient from whom the tumor was resected. Investigations which depend on tumor–stroma or tumor–immune system interactions may be better suited with an organotypic slice culture model or GEMM.

The PDX model is attractive for preclinical safety and efficacy testing of novel agents. It also allows for the testing of multimodality or combinational agent therapies. Advanced MRI and PET imaging can measure parametric blood flow, volumetrics, metabolic activity, and reveal macroscopic features such as regions of hypoxia or stromal invasion (Holzgreve et al. 2016). These data can be useful in determining responsiveness to therapy and efficacy of targeting in drug delivery. The maintenance of tumor diversity and complexity that are the highlights of the PDX model also translate to high biological variance within experimental arms. This can complicate experimental design and result in ballooning resource consumption and cost. It is best then to consider this model as "quality over quantity." The PDX model is well suited as an advanced preclinical model where one wishes to understand the pharmacodynamics, efficacy, and delivery of a novel agent. It is less suited for high-throughput drug screening. We have previously discussed how a PDX system can be used as either tumor avatars or proband models, depending on whether a PDX is intended to have a direct one-to-one correlation with the patient from which it is derived or whether a patient can be matched to an existing PDX, respectively (Willey et al. 2015). Furthermore, PDX are actively being developed for parallel or "co-clinical" trials in early phase therapeutic development.

5.3 Chick Chorioallantoic Membrane

The chick chorioallantoic membrane (CAM) assay is a well-documented, inexpensive, and highly effective in vivo basic science and preclinical model for the study of angiogenesis, tumorigenesis, metastasis, and invasion (Dohle et al. 2009; Lokman et al. 2012; Palmer et al. 2011). The CAM assay is typically carried out either in ovo, where a small window is cut into the shell exposing the CAM blood vessel network, or ex ovo, for which the egg content is completely removed from the shell and placed in a tissue culture dish (Dohle et al. 2009). Traditionally, the CAM assay was used by various industries to test the safety of consumer products (Debbasch et al. 2005). Test compounds can be placed on the membrane, away from any large vessels. The egg is then incubated, and the angiogenic potential of the compound is assessed by the formation of new vessels in and around the immediate area of application (Deryugina and Quigley 2008). The early chick embryo, being immune-deficient and nutrient-rich, is excellent for sustaining allo- and xenografts which can be readily injected intravenously into the large, accessible chorioallantoic vein (Palmer et al. 2011). Wnt pathway inhibition of human squamous cell carcinoma was found to reduce tumor growth and metastatic spread using the CAM model (Rudy et al. 2016). The use of Matrigel grafts containing tumor cells with or without the addition of novel compounds can be employed to study effects on migration and invasion in vivo (Lokman et al. 2012). It is possible to directly visualize many assays through the CAM, and multiple simultaneous tests of biomaterials can be conducted in the same CAM assay (Tufan and Satiroglu-Tufan 2005; Valdes et al. 2002).

This method is highly cost-effective as fertilized eggs are cheap to purchase and to maintain in large quantities though typically must be specially ordered. This is also an avian, embryonic model which in most cases requires less regulatory oversight and lower overhead cost for facilities and staff compared to murine models. The incubation time for this assay is short, on the order of days to weeks, so it is possible to screen hundreds of assays in a months' time. Studies which require tumor–immune cell or tumor–stromal interactions cannot be examined in this assay, as the early embryo is immune-deficient (Palmer et al. 2011). This is an appropriate model for researchers interested in understanding the basis of tumorigenesis, invasion, metastasis, and angiogenesis. It is also appropriate for high-throughput screening of novel compounds. This assay has been used to test inhibitors for 5-HT_7 in triple-negative breast cancer (Gautam et al. 2016), to test YF-452, an antiangiogenic compound suppressing VEGFR2 signaling (Liu et al. 2017), and to find that Cucurbitacin B reduces the metastatic and angiogenic potential of breast cancer cells via downregulation of VEGF/FAK/MMP-9 signaling (Sinha et al. 2016).

5.4 Liquid-Based 3D Models

The most common liquid-based 3D tumor culture model is spheroid culture. Spheroids recreate tumor characteristics such as hypoxia and necrosis, depending on the size of the spheroid. The hanging drop, rotary culture, and aqueous two-phase methods are most common for liquid-based 3D culture (Ham et al. 2016). The rotary orbital hydrodynamic culture system (RCCS) was described in 1999 as a system for growing anchorage-dependent or suspension cells (Mitteregger et al. 1999). This technique has been successfully employed to potentiate megakaryopoiesis and significantly improve the efficiency of platelet generation (Yang et al. 2016) pointing to the utility of this approach as a low-cost, efficient way to quickly produce high yields of cells, in a highly controlled way. This could have clinical implications for enriching blood transfusion products for patients or to produce stem/progenitor cell aggregates for downstream application or study (Laundos et al. 2017). These RCCS devices are being used in 3D culture systems described below to improve performance.

Spheroid culture can be used for the enrichment and isolation of stem-like, tumor-initiating cell populations (Ishiguro et al. 2017). Spheroid formation and limited dilution assays can be useful not only in identifying the tumor initiating potential of a population of cells, but also in creating clonogenic libraries for drug screening (Shaheen et al. 2016). Gene-editing and molecular labeling can also be combined with these techniques to accomplish cell lineage tracing in vivo (Woodworth et al. 2017). In addition, these models are being prototyped for precision medicine purposes in combination with PDX (Chia et al. 2017; Pauli et al. 2017).

5.5 Scaffold-Based 3D Models

Scaffold-based 3D tumor explant culture provides a solid or gel-like matrix that is either synthetic as in Matrigel®, or organically derived as in Myogel (Salo et al. 2015). These products provide an analog for the complex extracellular matrix (ECM) found in situ. In the case of tumor tissues, these 3D models are sometimes referred to as microtumors. Normal tissues derived from pluripotent cells can form 3D organoid structures which replicate the microenvironment and function of organs to varying degrees of complexity. ECM is important for cell adhesion, migration, invasion, and signaling. The stromal-ECM provides the scaffolding that shapes the organs of the body. Simulating the in situ TME ECM preserves the molecular and phenotypic characteristics of these patient-derived explants. 3D microtumors grown on a human biomatrix of HuBiogel™ have been shown to provide physiologically relevant microenvironments sufficient to maintain the phenotypic characteristics and surface markers of brain tumor-initiating cells (Gilbert et al. 2017). Additionally, we have found that 3D microtumors produced using this human biogel culture system can provide a high-throughput drug screening system by generating physiologically relevant tumors that can be molecularly profiled for drug response testing (Gilbert et al. 2016).

5.6 "Tumor-on-a-Chip"

The combination of microfluidics and organotypic explants have created chip-based technologies, including "organ-on-a-chip," "tumor-on-a-chip," and ultimately building toward "human-on-a-chip" systems (Dehne et al. 2017; Esch et al. 2015; Tsai et al. 2017). These bioengineered microdevices provide a closed system of 3D organotypic cultures that can replicate both TME conditions and system-level biology integration. These systems can be designed to study metastasis and the role of TME in tissue specificity (Portillo-Lara and Annabi 2016). These devices are being developed for drug screening and precision medicine as an alternative or supplement to preclinical animal testing (Ozcelikkale et al. 2017). In theory, cells could be obtained from a patient and grown out under conditions to produce induced pluripotent stem cells (IPSCs). These cells could be split and separately manipulated to form syngeneic 3D organoids, representing different systems of the body (kidney, liver, lung, cardiovascular). Organoids in separate compartments of the chip can be linked with microfluidic channels for nutrient and small molecule exchange, analogous to vascular and lymphatic systems. The introduction of therapeutic agents into the system could then provide a personalized, preclinical screen to predict efficacy and safety for that patient specifically.

5.7 Transwell and Migration Assays

Transwell cell migration and invasion assays employ a two-chamber system (Boyden Chamber), where a membrane and/or organotypic ECM layer separates two populations of cells. Chemoattractants in the lower chamber can initiate the migration of cells in the upper chamber enabling the measure of chemotaxis and invasion through the ECM. This can help uncover mechanisms of invasion and metastasis such as the role of lncRNA HOXA11-AS in cell invasion and migration capacity in breast cancer (Li et al. 2017). Other novel systems to investigate migration, mechanosensitivity, chemotaxis have been designed using multiple chambers linked with microfluidics and culture wells containing nanogrooves or micropores (Dhawan et al. 2017). These microculture devices are highly specialized and represent the union of nanotechnology, biomolecular engineering, and cellular biology leading to novel insights in cell morphology and phenotype. Raft cultures are a similar method to the transwell assay and are useful for forming organotypic skin equivalents (Arnette et al. 2016).

5.8 Monolayer Culture

Traditional in vitro cell culture methods are too numerous to cover comprehensively in the scope of this chapter. These are generally less clinically relevant preclinical models, but their low cost, availability, scalability, and versatility make them invaluable biomedical research tools. These models are generally useful in mechanistic, functional, and cellular/biochemical profiling studies. Tumor cell motility can be evaluated using a "wound healing" or scratch assay. Suppression of the Snail-1 transcription factor by siRNA knock-down was shown in this way to reduce cell motility and migration potential in bladder cancer cells (Salehi et al. 2017).

In vitro culture conditions introduce biases and confounding effects that lead to poor replication of the in situ conditions or phenotypes. Cell–cell and cell–extracellular matrix (ECM) interactions through channels and transporters such as connexin-43 in pancreatic ductal adenocarcinoma (PDAC) can regulate cytoplasmic pH to promote growth (Dovmark et al. 2017); although, cells are negative for Cx43 in monolayer. In diffuse intrinsic pontine glioma (DIPG), in vitro culture conditions between adherent cells, neurospheres, or serum containing medium drive massive changes in gene expressions and signal transduction leading to culture condition-dependent response to targeted therapies (Meel et al. 2017). Improper culture conditions can reduce primary culture heterogeneity and induce culture condition-dependent phenotypic changes such as in therapy sensitivity. In GBM primary culture, it is recommended that cells be cultivated in neural stem cell media (NSC) (with physiological levels of EGF (10 ng/mL) and FGF-2 (10 ng/mL) as the only added growth factors), in PDGF-enriched media (OPC media), in astrocyte-specific media (with added HBEGF), or in serum-enriched media (though this produces a loss in stem cell component and promotes a mesenchymal phenotype) in order to preserve intratumoral heterogeneity (Ledur et al. 2017). This is somewhat in contrast to previous reporting that brain tumor-initiating cell (BTIC) expansion when grown in monolayer culture on laminin was equivalent to neurosphere culture (Rahman et al. 2015). It is generally accepted that adherent culture conditions promote cell differentiation, whereas suspensions or spheroid culture conditions enrich for more stem cell-like populations. This highlights the sensitivity of cellular phenotypes based on culture conditions and microenvironment.

6 Discussion

Cancer is a complex disease, and the study of cancer can be equally complex. There are advantages and disadvantages in selecting different sources of tumor material just as there are pros and cons to using different explant models. The selection of which source material and explant model to use should be guided by your specific research aims. Explant models can be generated that faithfully recapitulate the complexity of the in situ tumor. In translational or clinical studies evaluating drug efficacy and safety, this complexity may be desired. In other studies, this complexity may be unnecessary and prohibitive to the success of the study. The CAM assay is an excellent model for studying questions related to angiogenesis, invasion, and metastasis. Phuong et al. employed a combination of in vitro migration assays and the CAM assay to interrogate the 11,12-epoxyeicosatrienoic acid pathway in the context of tamoxifen-resistant breast cancer (Thuy Phuong et al. 2017). In this case, a transgenic mouse model or transplantable tumor model would have over-complicated the study and offered diminishing returns or even confounded the results through a reduction in power due to over-complexity.

As the model complexity increases, relevance to the in situ condition increases, but so does the biological variance leading to the requirement of more samples to achieve adequate statistical power. This leads to increased resource cost and technical skill requirement. There are also inherent biases that are specific to a given model as well as technical bias or variance introduced in the implementation of more complex models. Statistical modeling, nonlinear dynamics, network modeling, chaos theory, and power analysis are all beyond the scope of this chapter and are themselves entire fields of study which we recommend for you to explore (Mochizuki 2016; Resendis-Antonio and Diener 2017; Rickles et al. 2007; Sauro 2009). A number of tools are being developed to handle this complexity through informatics and high-performance computing that would otherwise be impossible for an individual to fathom (Chen et al. 2017; Yue et al. 2015). The number and complexity of cancer explant models is also growing thanks to improvements in biomaterials and bioengineering advances (Albritton and Miller 2017) (Godugu and Singh 2016) (Pradhan et al. 2017) (Peela et al. 2017) (Sabhachandani et al. 2016) (Portillo-Lara and Annabi 2016). There is also a broad body of literature with suggestions on selecting appropriate tumor models and research models in general (Borowsky 2011; Jackson and Thomas 2017; Resendis-Antonio and Diener 2017; Silk et al. 2014; White et al. 2016).

Figure 2 is a simplified, practical graphic to aid in the selection of a cancer explant model appropriate for a given research question. It is important to note that any of the listed cancer explant models in Fig. 2 can be used with any of the sources for tumor material described in Table 1 except for organotypic slice which can only come from a human or animal source. The decision regarding which source material to use will depend primarily on accessibility and resource availability.

We qualitatively reduced the dimensionality of the model selection process to three features: (i) model complexity, (ii) variance, and (iii) and "precision vs.

throughput." Model complexity accounts for the biological complexity as well as the technical and logistical difficulties in utilizing the model. The variance refers to both the technical and biological variances. As the complexity of the model increases, the total variance or noise also increases. This increases the risk of type II errors and thus decreases the statistical power at a given level. More complex models will, therefore, require much higher numbers of replicates to achieve sufficient power. These trends are shown linearly for the sake of simplicity, but are likely much more complicated and exceptions to them do exist. This graphic is subjective, but it also captures the essence of the general trends.

Each model is binned under the trendline, representing the positive correlation between complexity and variance. The models to the left, mostly in vitro models, are better suited for basic, molecular science or functional studies. Employing lentiviral manipulated lung cancer cell lines, Rohrbach et. al. were able to uncover a positive correlation between MARCKS phosphorylation and increased radiation sensitivity (Rohrbach et al. 2017). Models increasing in complexity to the right begin to incorporate higher order interactions and are better suited for studying networks and systems. An exploration of GBM resistance to temozolomide (TMZ) in a GBM PDX model by kinomic profiling revealed global trends in signaling networks related to TMZ resistance (Anderson et al. 2014).

A third feature to be considered represents a spectrum from suitability for precision medicine to aptness for high-throughput screening (Fig. 2). This dimension is somewhat analogous to the concepts of "Breadth vs. Depth" and primarily refers to the clinical utility of the models. Models with less heat (blue) are readily accessible, reproducible, inexpensive, and well suited for high-throughput applications. Models with more heat (red) are more expensive, not readily scalable, but recapitulate much of the complexity of the in situ tumor conditions. These models could be applied to novel therapeutic development and preclinical trials, and could be adapted for personalized, precision medicine applications. The three prevailing cancer explant models for this purpose are PDX, GEMM, and chip-based technologies. There is a debate over the superiority of these models (Kersten et al. 2017), but the pace of scientific and technological advancement makes this difficult to judge. We contend that all models have strengths, weaknesses, and inherent biases which must be weighed in the context of a specific research aim. We hope researchers will use our recommendations as a starting point in their experimental design process, and we encourage them to investigate prospective models thoroughly to improve their odds of success.

7 Conclusions

The success of any study depends on thoughtful design and the selection of an appropriate model to address the research question. In this chapter, we have presented the challenges in cancer research stemming largely from the complexity and diversity of in situ tumors. Rarely, will any model accurately recapitulate the complexity of an in situ tumor, and in almost every case several models will be needed to address all aspects that may affect the responses to precision medicine therapies. We have reviewed different sources of tumor materials that can be obtained or generated for cancer research. We have also described several cancer explant models and summarized their suitability in addressing key areas of tumor biology. Finally, we have synthesized a guide to help in the selection of a cancer explant model based on the research aims of a given study.

Acknowledgements Thank you to Drs. G. Yancey Gillespie and Joshua C. Anderson for their help in editing this manuscript.

References

Albritton JL, Miller JS (2017) 3D bioprinting: improving in vitro models of metastasis with heterogeneous tumor microenvironments. Dis Model Mech 10(1):3–14. https://doi.org/10.1242/dmm.025049

Anderson JC, Duarte CW, Welaya K, Rohrbach TD, Bredel M, Yang ES, Willey CD (2014) Kinomic exploration of temozolomide and radiation resistance in Glioblastoma multiforme xenolines. Radiother Oncol 111(3):468–474. https://doi.org/10.1016/j.radonc.2014.04.010

Arnette C, Koetsier JL, Hoover P, Getsios S, Green KJ (2016) In vitro model of the epidermis: connecting protein function to 3D structure. Methods Enzymol 569:287–308. https://doi.org/10.1016/bs.mie.2015.07.015

Bagby S, Messersmith WA, Pitts TM, Capasso A, Varella-Garcia M, Klauck PJ, … Arcaroli J (2016) Development and maintenance of a preclinical patient derived tumor xenograft model for the investigation of novel anti-cancer therapies. J Vis Exp (115) https://doi.org/10.3791/54393

Borowsky AD (2011) Choosing a mouse model: experimental biology in context—the utility and limitations of mouse models of breast cancer. Cold Spring Harb Perspect Biol, 3

Braekeveldt N, Bexell D (2017) Patient-derived xenografts as preclinical neuroblastoma models. Cell Tissue Re.s https://doi.org/10.1007/s00441-017-2687-8

Brown KM, Xue A, Mittal A, Samra JS, Smith R, Hugh TJ (2016) Patient-derived xenograft models of colorectal cancer in pre-clinical research: a systematic review. Oncotarget, 7:66212–66225

Cao M, Mao J, Duan X, Lu L, Zhang F, Lin B, … Shen J (2017) In vivo tracking of the tropism of mesenchymal stem cells to malignant gliomas using reporter gene-based MR imaging. Int J Cancer. https://doi.org/10.1002/ijc.31113

Cermeño EA, García AJ (2016) Tumor-initiating cells: emerging biophysical methods of isolation. Curr Stem Cell Rep 2(1):21–32. https://doi.org/10.1007/s40778-016-0036-6

Chen JY, Pandey R, Nguyen TM (2017a) HAPPI-2: a comprehensive and high-quality map of human annotated and predicted protein interactions. BMC Genom 18(1):182. https://doi.org/10.1186/s12864-017-3512-1

Chen Q, Xu L, Chen J, Yang Z, Liang C, Yang Y, Liu Z (2017b) Tumor vasculature normalization by orally fed erlotinib to modulate the tumor microenvironment for enhanced cancer nanomedicine and immunotherapy. Biomaterials 148:69–80. https://doi.org/10.1016/j.biomaterials.2017.09.021

Chen W, Xu XK, Li JL, Kong, K. K., Li, H., Chen, C., . . . Li, F. C. (2017). MALAT1 is a prognostic factor in glioblastoma multiforme and induces chemoresistance to temozolomide through suppressing miR-203 and promoting thymidylate synthase expression *Oncotarget* (Vol. 8, pp. 22783–22799)

Chia S, Low JL, Zhang X, Kwang XL, Chong FT, Sharma A, DasGupta R (2017) Phenotype-driven precision oncology as a guide for clinical decisions one patient at a time. Nat Commun 8(1):435. https://doi.org/10.1038/s41467-017-00451-5

Choi SY, Lin D, Gout PW, Collins CC, Xu Y, Wang Y (2014) Lessons from patient-derived xenografts for better in vitro modeling of human cancer. Adv Drug Deliv Rev 79–80C:222–237. https://doi.org/10.1016/j.addr.2014.09.009

Choi YY, Lee JE, Kim H, Sim MH, Kim KK, Lee G, Cheong JH (2016) Establishment and characterisation of patient-derived xenografts as paraclinical models for gastric cancer. Sci Rep 6:22172. https://doi.org/10.1038/srep22172

Cooper DKC (2012) A brief history of cross-species organ transplantation. Bayl Univ Med Cent Proc, 25:49–57

Daher A, de Groot J (2017) Rapid identification and validation of novel targeted approaches for Glioblastoma: A combined ex vivo-in vivo pharmaco-omic model. Exp Neurol https://doi.org/10.1016/j.expneurol.2017.09.006

Das A, Monteiro M, Barai A, Kumar S, Sen S (2017) MMP proteolytic activity regulates cancer invasiveness by modulating integrins. Sci Rep 7(1):14219. https://doi.org/10.1038/s41598-017-14340-w

Davies AH, Wang Y, Zoubeidi A (2017) Patient-derived xenografts: a platform for accelerating translational research in prostate cancer. Mol Cell Endocrinol. https://doi.org/10.1016/j.mce.2017.03.013

Debbasch C, Ebenhahn C, Dami N, Pericoi M, Van den Berghe C, Cottin M, Nohynek GJ (2005) Eye irritation of low-irritant cosmetic formulations: correlation of in vitro results with clinical data and product composition. Food Chem Toxicol 43(1):155–165. https://doi.org/10.1016/j.fct.2004.09.004

Decker WK, da Silva RF, Sanabria MH, Angelo LS, Guimaraes F, Burt BM, Paust S (2017) Cancer immunotherapy: historical perspective of a clinical revolution and emerging preclinical animal models. Front Immunol 8:829. https://doi.org/10.3389/fimmu.2017.00829

Decrausaz L, Goncalves AR, Domingos-Pereira S, Pythoud C, Stehle JC, Schiller J, Nardelli-Haefliger D (2011) A novel mucosal orthotopic murine model of human papillomavirus-associated genital cancers. Int J Cancer 128(9):2105–2113. https://doi.org/10.1002/ijc.25561

Dehne EM, Hasenberg T, Marx U (2017) The ascendance of microphysiological systems to solve the drug testing dilemma. Future Sci OA, 3(2):Fso185. https://doi.org/10.4155/fsoa-2017-0002

Deryugina EI, Quigley JP (2008) CHAPTER TWO: chick embryo chorioallantoic membrane models to quantify angiogenesis induced by inflammatory and tumor cells or purified effector molecules. Methods Enzymol 444:21–41. https://doi.org/10.1016/s0076-6879(08)02802-4

Dhawan U, Pan HA, Shie MJ, Chu YH, Huang GS, Chen PC, Chen WL (2017) The spatiotemporal control of osteoblast cell growth, behavior, and function dictated by nanostructured stainless steel artificial microenvironments. Nanoscale Res Lett 12

Dobrolecki LE, Airhart SD, Alferez DG, Aparicio S, Behbod F, Bentires-Alj M, Lewis MT (2016) Patient-derived xenograft (PDX) models in basic and translational breast cancer research. Cancer Metastasis Rev 35(4):547–573. https://doi.org/10.1007/s10555-016-9653-x

Dohle DS, Pasa SD, Gustmann S, Laub M, Wissler JH, Jennissen HP, Dunker N (2009) Chick ex ovo culture and ex ovo CAM assay: how it really works. J Vis Exp 33. https://doi.org/10.3791/1620

Dovmark TH, Saccomano M, Hulikova A, Alves F, Swietach P (2017) Connexin-43 channels are a pathway for discharging lactate from glycolytic pancreatic ductal adenocarcinoma cells. Oncogene 36(32):4538–4550. https://doi.org/10.1038/onc.2017.71

Ellsworth DL, Blackburn HL, Shriver CD, Rabizadeh S, Soon-Shiong P, Ellsworth RE (2017) Single-cell sequencing and tumorigenesis: improved understanding of tumor evolution and metastasis. Clin Transl Med 6(1):15. https://doi.org/10.1186/s40169-017-0145-6

Esch EW, Bahinski A, Huh D (2015) Organs-on-chips at the frontiers of drug discovery. Nat Rev Drug Discov 14(4):248–260. https://doi.org/10.1038/nrd4539

Fehringer G, Kraft P, Pharoah PD, Eeles RA, Chatterjee N, Schumacher FR, … Hung RJ (2016) Cross-cancer genome-wide analysis of lung, ovary, breast, prostate, and colorectal cancer reveals novel pleiotropic associations. https://doi.org/10.1158/0008-5472.can-15-2980

Fiebig H-H, Burger AM (2002) Human tumor xenografts and explants. In: Teicher BA (ed) Tumor models in cancer research. Humana Press, Totowa, NJ, pp 113–137

Gao B, Long C, Lee W, Zhang Z, Gao X, Landsittel D, … Hara H (2017) Anti-Neu5Gc and anti-non-Neu5Gc antibodies in healthy humans PLoS One 12

Gautam J, Banskota S, Regmi SC, Ahn S, Jeon YH, Jeong H, Kim JA (2016) Tryptophan hydroxylase 1 and 5-HT7 receptor preferentially expressed in triple-negative breast cancer promote cancer progression through autocrine serotonin signaling. Mol Cancer 15(1):75. https://doi.org/10.1186/s12943-016-0559-6

Gilbert AN, Shevin RS, Anderson JC, Langford CL, Eustace N, Gillespie GY, … Willey CD (2016) Generation of microtumors using 3D human biogel culture system and patient-derived glioblastoma cells for kinomic profiling and drug response testing. J Vis Exp (112). https://doi.org/10.3791/54026

Gilbert AN, Walker K, Tran AN, Boyd NH, Gillespie GY, Singh R. K, Hjelmeland AB (2017) Modeling physiologic microenvironments in three-dimensional microtumors maintains brain tumor initiating cells. J Cancer Stem Cell Res 5. https://doi.org/10.14343/jcscr.2017.5e1004

Godugu C, Singh M (2016) AlgiMatrix-Based 3D cell culture system as an in vitro tumor model: an important tool in cancer research. Methods Mol Biol 1379:117–128. https://doi.org/10.1007/978-1-4939-3191-0_11

Ham SL, Joshi R, Thakuri PS, Tavana H (2016) Liquid-based three-dimensional tumor models for cancer research and drug discovery. Exp Biol Med (Maywood) 241(9):939–954. https://doi.org/10.1177/1535370216643772

Hanahan D, Weinberg RA (2000) The hallmarks of cancer. Cell 100(1):57–70

Healy LE (2017) Acquisition and reception of primary tissues, cells, or other biological specimens. Methods Mol Biol 1590:17–27. https://doi.org/10.1007/978-1-4939-6921-0_3

Heo EJ, Cho YJ, Cho WC, Hong JE, Jeon HK, Oh DY, Lee JW (2017) Patient-derived xenograft models of epithelial ovarian cancer for preclinical studies. Cancer Res Treat 49(4):915–926. https://doi.org/10.4143/crt.2016.322

Hoffman RM (2015) Patient-derived orthotopic xenografts: better mimic of metastasis than subcutaneous xenografts. Nat Rev Cancer 15(8):451–452

Holzgreve A, Brendel M, Gu S, Carlsen J, Mille E, Böning G, … Albert NL (2016) Monitoring of tumor growth with [18F]-FET PET in a mouse model of glioblastoma: SUV measurements and volumetric approaches. Front Neurosci 10. https://doi.org/10.3389/fnins.2016.00260

Huang C, Mezencev R, McDonald JF, Vannberg F (2017a) Open source machine-learning algorithms for the prediction of optimal cancer drug therapies. PLoS ONE 12(10):e0186906. https://doi.org/10.1371/journal.pone.0186906

Huang G, Yuan M, Chen M, Li L, You W, Li H, Ji G (2017b) Integrating multiple fitting regression and Bayes decision for cancer diagnosis with transcriptomic data from tumor-educated blood platelets. Analyst 142(19):3588–3597. https://doi.org/10.1039/c7an00944e

Humpel C (2015) Organotypic brain slice cultures: a review. Neuroscience 305:86–98. https://doi.org/10.1016/j.neuroscience.2015.07.086

Hung RJ, Fehringer G, Casey G, Gruber SB, Peters U, Goode E L, … Wilson MD (2017) Cross-cancer analysis reveals novel pleiotropic associations—response. https://doi.org/10.1158/0008-5472.can-17-1584

Ing N, Huang F, Conley A, You S, Ma Z, Klimov S, Knudsen BS (2017) A novel machine learning approach reveals latent vascular phenotypes predictive of renal cancer outcome. Sci Rep 7(1):13190. https://doi.org/10.1038/s41598-017-13196-4

Ishiguro T, Ohata H, Sato A, Yamawaki K, Enomoto T, Okamoto K (2017) Tumor-derived spheroids: relevance to cancer stem cells and clinical applications. Cancer Sci 108, 283–289

Jackson SJ, Thomas GJ (2017) Human tissue models in cancer research: looking beyond the mouse. Dis Model Mech 10, 939–942

Jiang Y, Qi X, Liu X, Zhang, J, Ji J, Zhu Z, … Yu Y (2017) Fbxw7 haploinsufficiency loses its protection against DNA damage and accelerates MNU-induced gastric carcinogenesis Oncotarget 8:33444–33456

Jimenez-Gonzalez M, Plaza-Garcia S, Arizeta J, Bianchessi S, Trigueros C, Reese T (2017) A longitudinal MRI study on lymph nodes histiocytosis of a xenograft cancer model. PLoS ONE 12(7):e0181043. https://doi.org/10.1371/journal.pone.0181043

Johansson-Percival A, He B, Li ZJ, Kjellen A, Russell K, Li J, Ganss R (2017) De novo induction of intratumoral lymphoid structures and vessel normalization enhances immunotherapy in resistant tumors. Nat Immunol 18(11):1207–1217. https://doi.org/10.1038/ni.3836

Kemper K, Krijgsman O, Kong X, Cornelissen-Steijger P, Shahrabi A, Weeber F, Peeper DS (2016) BRAF(V600E) kinase domain duplication identified in therapy-refractory melanoma patient-derived xenografts. Cell Rep 16(1):263–277. https://doi.org/10.1016/j.celrep.2016.05.064

Kersten K, de Visser KE, van Miltenburg MH, Jonkers J (2017) Genetically engineered mouse models in oncology research and cancer medicine EMBO Mol Med 9:137–153

Kohnken R, Porcu P, Mishra A (2017) Overview of the use of murine models in Leukemia and Lymphoma research. Front Oncol 7. https://doi.org/10.3389/fonc.2017.00022

Kumar SS, Rao MRK, Balasubramanian MP, Chaudhuri A (2012) Antiproliferative role of *Indigofera aspalathoides* on 20 methylcholanthrene induced fibrosarcoma in rats. Asian Pac J Trop Biomed 2:966–974

Lai Y, Wei X, Lin S, Qin L, Cheng L, Li P (2017) Current status and perspectives of patient-derived xenograft models in cancer research. J Hematol Oncol 10

Lannin DR, Wang S (2017) Are small breast cancers good because they are small or small because they are good? N Engl J Med 376(23):2286–2291. https://doi.org/10.1056/nejmsr1613680

Laundos TL, Silva J, Assuncao M, Quelhas P, Monteiro C, Oliveira C, Amaral IF (2017) Rotary orbital suspension culture of embryonic stem cell-derived neural stem/progenitor cells: impact of hydrodynamic culture on aggregate yield, morphology and cell phenotype. J Tissue Eng Regen Med 11(8):2227–2240. https://doi.org/10.1002/term.2121

Ledur PF, Onzi GR, Zong H, Lenz G (2017) Culture conditions defining glioblastoma cells behavior: what is the impact for novel discoveries? Oncotarget 8(40):69185–69197. https://doi.org/10.18632/oncotarget.20193

Lenting K, Verhaak R, ter Laan M, Wesseling P, Leenders W (2017) Glioma: experimental models and reality. Acta Neuropathol 133:263–282

Li W, Jia G, Qu Y, Du Q, Liu B (2017) Long non-coding RNA (LncRNA) HOXA11-AS promotes breast cancer invasion and metastasis by regulating epithelial-mesenchymal transition. Med Sci Monit 23:3393–3403

Liu S, Cheng C (2013) Alternative RNA splicing and cancer. Wiley Interdiscip Rev RNA 4(5):547–566. https://doi.org/10.1002/wrna.1178

Liu Y, He Y, Yang F, Cong X, Wang J, Peng S, Yi Z (2017) A novel synthetic small molecule YF-452 inhibits tumor growth through antiangiogenesis by suppressing VEGF receptor 2 signaling. Sci China Life Sci 60(2):202–214. https://doi.org/10.1007/s11427-016-0369-6

Lokman NA, Elder AS, Ricciardelli C, Oehler MK (2012) Chick chorioallantoic membrane (CAM) assay as an in vivo model to study the effect of newly identified molecules on ovarian cancer invasion and metastasis. Int J Mol Sci 13(8):9959–9970. https://doi.org/10.3390/ijms13089959

Martens-de Kemp SR, Brink A, van der Meulen IH, de Menezes RX, Te Beest DE, Leemans CR, Brakenhoff RH (2017) The FA/BRCA pathway identified as the major predictor of cisplatin response in head and neck cancer by functional genomics. Mol Cancer Ther 16(3):540–550. https://doi.org/10.1158/1535-7163.mct-16-0457

McLean DT, Strand DW, Ricke WA (2017) Prostate cancer xenografts and hormone induced prostate carcinogenesis. Differentiation 97:23–32. https://doi.org/10.1016/j.diff.2017.08.005

Meel MH, Sewing ACP, Waranecki P, Metselaar DS, Wedekind LE, Koster J, … Hulleman E (2017) Culture methods of diffuse intrinsic pontine glioma cells determine response to targeted therapies. Exp Cell Res. https://doi.org/10.1016/j.yexcr.2017.09.032

Melzer C, von der Ohe J, Lehnert H, Ungefroren H, Hass R (2017) Cancer stem cell niche models and contribution by mesenchymal stroma/stem cells. Mol Cancer 16

Mitteregger R, Vogt G, Rossmanith E, Falkenhagen D (1999) Rotary cell culture system (RCCS): a new method for cultivating hepatocytes on microcarriers. Int J Artif Organs 22(12):816–822

Mochizuki A (2016) Theoretical approaches for the dynamics of complex biological systems from information of networks. Proc Jpn Acad Ser B Phys Biol Sci 92(8):255–264. https://doi.org/10.2183/pjab.92.255

Morgan KM, Riedlinger GM, Rosenfeld J, Ganesan S, Pine SR (2017) Patient-derived xenograft models of non-small cell lung cancer and their potential utility in personalized medicine. Front Oncol 7:2. https://doi.org/10.3389/fonc.2017.00002

Na KJ, Choi H (2017) Tumor metabolic features identified by FDG PET correlates with gene networks of immune cell microenvironment in head and neck cancer. J Nucl Med. https://doi.org/10.2967/jnumed.117.194217

Naipal KAT, Verkaik NS, Sánchez H, van Deurzen CHM, den Bakker MA, Hoeijmakers JH, … van Gent DC (2016) Tumor slice culture system to assess drug response of primary breast cancer. BMC Cancer 16

Narkhede AA, Shevde LA, Rao SS (2017) Biomimetic strategies to recapitulate organ specific microenvironments for studying breast cancer metastasis. Int J Cancer 141(6):1091–1109. https://doi.org/10.1002/ijc.30748

Neve A, Kumar KS, Tripolitsioti D, Grotzer MA, Baumgartner M (2017) Investigation of brain tissue infiltration by medulloblastoma cells in an ex vivo model. Sci Rep 7

Otte A, Yang Y, von der Ohe J, Melzer C, Hillemanns P, Feuerhake F, Hass R (2016) SCCOHT tumors acquire chemoresistance and protection by interacting mesenchymal stroma/stem cells within the tumor microenvironment. Int J Oncol 49(6):2453–2463. https://doi.org/10.3892/ijo.2016.3735

Ozcelikkale A, Moon HR, Linnes M, Han B (2017). In vitro microfluidic models of tumor microenvironment to screen transport of drugs and nanoparticles. Wiley Interdiscip Rev Nanomed Nanobiotechnol 9(5). https://doi.org/10.1002/wnan.1460

Palmer TD, Lewis J, Zijlstra A (2011) Quantitative analysis of cancer metastasis using an avian embryo model. J Vis Exp (51). https://doi.org/10.3791/2815

Pauli C, Hopkins BD, Prandi D, Shaw R, Fedrizzi T, Sboner A, Rubin MA (2017) Personalized In Vitro and In Vivo cancer models to guide precision medicine. Cancer Discov 7(5):462–477. https://doi.org/10.1158/2159-8290.CD-16-1154

Pavlova NN, Thompson CB (2016) The emerging hallmarks of cancer metabolism. Cell Metab 23 (1):27–47. https://doi.org/10.1016/j.cmet.2015.12.006

Peela N, Truong D, Saini H, Chu H, Mashaghi S, Ham SL, Nikkhah M (2017) Advanced biomaterials and microengineering technologies to recapitulate the stepwise process of cancer metastasis. Biomaterials 133:176–207. https://doi.org/10.1016/j.biomaterials.2017.04.017

Perry CJ, Lawrence AJ (2017) Hurdles in basic science translation. Front Pharmacol 8. https://doi.org/10.3389/fphar.2017.00478

Portillo-Lara R, Annabi N (2016) Microengineered cancer-on-a-chip platforms to study the metastatic microenvironment. Lab Chip 16(21):4063–4081. https://doi.org/10.1039/c6lc00718j

Pradhan S, Clary JM, Seliktar D, Lipke EA (2017) A three-dimensional spheroidal cancer model based on PEG-fibrinogen hydrogel microspheres. Biomaterials 115:141–154. https://doi.org/10.1016/j.biomaterials.2016.10.052

Quail DF, Joyce JA (2017) The microenvironmental landscape of brain tumors. Cancer Cell 31 (3):326–341. https://doi.org/10.1016/j.ccell.2017.02.009

Qureshi-Baig K, Ullmann P, Haan S, Letellier E (2017) Tumor-Initiating Cells: a criTICal review of isolation approaches and new challenges in targeting strategies. Mol Cancer 16(1):40. https://doi.org/10.1186/s12943-017-0602-2

Rahman M, Reyner K, Deleyrolle L, Millette S, Azari H, Day BW, Reynolds BA (2015) Neurosphere and adherent culture conditions are equivalent for malignant glioma stem cell lines. Anat Cell Biol 48(1):25–35. https://doi.org/10.5115/acb.2015.48.1.25

Resendis-Antonio O, Diener C (2017) Editorial: systems biology and the challenge of deciphering the metabolic mechanisms underlying cancer. Front Physiol 8. https://doi.org/10.3389/fphys.2017.00537

Rickles D, Hawe P, Shiell A (2007) A simple guide to chaos and complexity. J Epidemiol Commun Health 61:933–937

Rivina L, Davoren M, Schiestl RH (2014) Radiation-induced myeloid leukemia in murine models. Hum Genomics 8:13

Rohrbach TD, Jones RB, Hicks PH, Weaver AN, Cooper TS, Eustace NJ, … Willey CD (2017) MARCKS phosphorylation is modulated by a peptide mimetic of MARCKS effector domain leading to increased radiation sensitivity in lung cancer cell lines. Oncol Lett 13:1216–1222

Rudy SF, Brenner JC, Harris JL, Liu J, Che J, Scott MV, … Carey TE (2016) In vivo Wnt pathway inhibition of human squamous cell carcinoma growth and metastasis in the chick chorioallantoic model. J Otolaryngol Head Neck Surg 45

Sabhachandani P, Motwani V, Cohen N, Sarkar S, Torchilin V, Konry T (2016) Generation and functional assessment of 3D multicellular spheroids in droplet based microfluidics platform. Lab Chip 16(3):497–505. https://doi.org/10.1039/c5lc01139f

Salehi S, Mansoori B, Mohammadi A, Davoudian S, Musavi Shenas SMH, Shajari N, Baradaran B (2017) An analysis of suppressing migratory effect on human urinary bladder cancer cell line by silencing of snail-1. Biomed Pharmacother 96:545–550. https://doi.org/10.1016/j.biopha.2017.10.044

Salo T, Sutinen M, Hoque Apu E, Sundquist E, Cervigne NK, de Oliveira CE, … Coletta RD (2015) A novel human leiomyoma tissue derived matrix for cell culture studies. BMC Cancer 15

Sauro HM (2009) Biological models of molecular network dynamics. In: Meyers RA (ed) Encyclopedia of complexity and systems science. Springer, New York, New York, NY, pp 554–570

Scott MC, Tomiyasu H, Garbe JR, Cornax I, Amaya C, O'Sullivan MG, Modiano JF (2016) Heterotypic mouse models of canine osteosarcoma recapitulate tumor heterogeneity and biological behavior. Dis Model Mech 9(12):1435–1444. https://doi.org/10.1242/dmm.026849

Shaheen S, Ahmed M, Lorenzi F, Nateri AS (2016) Spheroid-formation (Colonosphere) assay for in vitro assessment and expansion of stem cells in colon cancer. Stem Cell Rev 12:492–499

Silk D, Kirk PDW, Barnes CP, Toni T, Stumpf MPH (2014) Model selection in systems biology depends on experimental design. PLoS Comput Biol 10(6):e1003650. https://doi.org/10.1371/journal.pcbi.1003650

Sinha S, Khan S, Shukla S, Lakra AD, Kumar S, Das G, Meeran SM (2016) Cucurbitacin B inhibits breast cancer metastasis and angiogenesis through VEGF-mediated suppression of FAK/MMP-9 signaling axis. Int J Biochem Cell Biol 77(Pt A):41–56. https://doi.org/10.1016/j.biocel.2016.05.014

Thuy Phuong NT, Kim JW, Kim JA, Jeon JS, Lee JY, Xu W J, … Kang KW (2017) Role of the CYP3A4-mediated 11,12-epoxyeicosatrienoic acid pathway in the development of tamoxifen-resistant breast cancer. Oncotarget 8:71054–71069

Tsai HF, Trubelja A, Shen AQ, Bao G (2017) Tumour-on-a-chip: microfluidic models of tumour morphology, growth and microenvironment. J R Soc Interface 14

Tufan AC, Satiroglu-Tufan NL (2005) The chick embryo chorioallantoic membrane as a model system for the study of tumor angiogenesis, invasion and development of anti-angiogenic agents. Curr Cancer Drug Targets 5(4):249–266

Valdes TI, Kreutzer D, Moussy F (2002) The chick chorioallantoic membrane as a novel in vivo model for the testing of biomaterials. J Biomed Mater Res 62(2):273–282. https://doi.org/10.1002/jbm.10152

Verma B, Ritchie M, Mancini M (2017) Development and applications of patient-derived xenograft models in humanized mice for oncology and immune-oncology drug discovery. Curr Protoc Pharmacol 78:14.41.11–14.41.12. https://doi.org/10.1002/cpph.26

Viallard C, Larrivee B (2017) Tumor angiogenesis and vascular normalization: alternative therapeutic targets. Angiogenesis 20(4):409–426. https://doi.org/10.1007/s10456-017-9562-9

Viant C, Guia S, Hennessy RJ, Rautela J, Pham K, Bernat C, … Huntington ND (2017) Cell cycle progression dictates the requirement for BCL2 in natural killer cell survival. J Exp Med 214:491–510

Vidic I, Egnell L, Jerome NP, Teruel JR, Sjobakk TE, Ostlie A, … Goa PE (2017) Support vector machine for breast cancer classification using diffusion-weighted MRI histogram features: preliminary study. J Magn Reson Imaging. https://doi.org/10.1002/jmri.25873

White A, Tolman M, Thames HD, Withers HR, Mason KA, Transtrum MK (2016) The limitations of model-based experimental design and parameter estimation in sloppy systems. PLoS Comput Biol 12

Willey CD, Gilbert AN, Anderson JC, Gillespie GY (2015) Patient-derived xenografts as a model system for radiation research. Semin Radiat Oncol 25(4):273–280. https://doi.org/10.1016/j.semradonc.2015.05.008

William D, Mullins CS, Schneider B, Orthmann A, Lamp N, Krohn M, Linnebacher M (2017) Optimized creation of glioblastoma patient derived xenografts for use in preclinical studies. J Transl Med 15(1):27. https://doi.org/10.1186/s12967-017-1128-5

Winterhoff BJ, Maile M, Mitra AK, Sebe A, Bazzaro M, Geller MA, Starr TK (2017) Single cell sequencing reveals heterogeneity within ovarian cancer epithelium and cancer associated stromal cells. Gynecol Oncol 144(3):598–606. https://doi.org/10.1016/j.ygyno.2017.01.015

Woodworth MB, Girskis KM, Walsh CA (2017) Building a lineage from single cells: genetic techniques for cell lineage tracking [Review Article]. 18:230. https://doi.org/10.1038/nrg.2016.159

Yang Y, Liu C, Lei X, Wang H, Su P, Ru Y, Zhou J (2016) Integrated biophysical and biochemical signals augment megakaryopoiesis and thrombopoiesis in a three-dimensional rotary culture system. Stem Cells Transl Med 5(2):175–185. https://doi.org/10.5966/sctm.2015-0080

Yasui H, Kawai T, Matsumoto S, Saito K, Devasahayam N, Mitchell JB, Krishna MC (2017) Quantitative imaging of pO_2 in orthotopic murine gliomas: hypoxia correlates with resistance to radiation. Free Radic Res 51(9–10):861–871. https://doi.org/10.1080/10715762.2017.1388506

Yue Z, Kshirsagar MM, Nguyen T, Suphavilai C, Neylon MT, Zhu L, Chen JY (2015) PAGER: constructing PAGs and new PAG-PAG relationships for network biology. Bioinformatics 31 (12):i250–i257. https://doi.org/10.1093/bioinformatics/btv265

Zhang M, Lee AV, Rosen JM (2017) The cellular origin and evolution of breast cancer. Cold Spring Harb Perspect Med 7(3). https://doi.org/10.1101/cshperspect.a027128

Organotypic Models of Lung Cancer

Anna Pomerenke

Contents

Abstract Human lung cancer is one of the leading causes of death worldwide, with nearly 2 million of new cases diagnosed each year, often too late for a successful therapeutic intervention. In this chapter, organotypic models of lung cancer will be reviewed. Ex vivo tissue explants, spheroids, organoids, and novel bioengineering approaches are currently being used to study human lung cancer. Although there is no ideal method that will fully recapitulate the complex human lung architecture, the three-dimensional (3D) organotypic models described here represent a major advance from classical two-dimensional (2D) tissue culture models. Organotypic tissue cultures are better at predicting in vivo tissue responses to anticancer drugs or carcinogenic toxins. In addition, there is also a possibility to use these systems as part of personalized medicine (e.g., to assist clinicians in selecting appropriate therapeutic choices based on individual patient-derived tissue responses to drugs in vitro). There is still a need for improvement in the 3D culture systems, such as automation for high-throughput testing and reduction of costs. However, given the large number of failed clinical trials due to safety or efficacy reasons, more attention should be given to these more physiologically relevant 3D organotypic tissue culture models.

A. Pomerenke (✉)
Laboratory of Infectious Diseases, National Institute of Allergy and Infectious Diseases, National Institutes of Health, Bethesda, MD, USA
e-mail: anna.pomerenke@gmail.com

Current Topics in Microbiology and Immunology (2021) 430: 161–181
https://doi.org/10.1007/82_2017_79

Published Online: 9 February 2018

1 Introduction

Lung cancer is the most common cause of cancer-related death worldwide, with over 1.5 million deaths and almost 2 million new cases diagnosed per year (Ferlay et al. 2015). The major cause of lung cancer is chronic exposure to tobacco smoke, although other environmental factors, such as pollution, also play a role.

Classical models to study lung cancer include culture of cell lines or primary cells and mouse models. These systems have major limitations. Monolayers of cells do not allow to fully investigate the mechanisms governing the uncontrolled growth of tumors. When grown in monolayer, cells lose their normal phenotype (Baker and Chen 2012). Mouse models, although tremendously helpful in advancing our understanding of human disease, are not reliable in predicting human responses in vivo. For example, the failure rate of clinical trials is high even when safety and efficacy data from rodent models seem promising (Harrison 2016). However, advances in primary cell isolation, stem cell research, and bioengineering have resulted in establishing various three-dimensional (3D) models that can be used to study infections, carcinogenesis, and tissue responses to novel drugs in a more physiologically relevant setting.

The main role of the human respiratory system is air exchange that supplies oxygen to all the tissues and removes carbon dioxide from the circulation. The human respiratory system starts at the nose and comprises the pharynx, epiglottis, larynx, and trachea; the trachea then bifurcates into bronchi, which divide into smaller airways terminating in bronchioles with alveoli. There are approximately 16 generations of bronchi from the main bronchus. Gas exchange occurs in the 300 million alveoli present in the human lungs (Popper 2017b).

Many cell types are present along the airways. However, the composition and thickness of the airway are different depending on the location. For example, large bronchi comprise cartilage tissue, basal cells, and differentiated cells such as ciliated epithelial cells, goblet cells, and a limited number of club cells. Smaller bronchi have less cartilage and are lined with a simple columnar epithelial layer. Bronchioles, however, have no cartilage and are lined with simple non-ciliated cuboidal epithelial layer, including club cells and bronchialveolar stem cells. In the alveoli, there are two types of specialized epithelial cells: type I and type II pneumocytes. Type I cells are the main functional cells in the lung and are responsible for gas exchange, whereas type II cells secrete surfactants required for proper alveolar function. In addition, veins and arteries are near the bronchi and small capillaries surround the alveolar sacs. This arrangement allows for the efficient gas exchange in the alveoli. Also, the mechanical process of ventilation is crucial for air exchange, and breathing movement is controlled by the diaphragm. This mechanism is also used for the delivery of drugs via inhalers or nebulizers (Chen et al. 2014; Popper 2017b).

Cancer is one of the most devastating diseases that can affect the lung. There are many types of lung tumors, including small-cell lung cancer and non-small-cell lung cancer (NSCLC). NSCLC is the most prevalent form and includes squamous cell carcinomas and adenocarcinomas (Gridelli et al. 2015). The many forms of lung tumors have recently been reviewed by Popper (2017a). Different cells of the human lung may give rise to different types of lung carcinomas; for example, the consensus is that squamous cell carcinomas have origin in the proximal lung, whereas adenocarcinomas start in the distal lung (Chen et al. 2014).

Human tumors are not two-dimensional; they have their own microenvironment with blood-supplying capillaries and infiltration of cells of the immune system (Chen et al. 2014; Nagarsheth et al. 2017). Tumor cells interact with the surrounding healthy environment by various means, including the release of cytokines. There is also significant intra- and inter-tumor heterogeneity that cannot be recapitulated in the 2D cell line monolayers grown in a culture dish.

Lung cancer is a major health concern worldwide, and thus, it is imperative to study the mechanisms of carcinogenesis as well as tissue responses to novel drugs in reliable in vitro systems, whose results can be directly translated into the clinic. Recently, researchers have shifted toward using more physiologically relevant 3D organotypic models to study infections, disease, and cancer. In recent years, many comprehensive reviews have been published on the organotypic models used in biomedical research (Barrila et al. 2010; Bhatia and Ingber 2014; Horvath et al. 2016; Morgan et al. 2016; Shamir and Ewald 2014). The focus of this chapter is to review studies and applications related only to human lung cancer.

2 Organotypic Models of Lung Cancer

Organotypic three-dimensional models of human development and disease are being rapidly developed. These organotypic models incorporate the 3D spatial structure of tissue that includes multiple cell types, thereby allowing for cell-to-cell interactions and communications in vitro. Various 3D models have been used thus far to study lung development and disease. It is not possible here to review all the studies that used organotypic tissue culture to study lung cancer; however, the subsections below aim at summarizing the most commonly utilized and the most promising current models (also summarized in Table 1).

2.1 Ex Vivo Tissue Culture Models

2.1.1 Precision-Cut Lung Slices (PCLSs)

Precision-cut lung slices (PCLSs) are one of the most commonly used organotypic model to study human lung tissue ex vivo. It has been used to study drug toxicity, airway, and whole tissue responsiveness to various stimuli, including new small molecules or even vaccines (Abdull Razis et al. 2011; Behrsing et al. 2013; Cooper et al. 2009, 2010; Harrigan et al. 2004; Henjakovic et al. 2008a, b; Herbert et al.

Table 1 Overview of organotypic models used to study human lung cancer

Model	Advantages	Disadvantages	Current applications	References
Precision-cut lung slices (PCLSs)	– Serial lung slices from the same specimen can be used to study multiple conditions – Bronchi maintain their constricting properties – Give reproducible results	– Short viability – Difficulty in obtaining human specimens – They are only a snapshot of the lung architecture – Require processing in agarose	– Immunomodulation – Toxicity studies – Effects of allergen and pollutant exposure – Investigating mechanisms of chronic disease – Investigating mechanisms of carcinogenesis	Abdull Razis et al. (2011), Behrsing et al. (2013), Cooper et al. (2009, 2010, 2011a, b), Cooper and Panettieri (2008), Donovan et al. (2015, 2016), Ebsen et al. (2002), Harrigan et al. (2004), Henjakovic et al. (2008a, b), Herbert et al. (2017), Hess et al. (2016), Hofmann et al. (2015), Kim et al. (2015), Kistemaker et al. (2017), Koziol-White et al. (2016a, b), Lauenstein et al. (2014), Martin et al. (2001), Morin et al. (2013), Muller-Redetzky et al. (2012), Neuhaus et al. (2013, 2017), Ressmeyer et al. (2006), Rieg et al. (2014), Schleputz et al. (2012), Seehase et al. (2012), Switalla et al. (2010), Temann et al. (2017), van den Bosch et al. (2017), Van Dijk et al. (2016), Waypa et al. (2013), Wijeweera et al. (2001), Wohlsen et al. (2001, 2003), Wright et al. (2011)
Lung whole tissue explants (WTE)	– Easy to prepare once lung sample has been obtained – Can test multiple parameters on the same lung sample – Give reproducible results	– Short viability – Difficulty in obtaining human specimens – They are only a snapshot of the lung architecture – Not suitable for airway responsiveness studies – Data normalization required due to heterogeneity of each explant	– Immunomodulation – Toxicity studies – Effects of allergen and pollutant exposure – Investigating mechanisms of chronic disease – Investigating mechanisms of carcinogenesis	Hackett et al. (2008), Jacquet et al. (1997), Karekla et al. (2017), Pomerenke et al. (2016), Rimington et al. (2017)
Lung tumor xenografts	– Use of primary human tumors – Can be used to guide therapeutic approach and/or predict therapy outcome – Possibility of studying long-term effects	– Cost of maintaining animal facility – Technically challenging	– Testing personalized therapeutic outcomes – Toxicity studies	Chen et al. (2017a), Jung (2014), Kellar et al. (2015), Morgan et al. (2017), Russo et al. (2015)

(continued)

Table 1 (continued)

Model	Advantages	Disadvantages	Current applications	References
Spheroids	– Relatively easy to culture and cheap to maintain, depending on the source of the cells: cell lines vs primary cells – Reproducible results – Culture can be adapted for high-throughput testing – If formed using multiple cell types, they can recapitulate tumor 3D environment better than 2D systems	– If formed using single cell type, they hardly recapitulate heterogeneity of tumors – Relatively short culture time (days to weeks) – Lack of "airway" side where airway responses or drug delivery could be tested	– Investigating mechanisms of chronic disease – Investigating mechanisms of carcinogenesis – Investigating mechanisms of resistance of tumors to therapy – Testing personalized therapeutic outcomes	Amann et al. (2014), Barrera-Rodríguez and Fuentes (2015), Chimenti et al. (2017), Deslee et al. (2007), Dinh et al. (2017), Ekert et al. (2014), Endo et al. (2013), Eramo et al. (2007), Fessart et al. (2013), Henry et al. (2015), Huber et al. (2016), Lama et al. (2013), Leung et al. (2017), Onion et al. (2016), Pisanu et al. (2017), Razian et al. (2013), Sato et al. (2006), Weiswald et al. (2015)
Organoids	– Form tissue-like structures containing multiple cell types – Can be used for long-term study – Can be manipulated using CRIPR/ Cas9 technology – Possibility to study organoids resembling different areas of the respiratory tract	– Technically challenging – High cost and time consuming – Not yet available for high-throughput screening – Lack of "airway" side, but organoids form enclosed hollow structures	– Studying human lung development – Investigating mechanisms of chronic disease – Investigating mechanisms of carcinogenesis – Possible future uses in organ transplants	Barkauskas et al. (2017), Chen et al. (2017b), Danahay et al. (2015), Fatehullah et al. (2016), Kumar et al. (2011), Nikolić and Rawlins (2017), Rock et al. (2011), van den Bosch et al. (2017)
Air–liquid interface	– "Airway" side exposed on the surface – Long viability – Possibility of co-culturing tumor cells with normal respiratory epithelium – Possibility of studying airway and systemic drug delivery	– Differentiation of cells is time consuming, but predifferentiated cells are available commercially – Technically challenging and time consuming, but automation efforts are in progress	– Immunomodulation – Toxicity studies – Effects of allergen and pollutant exposure – Investigating mechanisms of chronic disease – Investigating mechanisms of carcinogenesis	Aufderheide et al. (2016), Balharry et al. (2008), Dwivedi et al. (2017), Geles et al. (2016), Gray et al. (1996), Konishi et al. (2016), Lenz et al. (2014), Mas et al. (2015), Mathis et al. (2013), Meenach et al. (2016), Pezzulo et al. (2011), Salar-Behzadi et al. (2017), Schamberger et al. (2015), Zanoni et al. (2016)

(continued)

Table 1 (continued)

Model	Advantages	Disadvantages	Current applications	References
Lung-on-a-chip	– Comprises epithelial and endothelial layers – Tight control over experimental parameters – Possibility to study immune cell invasion in response to various stimuli – Recapitulates human lung architecture and breathing mechanism – Can be part of the bodies-on-chips	– High cost – Bioengineering experts required for production and maintenance – Not suitable for long-term studies	– Immunomodulation – Toxicity studies – Effects of allergen and pollutant exposure – Investigating mechanisms of chronic disease – Investigating mechanisms of carcinogenesis – Modeling infection of the lung – Testing personalized therapeutic outcomes	Ahn et al. (2017), Bhatia and Ingber (2014), Esch et al. (2015), Hsu et al. (2011), Huh et al. (2012), Huh et al. (2010), Xu et al. (2013)
Lung 3D bioprinting	– Printed tissues highly recapitulate 3D structure of the organ – Tight control over cell types location and numbers	– High cost – Specialized equipment and bioengineering expertise required – Cells are printed on a scaffold	– Possible future uses in organ transplants – Other applications possible but not yet used in the 3D-printed lung	Horvath et al. (2015), Jin Woo et al. (2016), Munoz-Abraham et al. (2016), Murphy and Atala (2014)

2017; Hofmann et al. 2015; Koziol-White et al. 2016a, b; Martin et al. 2001; Morin et al. 2013; Neuhaus et al. 2013; Rieg et al. 2014; Seehase et al. 2012; Switalla et al. 2010; Temann et al. 2017; Wijeweera et al. 2001). In addition to human tissues, animal tissues, i.e., mouse, rat, and guinea pig, have also been used to study human disease (Harrigan et al. 2004; Henjakovic et al. 2008a; Ressmeyer et al. 2006). In order to obtain PCLSs, the tissue is first stabilized in low-melting agarose and then cut using precision slicers to the desired thickness (Morin et al. 2013). Slices are placed in tissue culture dishes and submerged in appropriate medium. After an initial incubation period, slices are exposed to stimuli such as *toll*-like receptor (TLR) agonists, allergens, or tobacco smoke and to test the efficacy and toxicity of novel small molecules (Cooper et al. 2009, 2010; Cooper and Panettieri 2008; Harrigan et al. 2004; Henjakovic et al. 2008a; Lauenstein et al. 2014; Schleputz et al. 2012; Seehase et al. 2012; Switalla et al. 2010).

One of the advantages of using lung slices is that the same lung specimen can be used to study multiple conditions to compare either the immune responses to various stimuli or the effect of a drug. This is important to ensure the reproducibility of the results. At the same time, PCLS retains mechanical properties of an airway, e.g., they can constrict in response to cytokines (Martin et al. 2001).

Studies using PCLS as a model aim at elucidating the mechanisms of chronic diseases, such as asthma or chronic obstructive pulmonary disease (Cooper et al. 2010, 2011a, b; Cooper and Panettieri 2008; Donovan et al. 2015; Kim et al. 2015; Kistemaker et al. 2017; Koziol-White et al. 2016b; Lauenstein et al. 2014; van den Bosch et al. 2017; Van Dijk et al. 2016; Wright et al. 2011). Others examine human or animal tissue responses to allergens, hypoxia, or infection with either bacterial or viral pathogens (Donovan et al. 2016; Ebsen et al. 2002; Muller-Redetzky et al. 2012; Waypa et al. 2001, 2013; Wohlsen et al. 2003; Wright et al. 2011). Not many studies have used lung slices to investigate human lung cancer. However, several groups studied the effects of known human carcinogens. For example, Harrigan et al. (2004) used rat PCLS model to demonstrate that in vitro exposure to benzo[a]pyrene (one of the major carcinogenic components of cigarette smoke) induces DNA adduct formation in PCLSs. Another group, also using rat PCLS model, found that arsenic caused activation of the pro-inflammatory transcription factors AP-1 and NF-κB (Wijeweera et al. 2001). Human PCLS from tumor samples could be used to study the tissue responses in preclinical models of anticancer therapy. In fact, animal PCLSs have already been used in this setting. For example, Behrsing et al. (2013) used rat PCLS model to test the effects of an anticancer drug on cytokine release. In addition, efforts are now underway to validate PCLS as a model for drug toxicity testing in preclinical settings (Hess et al. 2016; Neuhaus et al. 2017).

In summary, PCLSs are a robust model to study the effects of various stimuli in a highly controlled ex vivo environment. They show high reproducibility and can be maintained viable in vitro even for several weeks. In addition, because PCLS are very thin, it is possible to test multiple conditions using a series of slices derived from the same donor.

2.1.2 Lung Whole Tissue Explants (WTE)

Another three-dimensional approach to study the lung environment is represented by lung whole tissue explants (WTE). This approach is similar to PCLS; however, the tissue here can be cut into cubes rather than slices. WTE may not be the best model to study airway responsiveness as the airway 3D structure may not be fully preserved. Nevertheless, this approach has been successfully used to study responses of various stimuli, including TLR agonists (Hackett et al. 2008; Jacquet et al. 1997; Pomerenke et al. 2016; Rimington et al. 2017). It is also a good model to study lung tumors. Recently, Karekla et al. (2017) adapted this method to study human lung cancer. They used WTE to study tissue responses to cisplatin and TNF-related apoptosis-inducing ligand (TRAIL), finding the model to be robust and to yield reproducible results. This indicates that WTE can be successfully used to study tissue responses to cancer treatment.

This model requires less processing when compared to PCLS. Here, tissue does not need to be stabilized in agarose prior to cutting. There is, however, a question of how to compare different experimental conditions where every piece of tissue potentially has a different ratio of cell types. It is common to normalize the data by using the same number of explants per condition and to express data per weight of tissue. Other advantages of this model include those already mentioned above for PCLS, e.g., the possibility to use lung from the same patient in multiple experimental conditions.

2.1.3 Summary

In general, ex vivo human lung explants are a good model to study the effect of novel anticancer drugs in the 3D organotypic setting. Since human lung tissue is hard to obtain, it should be used in the later stages of preclinical development when lead compounds are already established. These models do not allow to study the mechanisms of carcinogenesis in vitro, as tissue will only be viable for a limited time. However, they can be used to study the effects of known carcinogens, such as cigarette smoke, and examining how they alter pathways that are important in cancer development.

2.2 Lung Tumor Xenografts

Human tumor xenografts in rodent animal models have also been employed to study human lung cancer. Several approaches have been explored, ranging from injection of human tumor cell lines to engraftment of primary human tumors. For example, Chen et al. (2017a) used human alveolar cell line, A549, mixed with Matrigel to produce xenografts in mice, while Russo et al. (2015) developed a model using patient-derived samples of NSCLC and implanted them into mice.

The latter study showed that these xenografts can be studied for up to 6 months without losing the distinctive markers of primary tumors.

Mouse xenograft models are not an organotypic model per se, and therefore, they are beyond the scope of this chapter. However, it is important to mention this approach as it is very popular for testing anticancer drugs; in 2016 alone, over 400 articles that used this system have been published. Mouse xenograft models in preclinical testing and their potential use in personalized medicine have been recently reviewed (Jung 2014; Kellar et al. 2015; Morgan et al. 2017).

2.3 Spheroids

Another widely used model for culturing cells in 3D is spheroid. Spheroids can be formed by seeding cells in ultra-low attachment plates by using the so-called "hanging-drop" method or by seeding cells into an ECM-like biomaterial. When cultured in these conditions, cells spontaneously aggregate into spheroids (Weiswald et al. 2015). Cells from various sources can be used to culture spheroids, including cell lines, primary cells, and even primary human tissue, both normal and tumorous (Endo et al. 2013; Eramo et al. 2007; Fessart et al. 2013; Henry et al. 2015; Leung et al. 2017; Pisanu et al. 2017; Weiswald et al. 2015).

There are many studies that used spheroids in the context of lung tissue and lung cancer. For example, spheroids can be formed using primary bronchial epithelial cells cultured in Matrigel. This system allows for acini formation that is hollow on the inside, thus reliably representing the 3D tissue structure as observed in vivo (Fessart et al. 2013). Spheroids produced from healthy primary human tissue samples were shown to contain lung progenitor cells which may help in the treatment of tissue fibrosis and other lung diseases (Dinh et al. 2017; Henry et al. 2015). Chimenti et al. (2017) investigated the behavior of lung progenitor cells in both 2D culture and in 3D spheroids and found marked differences in gene expression between the two models. In addition, they reported that when cultured as spheroids, lung progenitor cells started to show microtissue-like phenotype. Endo et al. (2013) used over 100 samples of patient-derived tumors to generate lung cancer spheroids. They have confirmed that these spheroids maintained primary tumor phenotypes and were sensitive to anticancer treatment. Eramo et al. (2007) also used patient-derived tumor tissue to form tumor spheroids and showed that $CD133^+$ cells could grow indefinitely and had tumorigenic properties when injected into immunocompromised mice and thus named these cells cancer stem cells (CSCs).

Spheroids are increasingly used in research studies that examine drug responses in a 3D lung environment (Ekert et al. 2014; Huber et al. 2016; Lama et al. 2013; Onion et al. 2016). Moreover, lung cancer spheroids can be co-cultured with stromal cells, such as fibroblasts, to try to recapitulate the tumor microenvironment (Amann et al. 2014). Sato et al. (2006) introduced knockdowns of multiple genes important in tumor development, such as *KRAS*, *TP53,* and epidermal growth factor receptor (*EGFR*), and showed that bronchial spheroids started to show signs of transformation, although they were not fully malignant. In addition, spheroid cultures are used in studies that try to explain the mechanisms of resistance of lung tumor to current treatments, e.g., cisplatin, and identify new targets for combination therapy (Barrera-Rodríguez and Fuentes 2015; Leung et al. 2017; Pisanu et al. 2017).

There are several advantages in studying lung cancer using the 3D spheroid model. It is a reproducible model, as seeding equal number of cells gives comparable spheroid sizes with similar responses to challenges in vitro, it is relatively inexpensive, and it mimics the 3D tumor microenvironment found in vivo. In addition, efforts are underway to adapt spheroid cultures into high-throughput assays (Razian et al. 2013). The disadvantages of this model include the relatively short culture time, i.e., days, and the lack of an "air side" that would allow to test airway-delivered drugs, and although some spheroids form hollow lumens, they are usually inside the spheroid (Deslee et al. 2007).

2.4 *Organoids*

Organoids are small 3D tissue-like specimens that can be cultured in vitro on their own or in Matrigel or other analogues of the extracellular matrix. They can be formed from stem cells of multiple sources, such as primary tissue or embryonic stem cells or induced pluripotent stem cells, and they differentiate and proliferate on their own. Many organs have now been successfully studied in vitro using organoids as a model (reviewed by Fatehullah et al. 2016). Sometimes, the terms "spheroids" and "organoids" are used interchangeably. For the purpose of this chapter, organoids are defined as 3D tissue-like models derived from a stem cell, whereas organoids are more complex than spheroids as the original stem cell gives rise to the multiple types of cells present in the tissue.

Organoids have not only been used to study organ development, but also adapted to model human disease and cell-to-cell interactions (Nikolić and Rawlins 2017). Various uses of human and murine lung organoids have been recently reviewed by Barkauskas et al. (2017). There are several subtypes of lung organoids, which derive their names from the original cells, e.g., tracheospheres (tracheal cell), bronchospheres (bronchial cell), or alveolospheres (alveolar cell). Spheroids derived from trachea or bronchi can be studied in vitro on their own; however, alveolar spheroids require support from mesenchymal cells (Barkauskas et al. 2017). Lung basal cells can give rise to more specialized cell types, such as ciliated cells, secreting cells, or alveolar cells (Danahay et al. 2015; Kumar et al. 2011; Rock et al. 2011). Bronchospheres comprising ciliated, basal, and goblet cells have recently been adapted into a high-throughput screening system by Danahay et al. (2015). In this interesting study, lung organoids were used to establish a disease-like state by exposing organoids to interleukin (IL)-13—a cytokine known to play an important role in airway remodeling—and this resulted in goblet-cell metaplasia. This effect was inhibited by antibodies specific to Notch2, thus confirming findings from an earlier study by Rock et al. (2011) who also reported on the importance of Notch signaling in cell differentiation in human and murine tracheospheres.

In a recent report, Chen et al. (2017b) used human pluripotent stem cells, both embryonic and inducible, and created budding lung organoids. They reported growing lung organoids for up to 170 days at which point their gene expression matched the profile found in vivo in human lung late in the second trimester of gestation. They have successfully grown organoids from many cell donors, including one with mutated interferon regulatory factor 7 (*IRF7*) gene. In addition, infection challenge with respiratory syncytial virus (RSV) closely resembled the infection in vivo. Also, they used CRISPR/Cas9 technology to delete the *HPS1* gene to model pulmonary fibrosis. HPS gene defects cause Hermansky–Pudlak syndrome, a rare autosomal recessive disorder that leads, inter alia, to pulmonary fibrosis (Vicary et al. 2016); indeed, lung organoids showed signs of fibrosis after 40 days. The detailed protocol used in this pivotal study was shared by the group (see Chen et al. 2017b).

In summary, lung organoids are a very promising tool in the field of human lung research, and soon scientists may be able to study the development of human lung with defective genes in vitro. In addition, lung organoids seem to be an appropriate model to study the mechanism that drives carcinogenesis. They are suitable for gene knockdown and gene editing using CRISPR/Cas9 and thus can help to identify genes that drive malignant transformation and metastasis. Moreover, the same lung organoid can be followed for months or perhaps even longer, and therefore any long-term changes can be monitored in real time. Disadvantages of this model include time and costs associated with the long-term organoid culture; this suggests that lung organoids may not be the best model to study drug efficacy and safety where high-throughput assays are utilized.

To date, there are no reports using true organoids to model human lung cancer. In the future, however, lung organoids could become fundamental tools in lung cancer therapy, i.e., the patient's own induced pluripotent stem cells could be used to grow a "replacement" lung for transplantation. However, the road ahead is still very long, as currently lung organoids take the same amount of time to develop as normal human lung during gestation. Nevertheless, if at some point scientists can accelerate their growth in vitro, the field of organ transplantation will be revolutionized.

2.5 Air–Liquid Interface

Air–liquid interface (ALI) is a physiologically relevant approach to study the blood-air barrier in vitro. In the in vivo pulmonary environment, air exchange is the most important process, and an adequate representation is imperative for successful investigation of lung cancer. Utilization of air–liquid interface is common in various models, e.g., in vitro cell culture on well inserts or in technologically advanced models including organs-on-chips (see section below). Cell lines, primary cells, and tumor cells have been used in well inserts in vitro to generate the 3D architecture of a human lung (Dwivedi et al. 2017; Geles et al. 2016; Gray et al. 1996; Lenz et al. 2014; Pezzulo et al. 2011; Salar-Behzadi et al. 2017). The process of differentiation of cells into the air-blood barrier may take weeks, but predifferentiated lung epithelial layers are now available commercially.

Under ALI conditions, cells are differentiated into distinct subtypes, including ciliated epithelial cells and mucus-producing cells (Gray et al. 1996). Studies have shown that the ALI model of respiratory tract responds to known respiratory toxins in a similar manner as reported in vivo (Balharry et al. 2008), and primary bronchial cells challenged with cigarette smoke are phenotypically similar to those found in smokers (Mathis et al. 2013; Schamberger et al. 2015). As such, pulmonary ALI cultures can be used to study the mechanisms of carcinogenesis due to, for example, exposure to cigarette smoke. In fact, recent studies reported the successful culture of cell line- and stem cell-derived lung spheroids using the ALI model, which has also been used in the context of anticancer therapeutics (Konishi et al. 2016; Meenach et al. 2016; Zanoni et al. 2016). Meenach et al. (2016) showed that cell responses using this 3D system were significantly different when compared to the 2D model, again suggesting that 3D systems are more physiologically relevant for preclinical testing. Moreover, novel ALI systems comprising normal, differentiated bronchial epithelium, fibroblasts, and lung tumor cells have been developed and used to assess the efficacy of anticancer therapeutics. Mas et al. (2015) used this system to evaluate MEK inhibitors and found that the growth of tumor cells with *KRAS* mutations was significantly inhibited in agreement with preclinical data obtained using other models.

The major advantage of ALI cultures is that they allow for studying the effects of drugs or toxins when applied onto the semidry airway site, thus mimicking the way drugs can be delivered in vivo. In addition, ALI models can be cultured for weeks. However, handling multiple dishes of ALI cultures is difficult and time consuming. To address this issue, automated computer-controlled systems are being developed (Aufderheide et al. 2016; Lenz et al. 2014). By reducing human error, automation should also increase the reproducibility of the results. In addition, co-culture of tumor cells with its normal surrounding environment could be employed for the evaluation of anticancer therapy using either airway (air side) or systemic (liquid side) delivery systems.

2.6 Bioengineering Approaches

2.6.1 Lung-on-a-Chip

Recent advances in bioengineering have resulted in astounding organ-on-a-chip models, and researchers have developed models for multiple organs, including human lung. Bhatia and Ingber (2014) have recently reviewed organs-on-chips models in more detail. Initially, bioengineers have created "simple" organs-on-chips consisting of only one cell type. But with time, the design of organs-on-chips has evolved to incorporate extracellular matrix and multiple cell types, allowing for cell–cell communications.

In pivotal reports, Huh et al. have established a "biologically inspired lung-on-a-chip" model that recapitulates the behavior of a human lung in vivo (Huh et al. 2010, 2012). They incorporated the epithelial and endothelial cells into a chip designed to support air–liquid interphase and mechanical strain to mimic the human airway and the breathing mechanism. In addition, the microfluidic system utilized in organs-on-chips constantly perfuses the chip, making it useful to study angiogenesis and immune cell infiltration. This may be particularly interesting when investigating infections or tumors associated with infections and inflammation. In fact, in the same reports, Huh et al. showed that introduction of the pro-inflammatory cytokine TNF-α led to CXCL8 induction in the lung-on-a-chip model, leading to neutrophil recruitment and migration through the endothelial and epithelial layers. This suggests that the lung-on-a-chip may be a good model to study the interactions of parenchymal lung cells with the cells of the immune system.

More recently, this bioengineering approach has been applied to human cancer research, termed "cancer-on-a-chip." Several cancers-on-chips have been engineered, including lung cancer (Ahn et al. 2017; Hsu et al. 2011; Xu et al. 2013). Xu et al. (2013) have created a lung cancer-on-a-chip to test drug toxicity in 3D using cell lines and, more importantly, patient-derived primary cancer cells. They tested multiple anticancer drug combinations to identify the best outcome for each patient. Hsu et al. (2011) have created a chip comprising lung tumor cells and fibroblasts, showing that tumor cells cause fibroblast activation via TGF-β secretion, whereas conditioned media from activated fibroblasts caused an enhanced migratory behavior of the lung tumor. This paracrine loop was abolished when the TGF-β receptor was inhibited. Tumors-on-chips have also been used to study the interactions of tumor cells with components of the tumor environment (e.g., immune cells or cytokines) in other cancer types (recently reviewed by Ahn et al. 2017). These approaches could also be used in the setting of lung cancer.

The lung-on-a-chip model has many advantages, including the possibility to tightly control the experimental parameters. Since organ chips are relatively small, it is possible to test multiple drugs and concentrations with minimal amounts of patient-derived material. Results from such personalized tests could be used as guidance to clinicians for the appropriate choice of treatment. In addition, there are no limits to the use of chips as models to study carcinogenesis. Exposure of lungs-on-chips to known carcinogens and carefully planned output measurements could lead to the identification of early biomarkers of lung cancer and thus may help in early diagnosis. In addition, the fact that findings observed in vitro using lungs-on-chips are reproducible in a mouse model in vivo underscores the value of organs-on-chips as a preclinical research tool (Huh et al. 2012).

There are also limitations to organs-on-chips (Esch et al. 2015). Particularly, chips are not suitable for long-term drug effect studies, and the use of a single organ-on-a-chip to test novel drugs hardly reflects drug metabolism in vivo. This is being addressed by connecting multiple organs-on-chips to form "bodies-on-chips." Another disadvantage is the high cost and the necessity for specialized bioengineers to build the chips. However, with the advancements and automation of technology, the costs of chips should decrease, and automated systems will allow non-engineers to use this model on a greater scale.

2.6.2 Lung 3D Printing

Another bioengineering approach to create three-dimensional tissue models is tissue 3D printing. There are several ways in which tissue printing can be performed, which have recently been reviewed in detail (Murphy and Atala 2014). To recapitulate human organ architecture in vitro, cells are printed onto a scaffold (usually made from natural polymers or synthetic materials, e.g., collagen or PEG, respectively). One of the greatest potentials of 3D tissue printing is to print organs for transplantation using the patient's own healthy or genome-edited cells. This application, however, is beyond the scope of this chapter (reviewed by Munoz-Abraham et al. (2016)).

Tissue 3D printing in the context of the lung has not been widely reported thus far. Horvath et al. (2015) used endothelial and epithelial cell lines to print air-blood barrier in vitro onto a Matrigel base. They showed that 3D bioprinting results in more uniform and thinner layers than using manual cell seeding. This could be important to increase the reproducibility of drug efficacy and toxicity assays. 3D bioprinting of other tissues, such as liver, have been reported by Jin Woo et al. (2016) who showed that the multicellular environment promoted angiogenesis in vitro. Bioengineering is a dynamic and fast-paced field, and one can already imagine personalized 3D printing of lung cancers that may recapitulate the individual patient's multicellular tumor environment.

3 Conclusions

This chapter has discussed models that are or could be used to study human lung cancer (see also Table 1). As mentioned in the Introduction, the major challenge for lung cancer research is the difficulty to recapitulate the pulmonary heterogenous three-dimensional environment. Despite having led to major advancements in science, monolayers of transformed cells are not a good model to study the mechanisms of lung carcinogenesis or lung cancer responses to novel drugs. There is a need to reconstruct the 3D pulmonary architecture ex vivo and to use it as a platform to address unresolved questions in lung cancer research, both to enhance the pace at which novel cancer treatments are developed and to understand how normal lung tissue undergoes cancerous transformation.

There is no ideal model to study lung cancer, and thus, researchers must carefully choose the right system that will be appropriate to address their specific question. For example, to study the long-term effects of a drug, they should not choose a model that has short viability. Most models described in this chapter are suitable for efficacy and toxicity studies associated with the drug development process. There is hope that the use of human organotypic 3D models will generate data that can increase the success rate of clinical trials. Using organotypic systems, we can also study the mechanisms of carcinogenesis (e.g., exposing lung tissue to cigarette smoke or other known carcinogens). Also, the generation of patient-derived tumor tissue samples could be used in personalized medicine, e.g., by providing individualized treatment based on the tissue responses observed ex vivo.

With the exponential advancement of technology and the development of novel bioengineering tools and biocompatible materials, one can envision a promising future for 3D systems to study lung cancer. It must be remembered that lung cancer is the leading cause of cancer-related death in the world. Although many efforts are undertaken to limit pollution and educate the population about risks associated with the use of tobacco products, the rate of new lung cancer cases is expected to remain the same or even rise over the next decade (Ferlay et al. 2013). Early diagnosis and effective treatments are thus paramount for reducing lung cancer mortality worldwide.

Acknowledgements This research was supported [in part] by the Intramural Research Program of the NIH, NIAID.

References

Abdull Razis AF, Bagatta M, De Nicola GR, Iori R, Ioannides C (2011) Up-regulation of cytochrome P450 and phase II enzyme systems in rat precision-cut rat lung slices by the intact glucosinolates, glucoraphanin and glucoerucin. Lung Cancer 71:298–305. https://doi.org/10.1016/j.lungcan.2010.06.015

Ahn J, Sei Y, Jeon N, Kim Y (2017) Tumor microenvironment on a chip: the progress and future perspective. Bioengineering 4:64

Amann A et al (2014) Development of an innovative 3D cell culture system to study tumour—stroma interactions in non-small cell lung cancer cells. PLoS ONE 9:e92511. https://doi.org/10.1371/journal.pone.0092511

Aufderheide M, Förster C, Beshay M, Branscheid D, Emura M (2016) A new computer-controlled air–liquid interface cultivation system for the generation of differentiated cell cultures of the airway epithelium. Exp Toxicol Pathol 68:77–87. https://doi.org/10.1016/j.etp.2015.10.001

Baker BM, Chen CS (2012) Deconstructing the third dimension: how 3D culture microenvironments alter cellular cues. J Cell Sci 125:3015–3024. https://doi.org/10.1242/jcs.079509

Balharry D, Sexton K, BeruBe KA (2008) An in vitro approach to assess the toxicity of inhaled tobacco smoke components: nicotine, cadmium, formaldehyde and urethane. Toxicology 244:66–76. https://doi.org/10.1016/j.tox.2007.11.001

Barkauskas CE, Chung MI, Fioret B, Gao X, Katsura H, Hogan BL (2017) Lung organoids: current uses and future promise Development (Cambridge, England) 144:986–997. https://doi.org/10.1242/dev.140103

Barrera-Rodríguez R, Fuentes JM (2015) Multidrug resistance characterization in multicellular tumour spheroids from two human lung cancer cell lines. Cancer Cell Int 15:47. https://doi.org/10.1186/s12935-015-0200-6

Barrila J, Radtke AL, Crabbé A, Sarker SF, Herbst-Kralovetz MM, Ott CM, Nickerson CA (2010) Organotypic 3D cell culture models: using the rotating wall vessel to study host–pathogen interactions. Nat Rev Micro 8:791–801

Behrsing HP, Furniss MJ, Davis M, Tomaszewski JE, Parchment RE (2013) In vitro exposure of precision-cut lung slices to 2-(4-amino-3-methylphenyl)-5-fluorobenzothiazole lysylamide dihydrochloride (NSC 710305, Phortress) increases inflammatory cytokine content and tissue damage. Toxicol Sci 131:470–479. https://doi.org/10.1093/toxsci/kfs319

Bhatia SN, Ingber DE (2014) Microfluidic organs-on-chips. Nat Biotech 32:760–772. https://doi.org/10.1038/nbt.2989

Chen Z, Fillmore CM, Hammerman PS, Kim CF, Wong K-K (2014) Non-small-cell lung cancers: a heterogeneous set of diseases. Nat Rev Cancer 14:535. https://doi.org/10.1038/nrc3775

Chen XW, Sun JG, Zhang LP, Liao XY, Liao RX (2017a) Recruitment of CD11b+ Ly6C + monocytes in non-small cell lung cancer xenografts challenged by anti-VEGF antibody Oncol Lett 14:615–622. https://doi.org/10.3892/ol.2017.6236

Chen Y-W et al (2017b) A three-dimensional model of human lung development and disease from pluripotent stem cells. Nat Cell Biol 19:542–549. https://doi.org/10.1038/ncb3510

Chimenti I et al (2017) Human Lung Spheroids as In Vitro Niches of Lung Progenitor Cells with Distinctive Paracrine and Plasticity Properties. STEM CELLS Transl Med 6:767–777. https://doi.org/10.5966/sctm.2015-0374

Cooper PR, Panettieri RA Jr (2008) Steroids completely reverse albuterol-induced beta(2)-adrenergic receptor tolerance in human small airways. J Allergy Clin Immunol 122:734–740. https://doi.org/10.1016/j.jaci.2008.07.040

Cooper PR et al (2009) TLR3 activation stimulates cytokine secretion without altering agonist-induced human small airway contraction or relaxation. Am J Physiol Lung Cell Mol Physiol 297:L530–537. https://doi.org/10.1152/ajplung.00133.2009

Cooper PR, Poll CT, Barnes PJ, Sturton RG (2010) Involvement of IL-13 in tobacco smoke-induced changes in the structure and function of rat intrapulmonary airways. Am J Respir Cell Mol Biol 43:220–226. https://doi.org/10.1165/rcmb.2009-0117OC

Cooper PR, Kurten RC, Zhang J, Nicholls DJ, Dainty IA, Panettieri RA (2011a) Formoterol and salmeterol induce a similar degree of beta2-adrenoceptor tolerance in human small airways but via different mechanisms. Br J Pharmacol 163:521–532. https://doi.org/10.1111/j.1476-5381.2011.01257.x

Cooper PR, Zhang J, Damera G, Hoshi T, Zopf DA, Panettieri RA Jr (2011b) C-027 inhibits IgE-mediated passive sensitization bronchoconstriction and acts as a histamine and serotonin antagonist in human airways. Allergy Asthma Proc 32:359–365. https://doi.org/10.2500/aap.2011.32.3460

Danahay H et al (2015) Notch2 Is required for inflammatory cytokine-driven goblet cell metaplasia in the lung. Cell Reports 10:239–252. https://doi.org/10.1016/j.celrep.2014.12.017

Deslee G et al (2007) Bronchial epithelial spheroids: an alternative culture model to investigate epithelium inflammation-mediated COPD. Respir Res 8:86. https://doi.org/10.1186/1465-9921-8-86

Dinh P-UC et al (2017) Derivation of therapeutic lung spheroid cells from minimally invasive transbronchial pulmonary biopsies. Respir Res 18:132. https://doi.org/10.1186/s12931-017-0611-0

Donovan C et al (2015) Rosiglitazone elicits in vitro relaxation in airways and precision cut lung slices from a mouse model of chronic allergic airways disease. Am J Physiol Lung Cell Mol Physiol 309:L1219–1228. https://doi.org/10.1152/ajplung.00156.2015

Donovan C, Seow HJ, Bourke JE, Vlahos R (2016) Influenza A virus infection and cigarette smoke impair bronchodilator responsiveness to beta-adrenoceptor agonists in mouse lung. Clin Sci (Lond) 130:829–837. https://doi.org/10.1042/cs20160093

Dwivedi AM, Upadhyay S, Johanson G, Ernstgard L, Palmberg L (2017) Inflammatory effects of acrolein, crotonaldehyde and hexanal vapors on human primary bronchial epithelial cells cultured at air-liquid interface Toxicol. In Vitro. https://doi.org/10.1016/j.tiv.2017.09.016

Ebsen M, Mogilevski G, Anhenn O, Maiworm V, Theegarten D, Schwarze J, Morgenroth K (2002) Infection of murine precision cut lung slices (PCLS) with respiratory syncytial virus (RSV) and chlamydophila pneumoniae using the Krumdieck technique. Pathol Res Pract 198:747–753. https://doi.org/10.1078/0344-0338-00331

Ekert JE, Johnson K, Strake B, Pardinas J, Jarantow S, Perkinson R, Colter DC (2014) Three-dimensional lung tumor microenvironment modulates therapeutic compound responsiveness in vitro—implication for drug development. PLoS ONE 9:e92248. https://doi.org/10.1371/journal.pone.0092248

Endo H et al (2013) Spheroid culture of primary lung cancer cells with neuregulin 1/HER3 pathway activation. J Thoracic Oncol Official Publ Int Assoc Study Lung Cancer 8:131–139. https://doi.org/10.1097/JTO.0b013e3182779ccf

Eramo A et al (2007) Identification and expansion of the tumorigenic lung cancer stem cell population. Cell Death Differ 15:504–514

Esch EW, Bahinski A, Huh D (2015) Organs-on-chips at the frontiers of drug discovery. Nat Rev Drug Discov 14:248–260. https://doi.org/10.1038/nrd4539

Fatehullah A, Tan SH, Barker N (2016) Organoids as an in vitro model of human development and disease. Nat Cell Biol 18:246–254. https://doi.org/10.1038/ncb3312

Ferlay J, Soerjomataram I, Ervik M, Dikshit R, Eser S, Mathers C, Rebelo M, Parkin DM, Forman D, Bray, F (2013) GLOBOCAN 2012 v1.0, cancer incidence and mortality worldwide: IARC CancerBase No. 11. International Agency for Research on Cancer. http://globocan.iarc.fr . Accessed 04/09/2017

Ferlay J et al (2015) Cancer incidence and mortality worldwide: sources, methods and major patterns in GLOBOCAN 2012. Int J Cancer 136:E359–386. https://doi.org/10.1002/ijc.29210

Fessart D, Begueret H, Delom F (2013) Three-dimensional culture model to distinguish normal from malignant human bronchial epithelial cells. Eur Respir J 42:1345–1356. https://doi.org/10.1183/09031936.00118812

Geles KG, Zhong W, O'Brien SK, Baxter M, Loreth C, Pallares D, Damelin M (2016) Upregulation of RNA processing factors in poorly differentiated lung cancer cells. Transl Oncol 9:89–98. https://doi.org/10.1016/j.tranon.2016.01.006

Gray TE, Guzman K, Davis CW, Abdullah LH, Nettesheim P (1996) Mucociliary differentiation of serially passaged normal human tracheobronchial epithelial cells. Am J Respir Cell Mol Biol 14:104–112. https://doi.org/10.1165/ajrcmb.14.1.8534481

Gridelli C et al (2015) Non-small-cell lung cancer. Nat Rev Dis Primers 1:15009. https://doi.org/10.1038/nrdp.2015.9

Hackett TL, Holloway R, Holgate ST, Warner JA (2008) Dynamics of pro-inflammatory and anti-inflammatory cytokine release during acute inflammation in chronic obstructive pulmonary disease: an ex vivo study. Respir Res 9:47. https://doi.org/10.1186/1465-9921-9-47

Harrigan JA, Vezina CM, McGarrigle BP, Ersing N, Box HC, Maccubbin AE, Olson JR (2004) DNA adduct formation in precision-cut rat liver and lung slices exposed to benzo[a]pyrene. Toxicol Sci 77:307–314. https://doi.org/10.1093/toxsci/kfh030

Harrison RK (2016) Phase II and phase III failures: 2013–2015. Nat Rev Drug Discov 15:817–818. https://doi.org/10.1038/nrd.2016.184

Henjakovic M et al (2008a) Ex vivo lung function measurements in precision-cut lung slices (PCLS) from chemical allergen-sensitized mice represent a suitable alternative to in vivo studies. Toxicol Sci 106:444–453. https://doi.org/10.1093/toxsci/kfn178

Henjakovic M et al (2008b) Ex vivo testing of immune responses in precision-cut lung slices. Toxicol Appl Pharmacol 231:68–76. https://doi.org/10.1016/j.taap.2008.04.003

Henry E et al (2015) Adult lung spheroid cells contain progenitor cells and mediate regeneration in rodents with bleomycin-induced pulmonary fibrosis. Stem Cells Transl Med 4:1265–1274. https://doi.org/10.5966/sctm.2015-0062

Herbert J, Thiermann H, Worek F, Wille T (2017) Precision cut lung slices as test system for candidate therapeutics in organophosphate poisoning. Toxicology 389:94–100. https://doi.org/10.1016/j.tox.2017.07.011

Hess A et al (2016) Prevalidation of the ex-vivo model PCLS for prediction of respiratory toxicity. Toxicol In Vitro 32:347–361. https://doi.org/10.1016/j.tiv.2016.01.006

Hofmann F, Blasche R, Kasper M, Barth K (2015) A co-culture system with an organotypic lung slice and an immortal alveolar macrophage cell line to quantify silica-induced inflammation. PLoS ONE 10:e0117056. https://doi.org/10.1371/journal.pone.0117056

Horvath L, Umehara Y, Jud C, Blank F, Petri-Fink A, Rothen-Rutishauser B (2015) Engineering an in vitro air-blood barrier by 3D bioprinting. Scientific Reports 5:7974. https://doi.org/10.1038/srep07974

Horvath P et al (2016) Screening out irrelevant cell-based models of disease. Nat Rev Drug Discov 15:751–769. https://doi.org/10.1038/nrd.2016.175

Hsu T-H, Xiao J-L, Tsao Y-W, Kao Y-L, Huang S-H, Liao W-Y, Lee C-H (2011) Analysis of the paracrine loop between cancer cells and fibroblasts using a microfluidic chip. Lab Chip 11:1808–1814. https://doi.org/10.1039/C1LC20090A

Huber JM et al (2016) Evaluation of assays for drug efficacy in a three-dimensional model of the lung. J Cancer Res Clin Oncol 142:1955–1966. https://doi.org/10.1007/s00432-016-2198-0

Huh D, Matthews BD, Mammoto A, Montoya-Zavala M, Hsin HY, Ingber DE (2010) Reconstituting organ-level lung functions on a chip. Science 328:1662–1668. https://doi.org/10.1126/science.1188302

Huh D et al (2012) A human disease model of drug toxicity-induced pulmonary edema in a lung-on-a-chip microdevice. Sci Transl Med 4:159ra147. https://doi.org/10.1126/scitranslmed.3004249

Jacquet M, Lambert V, Todaro A, Kremers P (1997) Mitogen-activated lymphocytes: a good model for characterising lung CYP1A1 inducibility. Eur J Epidemiol 13:177–183. https://doi.org/10.1023/a:1007354532547

Jin Woo L et al (2016) Development of a 3D cell printed construct considering angiogenesis for liver tissue engineering. Biofabrication 8:015007

Jung J (2014) Human tumor xenograft models for preclinical assessment of anticancer drug development. Toxicol Res 30:1–5. https://doi.org/10.5487/tr.2014.30.1.001

Karekla E et al (2017) Ex vivo explant cultures of non-small cell lung carcinoma enable evaluation of primary tumor responses to anticancer therapy. Cancer Res 77:2029–2039. https://doi.org/10.1158/0008-5472.can-16-1121

Kellar A, Egan C, Morris D (2015) Preclinical murine models for lung cancer: clinical trial applications. Biomed Res Int 2015:621324. https://doi.org/10.1155/2015/621324

Kim HJ et al (2015) Airway Smooth muscle sensitivity to methacholine in precision-cut lung slices (PCLS) from ovalbumin-induced asthmatic mice korean. J Physiol Pharmacol 19:65–71. https://doi.org/10.4196/kjpp.2015.19.1.65

Kistemaker LE et al (2017) The PDE4 inhibitor CHF-6001 and LAMAs inhibit bronchoconstriction-induced remodelling in lung slices. Am J Physiol Lung Cell Mol Physiol: ajplung 00069 02017. https://doi.org/10.1152/ajplung.00069.2017

Konishi S et al (2016) Directed induction of functional multi-ciliated cells in proximal airway epithelial spheroids from human pluripotent stem cells. Stem Cell Reports 6:18–25. https://doi.org/10.1016/j.stemcr.2015.11.010

Koziol-White CJ et al (2016a) Inhibition of spleen tyrosine kinase attenuates IgE-mediated airway contraction and mediator release in human precision cut lung slices. Br J Pharmacol 173:3080–3087. https://doi.org/10.1111/bph.13550

Koziol-White CJ et al (2016b) Inhibition of PI3 K promotes dilation of human small airways in a rho kinase-dependent manner. Br J Pharmacol 173:2726–2738. https://doi.org/10.1111/bph.13542

Kumar PA et al (2011) Distal airway stem cells yield alveoli in vitro and during lung regeneration following H1N1 influenza infection. Cell 147:525–538. https://doi.org/10.1016/j.cell.2011.10.001

Lama R, Zhang L, Naim JM, Williams J, Zhou A, Su B (2013) Development, validation and pilot screening of an in vitro multi-cellular three-dimensional cancer spheroid assay for anti-cancer drug testing. Bioorg Med Chem 21:922–931. https://doi.org/10.1016/j.bmc.2012.12.007

Lauenstein L et al (2014) Assessment of immunotoxicity induced by chemicals in human precision-cut lung slices (PCLS). Toxicol In Vitro 28:588–599. https://doi.org/10.1016/j.tiv.2013.12.016

Lenz AG et al (2014) Efficient bioactive delivery of aerosolized drugs to human pulmonary epithelial cells cultured in air-liquid interface conditions. Am J Respir Cell Mol Biol 51:526–535. https://doi.org/10.1165/rcmb.2013-0479OC

Leung AWY et al (2017) In vivo validation of PAPSS1 (3'-phosphoadenosine 5'-phosphosulfate synthase 1) as a Cisplatin-Sensitizing Therapeutic Target. Clin Cancer Res. https://doi.org/10.1158/1078-0432.ccr-17-0700

Martin C, Uhlig S, Ullrich V (2001) Cytokine-induced bronchoconstriction in precision-cut lung slices is dependent upon cyclooxygenase-2 and thromboxane receptor activation. Am J Respir Cell Mol Biol 24:139–145. https://doi.org/10.1165/ajrcmb.24.2.3545

Mas C, Boda B, CaulFuty M, Huang S, Wiszniewski L, Constant S (2015) Antitumour efficacy of the selumetinib and trametinib MEK inhibitors in a combined human airway-tumour-stroma lung cancer model. J Biotechnol 205:111–119. https://doi.org/10.1016/j.jbiotec.2015.01.012

Mathis C et al (2013) Human bronchial epithelial cells exposed in vitro to cigarette smoke at the air-liquid interface resemble bronchial epithelium from human smokers. Am J Physiol Lung Cell Mol Physiol 304:L489–L503. https://doi.org/10.1152/ajplung.00181.2012

Meenach SA, Tsoras AN, McGarry RC, Mansour HM, Hilt JZ, Anderson KW (2016) Development of three-dimensional lung multicellular spheroids in air- and liquid-interface culture for the evaluation of anticancer therapeutics. Int J Oncol 48:1701–1709. https://doi.org/10.3892/ijo.2016.3376

Morgan MM, Johnson BP, Livingston MK, Schuler LA, Alarid ET, Sung KE, Beebe DJ (2016) Personalized in vitro cancer models to predict therapeutic response: challenges and a framework for improvement. Pharmacol Ther 165:79–92. https://doi.org/10.1016/j.pharmthera.2016.05.007

Morgan KM, Riedlinger GM, Rosenfeld J, Ganesan S, Pine SR (2017) Patient-Derived Xenograft Models of Non-Small Cell Lung Cancer and Their Potential Utility in Personalized Medicine. Front Oncol 7:2. https://doi.org/10.3389/fonc.2017.00002

Morin JP, Baste JM, Gay A, Crochemore C, Corbiere C, Monteil C (2013) Precision cut lung slices as an efficient tool for in vitro lung physio-pharmacotoxicology studies. Xenobiotica 43:63–72. https://doi.org/10.3109/00498254.2012.727043

Muller-Redetzky HC et al (2012) Intermedin stabilized endothelial barrier function and attenuated ventilator-induced lung injury in mice. PLoS ONE 7:e35832. https://doi.org/10.1371/journal.pone.0035832

Munoz-Abraham AS, Rodriguez-Davalos MI, Bertacco A, Wengerter B, Geibel JP, Mulligan DC (2016) 3D printing of organs for transplantation: where are we and where are we heading? Curr Transplant R 3:93–99. https://doi.org/10.1007/s40472-016-0089-6

Murphy SV, Atala A (2014) 3D bioprinting of tissues and organs. Nat Biotechnol 32:773–785. https://doi.org/10.1038/nbt.2958

Nagarsheth N, Wicha MS, Zou W (2017) Chemokines in the cancer microenvironment and their relevance in cancer immunotherapy. Nat Rev Immunol 17:559–572. https://doi.org/10.1038/nri.2017.49

Neuhaus V et al (2013) Functional testing of an inhalable nanoparticle based influenza vaccine using a human precision cut lung slice technique. PLoS ONE 8:e71728. https://doi.org/10.1371/journal.pone.0071728

Neuhaus V et al (2017) Assessment of long-term cultivated human precision-cut lung slices as an ex vivo system for evaluation of chronic cytotoxicity and functionality. J Occup Med Toxicol 12:13. https://doi.org/10.1186/s12995-017-0158-5

Nikolić MZ, Rawlins EL (2017) Lung organoids and their use to study cell-cell interaction. Curr Pathobiol Rep 5:223–231. https://doi.org/10.1007/s40139-017-0137-7

Onion D et al (2016) 3-dimensional patient-derived lung cancer assays reveal resistance to standards-of-care promoted by stromal cells but sensitivity to histone deacetylase inhibitors. Mol Cancer Ther 15:753–763. https://doi.org/10.1158/1535-7163.mct-15-0598

Pezzulo AA et al (2011) The air-liquid interface and use of primary cell cultures are important to recapitulate the transcriptional profile of in vivo airway epithelia. Am J Physiol Lung Cell Mol Physiol 300:L25–L31. https://doi.org/10.1152/ajplung.00256.2010

Pisanu ME et al (2017) Blockade of Stearoyl-CoA-desaturase 1 activity reverts resistance to cisplatin in lung cancer stem cells. Cancer Lett 406:93–104. https://doi.org/10.1016/j.canlet.2017.07.027

Pomerenke A, Lea SR, Herrick S, Lindsay MA, Singh D (2016) Characterization of TLR-induced inflammatory responses in COPD and control lung tissue explants. Int J Chron Obstruct Pulmon Dis 11:2409–2417. https://doi.org/10.2147/copd.s105156

Popper H (2017a) Lung tumors. In: Pathology of lung disease: morphology—pathogenesis—etiology. Springer, Berlin, pp 353–575. https://doi.org/10.1007/978-3-662-50491-8_17

Popper H (2017b) Normal lung. In: Pathology of lung disease: morphology—pathogenesis—etiology. Springer Berlin, pp 7–20. https://doi.org/10.1007/978-3-662-50491-8_2

Razian G, Yu Y, Ungrin M (2013) Production of large numbers of size-controlled tumor spheroids using microwell plates. J Visualized Exp: JoVE:50665. https://doi.org/10.3791/50665

Ressmeyer AR, Larsson AK, Vollmer E, Dahlen SE, Uhlig S, Martin C (2006) Characterisation of guinea pig precision-cut lung slices: comparison with human tissues. Eur Respir J 28:603–611. https://doi.org/10.1183/09031936.06.00004206

Rieg AD et al (2014) Milrinone relaxes pulmonary veins in guinea pigs and humans. PLoS ONE 9: e87685. https://doi.org/10.1371/journal.pone.0087685

Rimington TL et al (2017) Defining the inflammatory signature of human lung explant tissue in the presence and absence of glucocorticoid. F1000Res 6:460. https://doi.org/10.12688/f1000research.10961.1

Rock JR, Gao X, Xue Y, Randell SH, Kong Y-Y, Hogan BLM (2011) Notch-dependent differentiation of adult airway basal stem cells. Cell Stem Cell 8:639–648. https://doi.org/10.1016/j.stem.2011.04.003

Russo MV et al (2015) A new mouse avatar model of non-small cell lung cancer Front. Oncol 5:52. https://doi.org/10.3389/fonc.2015.00052

Salar-Behzadi S, Wu S, Mercuri A, Meindl C, Stranzinger S, Frohlich E (2017) Effect of the pulmonary deposition and in vitro permeability on the prediction of plasma levels of inhaled budesonide formulation. Int J Pharm 532:337–344. https://doi.org/10.1016/j.ijpharm.2017.08.124

Sato M et al (2006) Multiple oncogenic changes (K-RASV12, p53 Knockdown, Mutant EGFRs, p16 Bypass, Telomerase) are not sufficient to confer a full malignant phenotype on human bronchial epithelial cells. Can Res 66:2116–2128. https://doi.org/10.1158/0008-5472.can-05-2521

Schamberger AC, Staab-Weijnitz CA, Mise-Racek N, Eickelberg O (2015) Cigarette smoke alters primary human bronchial epithelial cell differentiation at the air-liquid interface. Sci Rep 5:8163. https://doi.org/10.1038/srep08163

Schleputz M et al (2012) Neurally mediated airway constriction in human and other species: a comparative study using precision-cut lung slices (PCLS). PLoS ONE 7:e47344. https://doi.org/10.1371/journal.pone.0047344

Seehase S et al (2012) LPS-induced lung inflammation in marmoset monkeys—an acute model for anti-inflammatory drug testing. PLoS ONE 7:e43709. https://doi.org/10.1371/journal.pone.0043709

Shamir ER, Ewald AJ (2014) Three-dimensional organotypic culture: experimental models of mammalian biology and disease. Nat Rev Mol Cell Biol 15:647–664. https://doi.org/10.1038/nrm3873

Switalla S et al (2010) Natural innate cytokine response to immunomodulators and adjuvants in human precision-cut lung slices. Toxicol Appl Pharmacol 246:107–115. https://doi.org/10.1016/j.taap.2010.04.010

Temann A, Golovina T, Neuhaus V, Thompson C, Chichester JA, Braun A, Yusibov V (2017) Evaluation of inflammatory and immune responses in long-term cultured human precision-cut lung slices. Hum Vaccin Immunother 13:351–358. https://doi.org/10.1080/21645515.2017.1264794

van den Bosch T et al (2017) A 6-alkylsalicylate histone acetyltransferase inhibitor inhibits histone acetylation and pro-inflammatory gene expression in murine precision-cut lung slices. Pulm Pharmacol Ther 44:88–95. https://doi.org/10.1016/j.pupt.2017.03.006

Van Dijk EM, Culha S, Menzen MH, Bidan CM, Gosens R (2016) Elastase-induced parenchymal disruption and airway hyper responsiveness in mouse precision cut lung slices: toward an ex vivo COPD Model. Front Physiol 7:657. https://doi.org/10.3389/fphys.2016.00657

Vicary GW, Vergne Y, Santiago-Cornier A, Young LR, Roman J (2016) Pulmonary Fibrosis in Hermansky-Pudlak Syndrome. Ann Am Thorac Soc 13:1839–1846. https://doi.org/10.1513/AnnalsATS.201603-186FR

Waypa GB et al (2013) Superoxide generated at mitochondrial complex III triggers acute responses to hypoxia in the pulmonary circulation. Am J Respir Crit Care Med 187:424–432. https://doi.org/10.1164/rccm.201207-1294OC

Weiswald L-B, Bellet D, Dangles-Marie V (2015) Spherical cancer models in tumor biology. Neoplasia 17:1–15. https://doi.org/10.1016/j.neo.2014.12.004

Wijeweera JB, Gandolfi AJ, Parrish A, Lantz RC (2001) Sodium arsenite enhances AP-1 and NFkappaB DNA binding and induces stress protein expression in precision-cut rat lung slices. Toxicol Sci 61:283–294

Wohlsen A, Uhlig S, Martin C (2001) Immediate allergic response in small airways. Am J Respir Crit Care Med 163:1462–1469. https://doi.org/10.1164/ajrccm.163.6.2007138

Wohlsen A et al (2003) The early allergic response in small airways of human precision-cut lung slices. Eur Respir J 21:1024–1032

Wright JL et al (2011) Statin reverses smoke-induced pulmonary hypertension and prevents emphysema but not airway remodeling. Am J Respir Crit Care Med 183:50–58. https://doi.org/10.1164/rccm.201003-0399OC

Xu Z et al (2013) Application of a microfluidic chip-based 3D co-culture to test drug sensitivity for individualized treatment of lung cancer Biomaterials 34:4109–4117. https://doi.org/10.1016/j.biomaterials.2013.02.045

Zanoni M et al (2016) 3D tumor spheroid models for in vitro therapeutic screening: a systematic approach to enhance the biological relevance of data obtained. Scientific Reports 6:19103. https://doi.org/10.1038/srep19103

Pancreatic Cancer Organotypic Models

Abigail Coetzee, Richard Grose and Hemant Kocher

Contents

Abstract Pancreatic cancer has a dismal prognosis for patients due to late diagnosis and ineffective treatment options. There is a desperate need for more accurate disease models to enable improved therapies and diagnostic tests to reach the clinic. Pancreatic tumours have a high content of desmoplastic stroma, which forms a stiff, hypoxic tumour mass and contributes significantly to tumour development and metastatic spread. Therefore, 2D cell culture is not sufficient for understanding the complex biology of this disease. 3D in vitro models offer a more representative method of culturing cells for research in the laboratory. There are many different 3D models that can be used in research, organoids formed from patient or murine

---

A. Coetzee (✉) · R. Grose · H. Kocher
Barts Cancer Institute, London, UK
e-mail: a.s.coetzee@qmul.ac.uk

Current Topics in Microbiology and Immunology (2021) 430: 183–198
https://doi.org/10.1007/82_2019_155

Published Online: 22 February 2019

tumours are grown embedded in collagen or matrigel matrices, giving the potential for screening treatment options and personalised therapy in the future. Also, organotypic models using pancreatic cancer cell lines and stromal cells can be easily manipulated to study different aspects of pancreatic cancer and new therapeutic options in the laboratory. There are new emerging pancreatic cancer 3D models being developed, including microchip technology or synthetic scaffolds instead collagen and matrigel. All of these 3D culturing methods give an advantage over traditional 2D cell culture and could lead to improved understanding of this disease, translating to a better prognosis for patients in the clinic.

1 Introduction

1.1 Pancreatic Cancer

Pancreatic cancer has an extremely poor prognosis with less than 5% five-year survival rates. The most common form of pancreatic cancer is pancreatic ductal adenocarcinoma (PDAC), which is where most research is focused. PDAC patients tend to be diagnosed with either locally advanced or metastatic disease, meaning that curative surgery cannot be performed. Even patients with early-stage tumours that are treated with surgical resection can relapse with terminal metastatic disease (Siegel et al. 2018). There are many reasons for the poor outcomes of PDAC patients, such as late presentation of symptoms and difficulty in diagnosing this tumour. There is currently no effective early detection or screening test that can be used in the clinic. Also, PDAC tumours have a highly desmoplastic stroma, which forms a stiff, hypoxic tumour mass, rendering chemo- and radio-therapy ineffective. This is in contrast to most tumours, which tend to undergo angiogenesis to facilitate tumour growth and spread (Whatcott et al. 2015; Nielsen et al. 2016).

1.2 The Importance of 3D Models

The biology of pancreatic cancer is complex, and this has impeded the development of effective therapies for patients. As with many tumours, the responses seen in 2D cell culture in the laboratory do not translate to human disease. Many PDAC cell lines have been isolated and are used routinely for research; however, 2D cell culture is not representative of PDAC tumours in the clinic. Due to the nature of the disease, most of these cell lines are from metastatic tumours and therefore have a bias towards late-stage tumour growth (Deer et al. 2010). Moreover, cells passaged on plastic and grown over a long time can lead to genetic drift and selection of aggressive clones within the population. Also, monoculture of PDAC cancer cell lines excludes the interaction of key stromal cells, which can make up to 80% of the tumour mass (Neesse et al. 2011). It is well known that stromal cells enter into essential cross talk with cancer cells triggering tumour growth and invasion; therefore, it is necessary to consider these cells within research into the disease.

Genetically engineered mouse models (GEMMs) have been developed to mimic PDAC development in vivo. The most well-known PDAC GEMM model is the KPC mouse (LSL-Kras(G12D) (/+);LSL-Trp53(R172H) (/+);Pdx-1-Cre) (Olive and Tuveson 2006). However, these models tend to form multiple lesions within the pancreas, compared to the development of a single tumour within human patients. Also, most PDAC tumours form from spontaneous mutations, not germline mutations. As well as this, the development and maintenance of GEMM models is very expensive. Therefore, even though GEMMs have helped to further our knowledge of pancreatic cancer, there are still limitations to these models that must be considered (Bardeesy and DePinho 2002; Pérez–Mancera et al. 2012).

The most common cancer precursor lesions found in patients are pancreatic intraepithelial neoplasias (PanINs). We still do not fully understand the pathway from PanIN lesions to pancreatic cancer (Distler et al. 2014). The heterogeneous nature of the disease origin makes it difficult to develop appropriate models. 3D in vitro models give us a good chance to study pancreatic cancer in a more relevant environment, which could lead to improved understanding and therapeutic developments.

2 Pancreatic Organoids

2.1 Overview

Pancreatic ductal cells can now be cultured in a 3D matrix forming organoids. This allows PDAC cell lines to be grown in a more physiological environment (3D culture) or can be used to culture cells taken directly from patients or mice (3D organoids). In this case, the source tissue is digested mechanically and enzymatically. The small pieces of tissue are then embedded into a 3D matrix and supplemented with various growth factors. The epithelial cells form spheres embedded in the 3D collagen or matrigel matrix. There are a number of published methods of growing pancreatic organoids (Moreira et al. 2018; Baker et al. 2016; Boj et al. 2015; Huang et al. 2015; Walsh et al. 2016; Tsai et al. 2018; Li et al. 2014; Ootani et al. 2009).

2.2 Epithelial Organoids

A pancreatic cancer organoid model method that has been extremely successful was developed by Boj et al. This model uses PDAC cells from either murine or human pancreas and embeds the digested cells in a matrigel matrix (Fig. 1a). Many growth factors are then added to the culture to simulate the stromal signals, such as

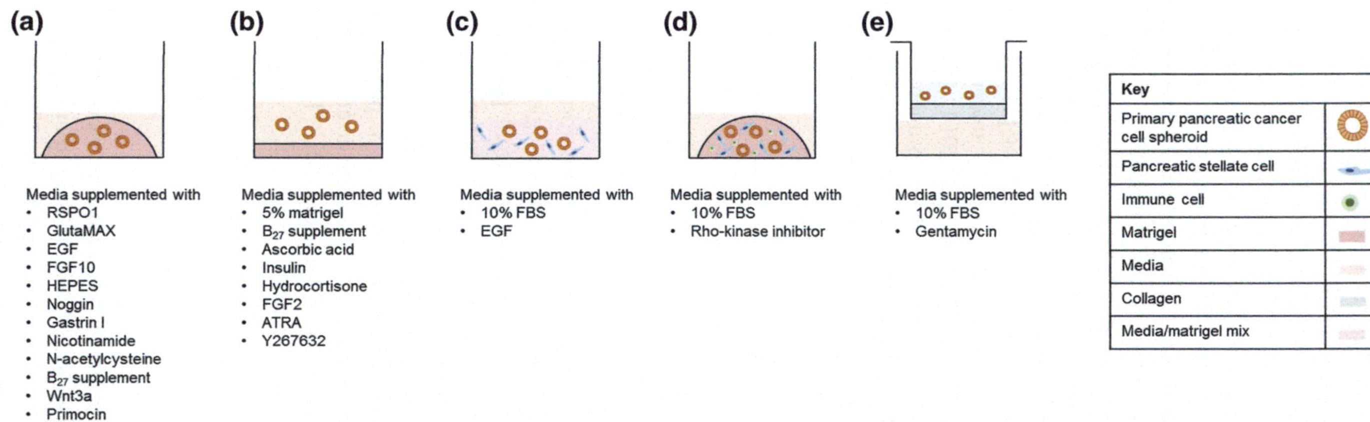

Fig. 1 Pancreatic cancer organoid methods. Pancreatic tumours from a human or murine pancreas are broken down into their cellular components. The cancer cells can then be embedded into a pure matrigel dome (**a**) with growth factors added to the media on top to stimulate cell growth. Another method is to culture these primary cells in media on top of a matrigel layer (**b**). The media must also be supplemented with many growth factors to stimulate cellular survival and growth. Alternatively, cancer cell spheroids can be co-cultured with stromal cells in a gel made up of 50% matrigel and 50% media, and drugs can be added to media on top of the gel (**c**). To make the organoid model more complex, immune cells can be added to stellate cells and cancer cell spheroids in a matrigel dome. Media can be added to the top of this dome, but less growth factors are needed due to the presence of stromal cells in the gels (**d**). Another method to culture cancer spheroids is to place them into collagen in a transwell insert. This gel can then be fed with media from below to create an air–liquid interface (**e**)

R-spondin-1 (RSPO1), GlutaMAX, epidermal growth factor (EGF), fibroblast growth factor 10 (FGF10), HEPES, Noggin, Gastrin I, nicotinamide, N-acetylcysteine, B_{27} supplement, Wnt3a and Primocin. These organoid cultures have been shown to recapitulate normal and tumour-associated pancreatic tissue and can also be implanted orthotopically into mice, creating PanIN and PDAC lesions. One major advantage of these organoids is that they can be cryogenically frozen and thawed, allowing continued study from a small original sample (Boj et al. 2015).

Another organoid method that has been developed uses resected patient PDAC tissue to isolate primary cells, which are then cultured in media on top of a matrigel layer (Fig. 1b). The media is supplemented with 5% matrigel, B_{27}, ascorbic acid, insulin, hydrocortisone, FGF2, all-trans retinoic acid (ATRA) and Y267632. Drug sensitivity was tested in these organoids with an EZH2 inhibitor. Response to the drug correlated with H3K27me3 in the organoids and matched patient tissue. Therefore, the organoids successfully mimicked the primary tumour in vitro. This could allow clinicians to screen potential therapies on a specific patient's tumour organoid to predict response and allow tailored personalised treatment options.

To avoid the need for fresh tissue, organoids can be established by inducing the differentiation of pluripotent stem cells into pancreatic exocrine cells (Huang et al. 2015). This is useful as it allows researchers to model pancreatic ducts without the difficulties of collecting and processing fresh tissue. Similar research has been carried out to develop models that can be used to study the endocrine pancreas, particularly for investigating diabetes (Sugiyama et al. 2013). The development of an in vitro method of pancreatic differentiation from progenitor cells makes it possible to understand the pathology of many diseases arising from both the exocrine and endocrine pancreatic compartments.

2.3 Co-culture

A benefit of using patient-derived organoids for research is that many organoids can be created from a small initial piece of tissue, such as a biopsy. Also these models have been shown to represent patient tumours, therefore making them invaluable for drug screening and investigation into the development of PDAC. One of the disadvantages of these models is that they can be expensive to set up, especially as lots of different growth factors must be added to the media. Many of these cultures only consider the epithelial compartment of the tumour, and therefore, the important signals from the stroma must be added in. A simpler organoid culture method of seeding fibroblasts and tumour cells together (1:2 respectively) in a matrigel/media mix could help to overcome this (Fig. 1c). Due to the presence of fibroblasts, the culture media only needs to be supplemented with 10% foetal bovine serum (FBS) and EGF (Walsh et al. 2016). Another advantage of these organoids is that they can be easily quantified using an optical imaging method to screen for metabolic changes within the cultures following drug treatment.

Complex multicellular organoid cultures derived from tumour resection, ascites or rapid autopsy have also been established. The initial tumour samples were dissected into the cellular components and cultured in matrigel domes (Fig. 1d). The tumour cells formed spheres in the matrigel, mimicking PDAC tumours, surrounded by cancer-associated fibroblasts to simulate the stroma. Human T-cells were added to the co-cultures and effectively mimicked infiltrating immune cells. This demonstrates that the classical tumour organoid model can be modulated to include stromal cells, allowing complex tumour-microenvironment interactions to be studied (Tsai et al. 2018).

The air–liquid interface model can also be used to grow pancreatic organoids using multiple cell types. In this protocol, pancreatic tissue is minced and segments are placed on top of a collagen gel (Fig. 1e). The media supplied to these models is replaced weekly. Following induction of mutant *Kras* or *p53*, the models reflect disease morphology within the pancreas, making this a useful way to study PDAC (Li et al. 2014; Ootani et al. 2009).

3 Pancreatic Organotypics

3.1 Overview

Pancreatic cancer organotypics are 3D in vitro models, which allow for interactions between different cells within the tumour niche to be investigated. The isolation and immortalisation of human pancreatic stellate cells, which are one of the major cells found within the stroma of PDAC tumours, has allowed cancer and stromal interactions to be explored (Froeling et al. 2009). Organotypic cultures provide a flexible 3D assay to study specific aspects of PDAC development and progression. An extracellular matrix (ECM) gel made up of collagen, matrigel, media and FBS can be used as the scaffold for these cultures.

3.2 PDAC Development Models

Pancreatic cancer cells in monoculture or in co-culture with stellate cells can be embedded within an organotypic gel to represent cells that have invaded into the surrounding pancreas. The cancer cells will form ducts in the gel, and the cultures can be fed with media placed on top of the gel (Fig. 2a). An alternative form of this organotypic is to place the stellate cells within the ECM gel, allow it to set and then seed the cancer cells on top. These gels can also be treated with media placed on top of the gel. This allows the researcher to investigate early-stage PDAC, and it has been shown that in the presence of stellate cells, Ezrin is translocated to the basal surface of the cancer cells, reflecting early events in PDAC invasion (Froeling et al. 2009).

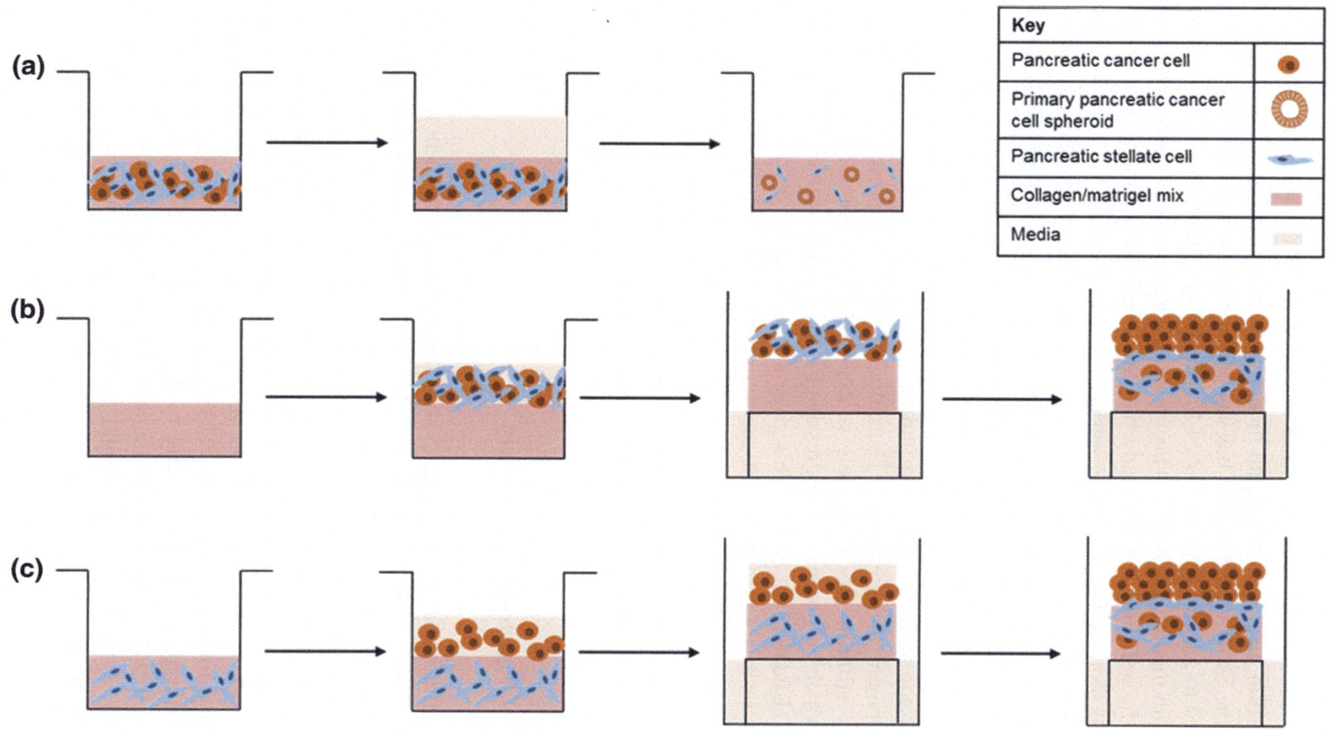

Fig. 2 Pancreatic cancer organotypic method. A mix of collagen and matrigel is set to form the organotypic gel. Cells can then either be seeded within the gel in mono- or co-culture (**a**) in co-culture on top of the gel (**b**) or the stellate cells can be pre-seeded into the organotypic gel, and the cancer cells can be placed on top of the gel (**c**). The organotypic gels can then either be fed with media on top or can be raised onto a metal grid and fed with media from below, creating an air–liquid interface. At the end of the experiment, the gels can be harvested and analysed

3.3 PDAC Invasion Models

Organotypic models have been modified to easily study the role of invasion in PDAC progression, using the air–liquid model system. Pancreatic stellate cells are seeded inside the ECM gel. Once this has set, pancreatic cancer cells are seeded on top and allowed to attach to the gel for 24 h. The gels are then placed onto a metal grid and fed with media from below. This creates a chemotactic gradient, triggering invasion into the gel (Fig. 2b). To study the direct cell–cell interaction of stellate and cancer cells, both cell types can be seeded on top of these ECM gels rather than inside. The gel is then raised onto the metal grid and fed from below as before (Fig. 2c). It is well established that stellate cells lead the way, allowing cancer cells to follow, during invasion in PDAC tumours (Koikawa et al. 2018; Apte et al. 2013). Therefore, this model more accurately represents this process (Coleman et al. 2014).

Stiffness of the tumour microenvironment can have a significant impact on PDAC development and treatment. Therefore, creating a representative desmoplastic stromal stiffness in 3D models is essential. The ratio of cells in these organotypics which best represents aggressive PDAC was determined as 2:1 stellate to cancer cells (Kadaba et al. 2013). This gives a representative model of pancreatic cancer that can be easily manipulated to answer specific questions in the laboratory. Organotypics can be fixed in formalin after culture and sections stained by H&E to look at invasion. Sections can also be stained by immunohistochemistry or immunofluorescence for other markers, such as cleaved caspase 3 to look at apoptosis or Ki67 to study proliferation (Coleman et al. 2014).

3.4 PDAC Organotypic Analysis

Organotypic cultures can be used to study the effects of drug treatment regimens on PDAC growth and invasion. Drugs can be added to the media below the cultures as often as necessary, and these will diffuse through the gel. The 3D gels can be left for 10 days to establish invasion before beginning treatment to mimic advanced PDAC. This is the stage that most patients are diagnosed, meaning the models are representative of human disease. Also, results from these assays have been reproduced in the KPC mouse model. This highlights the relevance of organotypic cultures as a physiomimetic model of PDAC (Carapuça et al. 2016).

PDAC organotypics provide a useful model to study disease progression and investigate new treatment options. The model can be easily manipulated to focus on specific aspects of the tumour and could be developed further to include other stromal cells, as well as stellate cells.

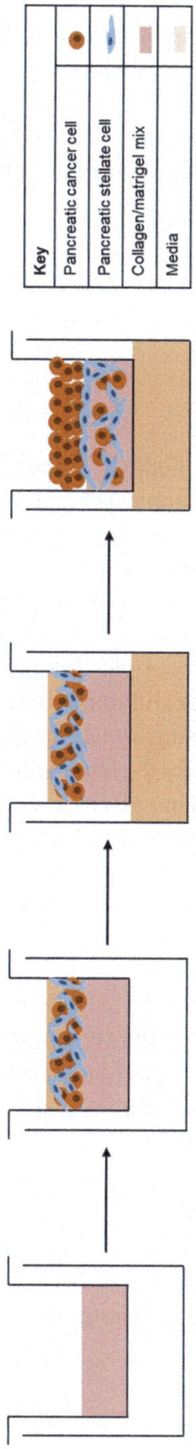

Fig. 3 Pancreatic cancer mini-organotypic method. A mix of collagen and matrigel is set to form the organotypic gel within a transwell insert. Cells can then be seeded in co-culture on top of the gel and fed with media from below, creating an air–liquid interface. At the end of the experiment, the gels can be harvested and analysed

3.5 Mini-organotypics

A smaller version of the traditional organotypic (mini-organotypic) has been developed. This involves placing gels into transwell inserts in a 24-well plate rather than on metal grids. These gels are much smaller than the original 3D models and therefore are ideal for use with primary pancreatic cells. The gels can also be fed with media from below, maintaining the air–liquid interface and promoting cellular invasion (Fig. 3) (Coleman et al. 2014).

4 Other Models

4.1 Spheroids

One of the drawbacks of many 3D culturing methods is that it is difficult to scale them up to a high-throughput assay. A more simple 3D culture method is to create spheroids of pancreatic cancer cells in a hanging drop of methylcellulose. This is inert and therefore exerts no ECM pressure on to the cells; it simply helps the cells to aggregate. The pancreatic cells can then form their own matrix, which recapitulates the situation in vivo, without exerting any signalling bias. This method can be used to easily screen drug responses in a more physiological setting, as well as examine molecular changes within cells under different conditions (Longati et al. 2013). Other models have been developed using synthetic 3D structures, such as hydrogels made of chitosan or chitosan and poly γ-glutamic acid, that allow PDAC cancer cells to attach and form 3D cultures. This gives an easy method to analyse tumour cell behaviour in a 3D in vitro environment (Chiellini et al. 2016).

4.2 Extracellular Matrix Models

One of the reasons behind the poor prognosis of PDAC patients is that tumours are often diagnosed at a late stage of disease, after the tumour has metastasised. Even patients that undergo surgery will often relapse with metastatic disease. Therefore, it is important to be able to effectively study PDAC invasion and metastasis in the laboratory. To correctly mimic this in 3D in vitro models, it is critical to have the correct ECM composition, desmoplastic reaction and stiffness. Synthetic hydrogels have been developed to study cellular interactions in ovarian and prostate cancer using modifiable 3D platforms (Loessner et al. 2014, 2016). A similar model that has been developed to investigate metastasis uses oligomer as a standard type 1 collagen, along with matrigel, to re-create tumour–stroma interactions that drive EMT within PDAC cells. By varying the concentrations of collagen in these models, different ECM stiffness can be achieved and therefore the PDAC tumour environment can be established (Puls et al. 2017).

4.3 Microfluidics

Another approach to 3D models that has recently been developed for use in cancer is using microfluidics and 'organ-on-chip' technology. A study using PDAC cell cultures grown on collagen-coated HepaChip® chambers showed that they resembled 3D cultures grown on agarose-coated plates. Cells were assembled in the chip using dielectrophoretic forces, and medium was perfused through for up to 9 days of culture. The advantage of using a microfluidic chip to culture PDAC cells is that they can be established quickly and can mimic the perfusion rates of patient tumours. This makes it easy to screen responses to different compounds, allowing the potential for a personalised therapy approach to treating patients in the clinic (Beer et al. 2017).

4.4 Tumour Slice Models

To understand the complex role of the stroma and immune system in pancreatic cancer, pancreatic tumour slice models can be used. One method for the growth of tumour slices is to transfer fresh resection specimens from patients to the laboratory in cold media (containing FBS and penicillin/streptomycin). The tumour is then placed into an extracellular salt solution, for example containing NaCl, KCl, $NaHCO_3$, NaH_2PO_4, $MgCl_2$, $CaCl_2$, glucose and HEPES (pH 7.4). The tumour is sliced in sterile conditions and placed onto membranes, which can be pre-coated with collagen. The tumour slices can be grown in media supplemented with FBS, GlutaMAX, $NaHCO_3$, HEPES, L-cystine and antibiotics. Tumour slices have been shown to maintain viability and structure in these ex vivo cultures for up to 6 days. This could be a useful method to screen personalised therapies for patients or to examine the stromal features of PDAC tumours in more detail, such as identifying the immune profile and allowing improved therapeutic targeting (Jiang et al. 2017; Vaira et al. 2010).

4.5 Mini-pancreas

In the development of human or murine pancreas in vitro models, it is important to consider the signals received from the mesenchyme, which allows both the endocrine and exocrine compartments to be established. Complex pancreatic organoids have been cultured from progenitor cells induced to create a pancreas in a dish. The cells were grown in matrigel, supplemented with FGF10, heparin, R-spondin-1, EGF and the ROCK inhibitor Y-27632. Progenitor cells differentiated and rearranged into pancreatic structures when exposed to these factors, particularly after inducing FGF and Notch signalling. The organoids mimicked pancreas development as the cells formed a polarised ductal network with acinar cells at the tips and endocrine cells in a central compartment (Greggio et al. 2013). A 3D model of a pancreas grown in vitro is an invaluable tool for studying PDAC development and treatment of pancreatic disease. These models can be taken even further to consider the roles of other cells, such as endothelial, immune and neuronal cells (Scavuzzo et al. 2017).

5 Conclusion

Pancreatic cancer 3D in vitro models provide a useful tool for studying this disease in a more relevant setting. Organoids and organotypics can be manipulated to study different aspects of pancreatic cancer development and progression. Patient-derived organoids also provide the potential for drug screening and personalised therapy to be used in the clinic. It is important to consider not only the role of stromal cells but also the ECM stiffness in PDAC development and treatment. The rapid growth of many different methods of studying pancreatic cancer in 3D models shows the relevance of this work to enable future research to be carried out (Table 1). Hopefully, these more representative in vitro models will translate into improved patient prognosis in the clinic.

Table 1 Comparison of pancreatic cancer 3D models

Models	Cancer cell organoids	Co-culture organoids	Organotypics	Spheroids	Microfluidics	Tumour slices	ECM models	Mini-pancreas
Summary	Cancer cells derived from patient tumours can be grown either embedded in or on top of matrigel or collagen to form organoid cultures	Cancer cells derived from patient samples can be cultured in or on top of matrigel with other stromal cells	Cancer cells are grown in or on top of a collagen/ matrigel mix. An air–liquid interface can be used to study invasion	Cells are grown in methylcellulose droplets, allowing them to form spheroids	Cells are assembled on a Chip in collagen-coated chambers, and these can be perfused at a similar rate to a pancreatic tumour	Tumours from patients are transferred to the laboratory and cut into slices. These are grown on top of collagen and covered with media for ex vivo tumour culture	Cancer development in tumours with different ECM stiffness can be modelled by using different oligomer collagen mixes	Progenitor cells are induced to develop into endocrine and exocrine cells, creating a pancreas in a dish
Advantage	Cells are derived from human patients, and therefore are representative of the tumour	Less factors need to be added due to the presence of stromal cells, and this is more representative of the whole tumour	Easy to set up and analyse. Can be manipulated to study different aspects of PDAC	Easy to set up and analyse. Reproducible due to no ECM—allows cells to interact on their own	Quick to establish cultures, could be used to screen different drugs quickly for personalised therapy	Includes all of the cells present in the tumour and allows the complex cellular interactions to take place	Easy to establish and manipulate these cultures. Also these are easy to analyse	This is a complex model that includes all of the structures within the pancreas
Disadvantage	It can be difficult to grow primary cells, and the cultures require a lot of extra growth factors to survive	Obtaining and processing patient tumours into all of the cell types can be difficult	Uses cells that have been in 2D culture. Air–liquid interface does not completely mimic the pancreas	Does not consider the stromal role in PDAC	Simple culture with no stromal cells included	Difficult to set up and maintain. Cultures can only be kept for 6 days, and it is complex to analyse	Simple model and does not assess the effect of stromal cell interactions	It difficult to set up and requires growth factors to induce differentiation into the different cell types

References

Apte MV, Wilson JS, Lugea A, Pandol SJ (2013) A starring role for stellate cells in the pancreatic cancer microenvironment. Gastroenterology 144:1210–1219

Baker LA, Tiriac H, Clevers H, Tuveson DA (2016) Modeling pancreatic cancer with organoids. Trends in cancer 2:176–190

Bardeesy N, Depinho RA (2002) Pancreatic cancer biology and genetics. Nat Rev Cancer 2:897–909

Beer M, Kuppalu N, Stefanini M, Becker H, Schulz I, Manoli S, Schuette J, Schmees C, Casazza A, Stelzle M, Arcangeli A (2017) A novel microfluidic 3D platform for culturing pancreatic ductal adenocarcinoma cells: comparison with in vitro cultures and in vivo xenografts. Sci Rep 7:1325

Boj SF, Hwang C-I, Baker LA, Chio IIC, Engle DD, Corbo V, Jager M, Ponz-Sarvise M, Tiriac H, Spector MS, Gracanin A, Oni T, Yu KH., van Boxtel R, Huch M, Rivera KD, Wilson JP, Feigin ME, Öhlund D, Handly-Santana A, Ardito-Abraham CM, Ludwig M, Elyada E, Alagesan B, Biffi G, Yordanov GN, Delcuze B, Creighton B, Wright K, Park Y, Morsink FHM, Molenaar IQ, Rinkes IHB, Cuppen E, Hao Y, Jin Y, Nijman IJ, Iacobuzio-Donahue C, Leach SD, Pappin DJ, Hammell M, Klimstra DS, Basturk O, Hruban RH, Offerhaus GJ, Vries RGJ, Clevers H, Tuveson DA (2015) Organoid models of human and mouse ductal pancreatic cancer. Cell 160, 324–338

Carapuça E F, Gemenetzidis E, Feig C, Bapiro TE, Williams MD, Wilson AS, Delvecchio FR, Arumugam P, Grose RP, Lemoine NR, Richards FM, Kocher HM (2016) Anti-stromal treatment together with chemotherapy targets multiple signalling pathways in pancreatic adenocarcinoma. The J Pathol

Chiellini F, Puppi D, Piras AM, Morelli A, Bartoli C, Migone C (2016) Modelling of pancreatic ductal adenocarcinoma in vitro with three-dimensional microstructured hydrogels. RSC Adv 6:54226–54235

Coleman SJ, Watt J, Arumugam P, Solaini L, Carapuca E, Ghallab M, Grose RP, Kocher HM (2014) Pancreatic cancer organotypics: High throughput, preclinical models for pharmacological agent evaluation. World J Gastroenterol 20:8471–8481

Deer EL, Gonzalez-Hernandez J, Coursen JD, Shea JE, Ngatia J, Scaife CL, Firpo MA, Mulvihill SJ (2010) Phenotype and Genotype of Pancreatic Cancer Cell Lines. Pancreas 39:425–435

Distler M, Aust D, Weitz J, Pilarsky C, Grützmann R, Tzmann R (2014) Precursor lesions for sporadic pancreatic cancer: PanIN, IPMN, and MCN. BioMed Res Int 2014, 11

Froeling FEM, Mirza TA, Feakins RM, Seedhar A, Elia G, Hart IR, Kocher HM (2009) Organotypic culture model of pancreatic cancer demonstrates that stromal cells modulate E-Cadherin, β-Catenin, and Ezrin expression in tumor cells. Am J Pathol 175:636–648

Greggio C, De Franceschi F, Figueiredo-Larsen M, Gobaa S, Ranga A, Semb H, Lutolf M, Grapin-Botton A (2013) Artificial three-dimensional niches deconstruct pancreas development in vitro. Development 140:4452–4462

Huang L, Holtzinger A, Jagan I, Begora M, Lohse I, Ngai N, Nostro C, Wang R, Muthuswamy LB, Crawford HC, Arrowsmith C, Kalloger SE, Renouf DJ, Connor AA, Cleary S, Schaeffer DF, Roehrl M, Tsao M-S, Gallinger S, Keller G, Muthuswamy SK (2015) Ductal pancreatic cancer modeling and drug screening using human pluripotent stem cell and patient-derived tumor organoids. Nat Med 21:1364–1371

Jiang X, Seo YD, Chang JH, Coveler A, Nigjeh EN, Pan S, Jalikis F, Yeung RS, Crispe IN, Pillarisetty VG (2017) Long-lived pancreatic ductal adenocarcinoma slice cultures enable precise study of the immune microenvironment. OncoImmunology 6:e1333210

Kadaba R, Birke H, Wang J, Hooper S, Andl CD, Di Maggio F, Soylu E, Ghallab M, Bor D, Froeling FEM, Bhattacharya S, Rustgi AK, Sahai E, Chelala C, Sasieni P, Kocher HM (2013) Imbalance of desmoplastic stromal cell numbers drives aggressive cancer processes. J Pathol 230:107–117

Koikawa K, Ohuchida K, Takesue S, Ando Y, Kibe S, Nakayama H, Endo S, Abe T, Okumura T, Horioka K, Sada M, Iwamoto C, Moriyama T, Nakata K, Miyasaka Y, Ohuchida R, Manabe T, Ohtsuka T, Nagai E, Mizumoto K, Hashizume M, Nakamura M (2018) Pancreatic stellate cells reorganize matrix components and lead pancreatic cancer invasion via the function of Endo180. Cancer Lett 412:143–154

Li X, Nadauld L, Ootani A, Corney DC, Pai RK, Gevaert O, Cantrell MA, Rack PG, Neal JT, Chan CWM, Yeung T, Gong X, Yuan J, Wilhelmy J, Robine S, Attardi LD, Plevritis SK, Hung KE, Chen C-Z, Ji HP, Kuo CJ (2014) Oncogenic transformation of diverse gastrointestinal tissues in primary organoid culture. Nat Med 20:769–777

Loessner D, Holzapfel BM, Clements JA (2014) Engineered microenvironments provide new insights into ovarian and prostate cancer progression and drug responses. Adv Drug Deliv Rev 79–80:193–213

Loessner D, Meinert C, Kaemmerer E, Martine LC, Yue K, Levett PA, Klein TJ, Melchels FP, Khademhosseini A, Hutmacher DW (2016) Functionalization, preparation and use of cell-laden gelatin methacryloyl-based hydrogels as modular tissue culture platforms. Nat Protoc 11:727–746

Longati P, Jia X, Eimer J, Wagman A, Witt M-R, Rehnmark S, Verbeke C, Toftgård R, Löhr M, Heuchel RL (2013) 3D pancreatic carcinoma spheroids induce a matrix-rich, chemoresistant phenotype offering a better model for drug testing. BMC Cancer 13:95

Moreira L, Bakir B, Chatterji P, Dantes Z, Reichert M, Rustgi AK (2018) Pancreas 3D organoids: current and future aspects as a research platform for personalized medicine in pancreatic cancer. Cell Mol Gastroenterol Hepatol 5:289–298

Neesse A, Michl P, Frese KK, Feig C, Cook N, Jacobetz MA, Lolkema MP, Buchholz M, Olive KP, Gress TM, Tuveson DA (2011) Stromal biology and therapy in pancreatic cancer. Gut 60:861–868

Nielsen MFB, Mortensen MB, Detlefsen S (2016) Key players in pancreatic cancer-stroma interaction: Cancer-associated fibroblasts, endothelial and inflammatory cells. World J Gastroenterol 22:2678–2700

Olive KP, Tuveson DA (2006) The use of targeted mouse models for preclinical testing of novel cancer therapeutics. Clin Cancer Res 12:5277–5287

Ootani A, Li X, Sangiorgi E, Ho QT, Ueno H, Toda S, Sugihara H, Fujimoto K, Weissman IL, Capecchi MR, Kuo CJ (2009) Sustained in vitro intestinal epithelial culture within a Wnt-dependent stem cell niche. Nat Med 15:701–706

Pérez-Mancera PA, Guerra C, Barbacid M, Tuveson DA (2012) What we have learned about pancreatic cancer from mouse models. Gastroenterology 142:1079–1092

Puls TJ, Tan X, Whittington CF, Voytik-Harbin SL (2017) 3D collagen fibrillar microstructure guides pancreatic cancer cell phenotype and serves as a critical design parameter for phenotypic models of EMT. PLoS ONE 12:e0188870

Scavuzzo MA, Yang D, Borowiak M (2017) Organotypic pancreatoids with native mesenchyme develop Insulin producing endocrine cells. Sci Rep 7:10810

Siegel RL, Miller KD, Jemal A (2018) Cancer statistics, 2018. CA Cancer J Clin 68:7–30

Sugiyama T, Benitez CM, Ghodasara A, Liu L, Mclean GW, Lee J, Blauwkamp TA, Nusse R, Wright CV, Gu G, Kim SK (2013) Reconstituting pancreas development from purified progenitor cells reveals genes essential for islet differentiation. Proc Natl Acad Sci U S A 110:12691–12696

Tsai S, Mcolash L, Palen K, Johnson B, Duris C, Yang Q, Dwinell MB, Hunt B, Evans DB, Gershan J, James MA (2018) Development of primary human pancreatic cancer organoids, matched stromal and immune cells and 3D tumor microenvironment models. BMC Cancer 18:335–335

Vaira V, Fedele G, Pyne S, Fasoli E, Zadra G, Bailey D, Snyder E, Faversani A, Coggi G, Flavin R, Bosari S, Loda M (2010) Preclinical model of organotypic culture for pharmacodynamic profiling of human tumors. Proc Natl Acad Sci U S A 107:8352–8356

Walsh AJ, Castellanos JA, Nagathihalli NS, Merchant NB, Skala MC (2016) Optical imaging of drug-induced metabolism changes in murine and human pancreatic cancer organoids reveals heterogeneous drug response. Pancreas 45:863–869

Whatcott CJ, Diep CH, Jiang P, Watanabe A, Lobello J, Sima C, Hostetter G, Shepard HM, Von Hoff DD, Han H (2015) Desmoplasia in primary tumors and metastatic lesions of pancreatic cancer. Clin Cancer Res 21:3561–3568

Breast Organotypic Cancer Models

Pilar Carranza-Rosales, Nancy Elena Guzmán-Delgado, Irma Edith Carranza-Torres, Ezequiel Viveros-Valdez and Javier Morán-Martínez

Contents

P. Carranza-Rosales (✉) · I. E. Carranza-Torres
Departamento de Biología Celular y Molecular, Instituto Mexicano del Seguro Social. Centro de Investigación Biomédica del Noreste, Monterrey, Nuevo León, Mexico
e-mail: carranza60@yahoo.com.mx

I. E. Carranza-Torres
e-mail: mitzba@hotmail.com

N. E. Guzmán-Delgado
Unidad Médica de Alta Especialidad # 34, División de Investigación, Instituto Mexicano del Seguro Social, Monterrey, Nuevo León, Mexico
e-mail: nancyegd@gmail.com

E. Viveros-Valdez
Departamento de Química Analítica, Ciudad Universitaria, Universidad Autónoma de Nuevo León, Facultad de Ciencias Biológicas, San Nicolás de los Garza, Nuevo León, Mexico
e-mail: ezequielviveros@hotmail.com

J. Morán-Martínez
Departamento de Biología Celular y Ultraestructura, Universidad Autónoma de Coahuila, Facultad de Medicina. Centro de Investigación Biomédica, Torreón, Coahuila, Mexico
e-mail: javmoran@yahoo.com

Current Topics in Microbiology and Immunology (2021) 430: 199–223
https://doi.org/10.1007/82_2018_86

Published Online: 20 March 2018

Abstract Breast cancer is the most common cancer type diagnosed in women, it represents a critical public health problem worldwide, with 1,671,149 estimated new cases and nearly 571,000 related deaths. Research on breast cancer has mainly been conducted using two-dimensional (2D) cell cultures and animal models. The usefulness of these models is reflected in the vast knowledge accumulated over the past decades. However, considering that animal models are three-dimensional (3D) in nature, the validity of the studies using 2D cell cultures has recently been questioned. Although animal models are important in cancer research, ethical questions arise about their use and usefulness as there is no clear predictivity of human disease outcome and they are very expensive and take too much time to obtain results. The poor performance or failure of most cancer drugs suggests that preclinical research on cancer has been based on an over-dependence on inadequate animal models. For these reasons, in the last few years development of alternative models has been prioritized to study human breast cancer behavior, while maintaining a 3D microenvironment, and to reduce the number of experiments conducted in animals. One way to achieve this is using organotypic cultures, which are being more frequently explored in cancer research because they mimic tissue architecture in vivo. These characteristics make organotypic cultures a valuable tool in cancer research as an alternative to replace animal models and for predicting risk assessment in humans. This chapter describes the cultures of multicellular spheroids, organoids, 3D bioreactors, and tumor slices, which are the most widely used organotypic models in breast cancer research.

1 Introduction

Cancer is the second leading cause of death worldwide. In 2015, 8.8 million deaths were attributed to this disease, of which 571,000 were due to breast cancer. Breast cancer is ranked fifth among the five major cancers causing the highest number of deaths in the world (World Health Organization 2017). Due to the alarming predictions of an increase in the number of cancer cases and mortality rates in the coming years, many efforts by health authorities are directed toward the prevention and treatment of this disease. In line with this, the International Agency for Research on Cancer has predicted that if the disease is not inhibited, and the resistance of tumor cells to current treatments increases, by 2035 the number of cancer cases and mortality rates will almost double worldwide (Ferlay et al. 2013, 2015). In the USA, the incidence and mortality estimated in 2017 were 252,710 new cases and 40,610 deaths (Siegel et al. 2017).

In the last 30 years, in vivo models and two-dimensional (2D) cell cultures have dominated in preclinical breast cancer research. However, current estimates suggest that >90% of drugs studied using these systems failed during clinical trials (Roberts et al. 2004; Kamb 2005; Pampaloni et al. 2009; Mak et al. 2014). For this reason, the development of more robust and predictive experimental models to evaluate antineoplastic therapies has become a priority. The actual trend is toward the

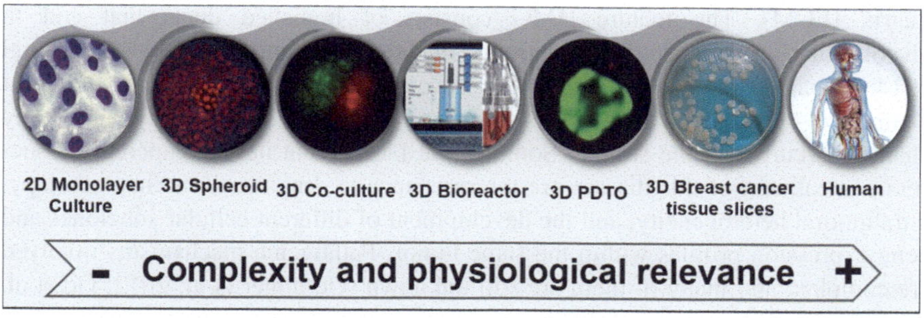

Fig. 1 Organotypic 3D models for breast cancer research. Several systems have been used to study the characteristics of a tumor and to search new therapeutic options. The 2D cancer cell lines are obtained by enzymatic digestion from tumors, they are easy to work with, but lack the tumor stroma and as a result of adaptation to in vitro culture conditions, they barely retain the phenotypic signature of the original tumor. The 3D breast organotypic cancer models are now widely used as they are a physiologically relevant system and closely remain the original tumor. Multicellular spheroids can be obtained as monocultures or cocultures with fibroblasts, stromal stem cells, or endothelial cells; they lack tumor–stroma interactions that involve the extracellular matrix or cells of the immune system. The cocultures can be performed with any of the 3D models described here (spheroids, bioreactors, PDTO, and mammary tumor slices). In cocultures, fibroblasts, endothelial cells, immune cells, or another cancer cells are exogenously added, which extend the predictive capabilities of these systems. Bioreactors are cell culture systems able to be monitored by software to study or to control physicochemical variables and nutrient supply affecting the tumor microenvironment. PDTO can grow from mechanically or enzymatically dissociated biopsies and keep tumor cells within the same tumor environment allowing the ex vivo propagation of individual tumors; however, microenvironment cannot be controlled because they consist mainly epithelial cells and are devoid of tumor vasculature, stromal and immune cells. Breast tumor slices contain virtually all the tumor cells from the original tumor where are obtained, the histological and three-dimensional structure of the tumor, elements of the immune system, extracellular matrix, inter- and extracellular interactions, and metabolic capacity are preserved. (Figure was adapted in part from Fong et al. 2016)

establishment of three-dimensional (3D) organotypic models such as multicellular spheroids, 3D cocultures, organoids, 3D bioreactors, and tissue slices. Interestingly, breast tumor slices are frequently considered as a system that more closely resembles situations taking place in vivo due to the characteristics and advantages that will be mentioned further. In Fig. 1, the most common 3D organotypic cultures used in the research and evaluation of new therapies for breast cancer and other types of cancer are shown.

1.1 Anatomy of the Human Mammary Gland (HMG) and Development of Breast Cancer

The HMG comprises different cell types, each one with specific functions, and a particular stroma or microenvironment. The development, proliferation, differentiation, and maintenance of the HMG are directly influenced by the extracellular matrix (ECM). The mature HMG consists of branched ducts that end in alveolar-like structures known as terminal ductal lobular units (Visvader and Stingl 2014; Inman et al. 2015). During the development of breast cancer, the most distinctive features are the loss of branched glandular morphology and many changes occurred in the stroma. Some of the factors that have hampered the successful treatment of this disease are the resistance to drugs used in chemotherapy, intratumoral heterogeneity, and the development of different cellular subclones and gene expression profiles within the same tumor. Pathogenic mechanisms involved are complex, and many of them are still unknown (Gerlinger et al. 2012; Ge et al. 2012; Marquette and Nabell 2012).

With regard to the etiology of breast cancer, it is a multifactorial disease and known risk factors are family history of breast cancer and aging, (Colditz et al. 2012), mutation of the BRCA1 and BRCA2 genes (Goodwin et al. 2012; Mavaddat et al. 2012), mammary tissue density (Razzaghi et al. 2012), personal history of benign proliferative breast disease (Goldacre et al. 2010; Kabat et al. 2010), early menarche/late menopause (Ritte et al. 2013), nulliparity, old age at the time of first birth, history of hormone replacement therapy, environmental factors such as the use and exposure to ionizing radiation, and lifestyle factors such as high-caloric diets, sedentary lifestyle, alcohol consumption, smoking, and obesity (Hiatt and Brody 2018).

Recent molecular and genetic studies have emphasized that breast cancer is a highly heterogeneous group of diseases that differ in their prognosis and response to treatment. The WHO Classification of Maternal Tumors (fourth edition) (Lakhani et al. 2012) recognizes more than 20 different subtypes. The most frequent are the carcinomas, which have epithelial origin; these tumors are subdivided as in situ carcinomas and invasive carcinomas (Schnitt and Lakhani 2014). The prognosis and selection of the treatment depends on (a) clinical characteristics such as menopausal status, and clinical stage of the disease according to the tumor size, axillary lymph node status, and presence of metastases (TNM classification), and (b) pathological type based on histological and immunohistochemical characteristics (Amin et al. 2017).

In addition to the TNM classification and histological type/grade, it is essential to complement the diagnosis of breast cancer with immunohistochemical biomarkers for alpha-estrogen receptor (ER), progesterone receptor (PR), Ki-67, and HER2. From results of RE, RP, and HER2 expression, breast cancer is classified into one of the following types: Luminal A [ER(+), PR (+), HER2 (-)], Luminal B [ER (+), PR(+), HER2 (+)], HER2 [ER (-), PR(-), HER2 (+)], and triple-negative or basal-like [RE(-), RP (-), HER2 (-)]. The expression of these biomarkers can help the prognosis of the disease and establish therapeutic options, including hormone therapy, chemotherapy, and anti-HER2 therapy (Ross et al. 2003; Colomer et al. 2017; Coates et al. 2015).

Additionally, gene profile tests represent a useful tool for the personalized choice of systemic treatment in a selected group of patients with invasive carcinoma. This allows greater certainty of the usefulness of the indicated treatment, avoiding the toxicity of therapies with limited benefit. Its use is recommended exclusively in patients with early disease and hormone-sensitive tumors (ER positive) or HER-2 negative (Gradishar et al. 2016; Senkus et al. 2015; Harris et al. 2016; Coates et al. 2015). In order to establish a prognostic category and decide whether adjuvant therapy may be limited to hormone therapy, one of the four available genetic prognostic platforms (MammaPrint®, Oncotype DX®, Prosigna® or EndoPredict®) can be used (Colomer et al. 2017).

1.2 2D Models in Breast Cancer Research

Due to the anatomical and structural complexities of the HMG, it is evident that the culture of human mammary epithelial cells under standard conditions does not allow the analysis of morphological aspects and limits the evaluation of functional characteristics. However, most preclinical research on breast cancer is based on the use of breast cancer cell lines, either as primary cultures or as cultures of established cell lines, which do not completely reflect the complex tissue architecture of tumors, thus limiting their predictive value (Van der Kuip et al. 2006). These limitations inherent to 2D cultures are mainly due to the fact that the most commonly used cell lines have been maintained in vitro for decades, and it is well documented that genetic and phenotypic transformations occur during long-term artificial cell propagation (Speirs et al. 1998; Joshi et al. 2011). Furthermore, given the known genetic and epigenetic heterogeneity of human cancers, it is probable that these cells represent clones that are more apt to grow in vitro (Williams et al. 2013); however, the most important limitation is that they lack the 3D microenvironment of the intact tissue and stroma. Stromal tissue plays an important role in human breast cancer development (Shekhar et al. 2001) since it represents more than 80% of the mammary volume of the resting MG (Drife 1986).

The most widely used breast cancer cells in the world are the MCF-7 cells, this is due to their specific hormone sensitivity through the expression of estrogen receptor (ER), which makes MCF-7 cells an ideal model to study hormone response (Levenson and Jordan 1997; Holliday and Speirs 2011). MCF-7, MDA-MB-231, and T47D breast cancer cell lines are used in more than two-thirds of related studies; however, this fact raises issues on how representative these cell lines are with regard to the different types of breast cancer, each of one with distinct behavioral and clinical implications (Lacroix et al. 2004; Dai et al. 2017).

1.3 In Vivo Models in Breast Cancer Research

Since early 1900s (Holen et al. 2017) and until 1980, in vivo-animal models were used for the discovery of new drugs; later, as the demand for drugs and high-throughput selection that was required increased, the use of in vivo models became very expensive and unethical (Ravi et al. 2015).

Mice are the most commonly used as in vivo preclinical models to study breast cancer. At present, there are dozens of different murine models which have been manipulated to reflect different characteristics and stages of breast cancer, in such a way that researchers can select the mouse model that answers their specific research questions. Preclinical mouse models more frequently cited in breast cancer studies are briefly described below:

Cell-derived xenografts (CDXs) are the most used murine models in breast cancer research; they consist of immunocompromised mice that are grafted with cancer cells of human origin (Neve et al. 2006). These mice have helped to study basic aspects of cancer genetics; despite their widespread use, they have not been used for predicting responses to clinical therapies (reviewed in Whittle et al. 2015). CDXs are followed in frequency of use by patient-derived xenografts (PDXs), these systems also use immunodeficient mice to which primary cultures of human cancer cells or fragments of tumors are grafted. Although this methodology is not new (reviewed in Hoffman 2015), they have now returned to great importance because new methodologies have managed to preserve basic aspects of human primary tumors (DeRose et al. 2011, 2013; Hidalgo et al. 2014). Although these models directly use human clinical samples they have the disadvantage that the generated xenografts possess scarce stroma, which is of murine origin (non-human), resulting in chimeric tumors whose biological behavior could be significantly different from those human tumors (Hahn and Weinberg 2002). Another model is the genetically modified mice (GEMM), which are based on the use of mice transfected with oncogenes targeting specific molecular pathways in breast cancer, and although these models have the disadvantage that the sequences directing the expression of their transgenes are not well defined, and do not necessarily reflect what happens in human tumors, they continue to be used in breast cancer research and have helped to understand early events of the tumor process (Blaas et al. 2016; Arun et al. 2016). A detailed description of the advantages and disadvantages of the models mentioned above can be reviewed in the publication of Holen et al. (2017).

On the other side, since the most important limitations in all CDX and PDX models are the use of immunocompromised mice, recently humanized models have been developed and they can generate functional immune systems with expansive capabilities that represent pioneer platforms to evaluate compounds for treatment of a variety of human diseases, including breast cancer (Zheng et al. 2015, 2016; Dallas et al. 2016).

All of these preclinical murine models have resulted from selective endogamy, tumor grafting, or genetic manipulations that have been aimed at improving the incidence and characteristics of specific human tumors; nonetheless, the genetic background of each strain has been profoundly altered (Kim et al. 2004). Also, it has been documented that mouse models do not faithfully replicate the HMG microenvironment (Koren and Bentires-Alj 2015). For example, human breast tissue has more fibrous connective tissue and fewer adipocytes than mouse mammary tissue (Bischel et al. 2015), and also humans react to drug toxicity in a different manner than animals (de Boo and Hendriksen 2005). These facts, together with the variations and differences between the physiology of humans and mice, defects and partialities in the experimental designs (Pound and Bracken 2014) and an inadequate selection of the murine model under study have not allowed to predict the efficacy and toxicity of therapeutic treatments.

Other issues are related to strong ethical concerns which are motivating governments and regulatory organizations to support the implementation of alternative methods for animal testing. The replacement of animal models in drug discovery and toxicity testing is one of the premises of the 3R's principle, which was described in 1959 by Russell and Burch in their book "The Principles of Humane Experimental Technique." The principle of the 3R's proposes the following: *Replace* the procedures that use animals with others that do not require them, *Reduce* the number of animals used, and *Refine* the methods used to improve their effectiveness and/or reduce suffering or pain inflicted on the animals (Knight 2008; Pampaloni et al. 2009; Griffin et al. 2014).

In accordance with the above mentioned, it is considered that preclinical models must take into account the heterogeneity and complexity of tumors to successfully identify new breast cancer treatment regimens since it has been reported that when the physiology of a solid tumor is modified or altered, the cytotoxic response to various antineoplastic agents may vary due to interactions between cancer cells and their microenvironment, impacting the results of chemotherapy (Netti et al. 2000; Sherman-Baust et al. 2003; Grantab et al. 2006). Recommendations are directed to increase the rate of clinical successes by using specific models of human diseases, which may help to eliminate problems due to species variations (Kamb 2005; Langley et al. 2017).

2 Organotypic Models for Breast Cancer Research

Organotypic 3D cultures are experimental models more frequently used in breast cancer research as they recapitulate the organization and functional differentiation of the HMG and allow the examination of complex interactions among different cellular populations, and the understanding of factors that determine the behavior of the tumor (Holliday et al. 2009). Compared with tests made in animals, organotypic models have the potential to improve drug detection and identify toxic and ineffective substances at an earlier stage in the research of new drugs, while simultaneously favoring a reduction in the number of in vivo tests, which are today regarded as ethically controversial and very expensive (Pampaloni et al. 2009; Ravi et al. 2015). The most commonly used organotypic cultures in breast cancer investigations include microfluidic chip-based scaffolds (Song et al. 2009; Sung et al. 2011), multicellular spheroids (mammospheres) (Gallardo Pérez et al. 2006; Ivascu and Kubbies 2007), organoids derived from cells or tissues (Nguyen-Ngoc et al. 2012; Sokol et al. 2016), cocultures in 3D matrices (Jaganathan et al. 2014), and ex vivo cultures of breast cancer slices or explants (Carranza-Torres et al. 2015;

Morrisey et al. 2015; Vesci et al. 2015; Naipal et al. 2016). The increasing use of these methodologies is because results obtained using them are similar to those observed using in vivo models (Caddeo et al. 2017).

2.1 Mammospheres: Multicellular 3D Spheroids of Mammary Tumors

Spheroids constitute an experimental 3D model of intermediate complexity between 2D cultures and tumors in vivo. Pioneering works were developed by Holtfreter (1944), Moscona (1952), Leighton (1951), and Sutherland et al. (1971). The behavior of tumor spheroids is very different from the one observed 2D cultures (Lin and Chang 2008), and the cellular heterogeneity in the 3D spheroid allows the imitation of the tumor microenvironment and favors the study of different aspects of cancer biology (Oudar 2000). During the development of the spheroids, several cells adhere to each other to form a spherical cell mass that allows cell–cell interactions. At the structural level, three layers of cell populations can be distinguished: an apoptotic/necrotic center, an intermediate layer of quiescent cells, and an outer layer of proliferating cells (Kim et al. 2004). Peripheral cells have an adequate and constant supply of oxygen and nutrients; therefore, they are able to proliferate (LaRue et al. 2004), and they also possess a high metabolic rate that can modify the levels of oxygen and nutrients between the different layers of cells (Walenta et al. 2000), which induces hypoxia in the inner center, resulting in the necrotic and apoptotic core (Riffle and Hegde 2017). When different types of human cancer cells are grown in optimal conditions, the distance from the periphery of the spheroid to the necrotic center varies between 50, 100, 220, and 300 μm, depending on the cell types, consumption and supply of nutrients, and the density of cellular packing in each layer of the spheroid. Similarly, radial measurements between vessels and necrotic areas in different tumors of rodents and humans vary from 50 to 250 μm (Sutherland 1988). For this reason, multicellular spheroids are considered as models of intermediate complexity between 2D cultures and tumors in vivo.

In recent years, this model system has been frequently used in cancer biology, since it is possible to obtain spheroids from different tissues. In accordance to this, distinct authors have designated the name of the spheroids depending on the cell line from which they were developed; in this way, the spheroids obtained from glioma tumor cells are called "neurospheres," "colonospheres" are spheroids from colon, "prostaspheres" from prostate, and "mammospheres" from breast (Ishiguro et al. 2017).

Mammospheres can be obtained from monocultures or cocultures with fibroblasts, stromal stem cells, or endothelial cells (Gauck et al. 2017). To form mammospheres, cells are grown in containers with a non-sticky and static surface, which allows the development of the primary nucleus in the spheroid. In a second step using orbital agitation, mammospheres are consolidated. Mammospheres are grown using liquid overlay and spinner flasks techniques, the difference lies in the use of adherent substrates in the first method and absence thereof in the latter method (Nagelkerke et al. 2013). In the liquid overlay method, the lower surface of a Petri dish is covered with agarose, and the breast cancer cell line is seeded at a confluence of 40–60%, afterward, the cells are incubated under standard conditions. Cells are then transferred to containers with fresh culture medium and are kept under constant agitation. Mammospheres are formed within 2–5 weeks influenced by serum-signaling factors such as carcinoembryonic antigen, interferon-γ, IGF-II, heregulin β1, plasmin, and E- and N-cadherin (Gallardo Pérez et al. 2006; Ivascu and Kubbies 2007). The most widely used cell lines for obtaining mammospheres are MCF-7, BT-474, T-47D, MDA-MB-361, MDA-MB-435S, MDA-MB-231, MDA-MB-468, and SK-BR-3 (Ivascu and Kubbies 2007).

2.2 *Applications of Mammospheres*

By virtue of their similar behavior to small avascular tumors, mammospheres are excellent models to study breast cancer (Ham et al. 2016). It has been reported that the tumor microenvironment created in mammospheres allows the expression of cellular markers other than those found when monolayer cells are cultured (Kenny et al. 2007). These markers are formed through interactions generated by the tumor microenvironment with stromal cells and are involved in the development and progression of breast cancer by inducing conditions that favors the growth and invasion of tumors, reduction in apoptosis, induction of angiogenesis, and metastasis (Kaur et al. 2011; Rodríguez et al. 2015). The similarity between the biological niche of spheroids and small viable tumors suggest that mammospheres are a realistic experimental model that can be used to optimize and develop antitumor therapies. Their use has already been reported in the study of drug resistance (He et al. 2015), and they also have been proposed as novel models in the systematic search for new drugs (Reid et al. 2014), including the investigation of natural products with antineoplastic potential (Bassa et al. 2016). Mammospheres have been used to study breast cancer stem cells (BCSCs) in different types of mammary tumors, for example, to study Hedgehog signaling pathway role in paclitaxel resistance and sensitivity to salinomycin (He et al. 2015), in the identification of deregulated proteins in BCSCs through proteomic studies (Li et al. 2013), or in the analysis of the effect of pomegranate extract on the modulation of factors present in BCSCs associated with the inhibition of epithelial–mesenchymal transitions (Nallanthighal et al. 2017).

The aforementioned aspects convert mammospheres into a model with great potential of biomedical and clinical applications as they combine the importance of the tissue arrangement in the tumor microenvironment with the practicality of maintaining the controlled conditions of in vitro tests. In turn, this could allow designing more precise experiments to conduct sensitivity tests to develop new and better therapies for breast cancer treatment.

2.3 Patient-Derived Tumor Organoids (PDTOs)

The term "organoid" reflects the ability of culture conditions to propel cells to self-organize themselves into structures that mimic the architecture of the organ from which they were derived while preserving cell–cell and cell–ECM interactions (Sato et al. 2009, 2011). The microenvironment in organoids resembles the original tumor microenvironment more accurately than that in traditional 2D cultures. Different organoid systems share the following features: are cultured and embedded within an ECM gel (collagen I or Matrigel); form 3D structures of epithelial cells generated from smaller multicellular units (such as enzymatically or mechanically dissociated primary tissue) or from single cells (embryonic stem or pluripotent cells differentiated in vitro). In terms of tumor heterogeneity, organoids are considered an intermediate model between cancer cell lines (in vitro) and xenografts (DeRose et al. 2013; Baker et al. 2016; Linnemann et al. 2017; Pauli et al. 2017). Patient-derived tumor organoids (PDTOs) are preclinical models that allow the ex vivo propagation of individual tumors and provide an excellent platform for studying tumor progression, invasion, and drug response (Walsh et al. 2016); they are constituted by 200–1000 adherent tumoral cells and reflect the cellular heterogeneity present in the primary tumor (Cheung et al. 2013).

2.4 Applications of PDTOs

Because PDTOs can grow from mechanically or enzymatically dissociated biopsies and keep tumor cells within the same tumor environment, they represent an attractive alternative to investigate therapeutic compounds and individual drug therapy for breast cancer (Campbell et al. 2011; Walsh et al. 2014; Francies et al. 2016). This 3D organotypic system has been used to evaluate the response to clinically relevant antitumor agents such as paclitaxel, anti-HER2 antibody, trastuzumab, the phosphatidylinositol-3-kinase pathway inhibitor XL147, and combinations thereof (Walsh et al. 2014).

Other applications include those mentioned in the report of Nguyen-Ngoc and colleagues (2012) that cultured the epithelium of human mammary carcinomas in different ECMs. They modeled the normal microenvironment of the mammary epithelium using Matrigel, and to mimic the stromal matrix of invasive mammary carcinomas, type I collagen was used. Their data suggest that changes in the ECM-induced signals that trigger invasion and local dissemination. They also analyzed cell adhesion genes and found that the removal of P-cadherin was sufficient to allow dissemination in type I collagen (Nguyen-Ngoc et al. 2012). Organoids derived from breast cancer cells have also been used to understand mechanisms underlying the invasiveness of tumor cells and metastasis (Chandhoke et al. 2017; Chanda et al. 2017). In 2013, Cheung and coworkers used PDTOs grown in 3D collagen type I gels to identify the most invasive cancer cells in primary breast tumors. They found that among the major subtypes of human breast cancer, collective invasion was promoted by specialized cancer cells expressing basal epithelial genes such as cytokeratin-14 and p63. According to their results, heterotypic interactions between epithelial subpopulations are fundamental for collective invasion (Cheung et al. 2013). Walsh and coworkers (2014) prepared organoids from six different samples from primary breast cancer tumors to predict therapeutic response. They evaluated the sensitivity to treatments using optical metabolic images that quantify fluorescence intensity and half-life of NADH and FAD. Compared with organoids derived from xenografts, the morphology of PDTOs differed between patients and within the subtypes of breast cancer and showed great heterogeneity within primary tumors. In another study, the cellular morphology, viability, and response to paclitaxel, tamoxifen, and trastuzumab of PDTO from frozen tissues were studied. It was found that it is possible to grow viable organoids from frozen tissues as long as the researcher is aware that the freezing process can introduce cellular differences and reduced response to drugs (Walsh et al. 2016). On the other side, Sokol and coworkers (2016) described a 3D culture system that supports the growth of primary epithelial cells of the human breast. This culture stimulates the growth of morphologically complex and hormone-sensitive breast tissues. Primary human epithelial cells self-organized and showed complex ductal and lobular morphologies observed in the breast. The ability to cultivate hormone-sensitive human breast tissue in hydrogels with defined components will enhance future studies based on the development and biology of human HMG, with potential implications for understanding the biology of breast cancer (Sokol et al. 2016).

Although organoid cultures can be prepared between 7 and 10 days (Sato et al. 2011), the propagation time required to obtain drug sensitivity data is from 1 to 3 months, depending on the size and origin of the tumor sample used. To obtain useful results for the treatment of patients using patient-derived xenografts (PDXs), 6–8 months are required. The cost of establishing and propagating an organoid culture system is more expensive than growing cell lines but less expensive than PDX (Weeber et al. 2017). As with multicellular spheroid cultures, PDTO cultures lack control of the tumor microenvironment because they are mainly epithelial cultures devoid of stromal cells and tumor vasculature (Junttila and de Sauvage 2013). Despite these disadvantages, several efforts are being made to improve these systems. One example of these efforts is the application of coculture techniques for both, organoids and mammospheres. Cocultures of organoids with the main cell types from breast tissue to mimic genetic and phenotypic characteristics of the microenvironment in vivo have been reported (Holliday et al. 2009; Kimlin et al. 2013). Moreover, there are methodologies of coculture of mammary tissue organoids with fibroblast spheroids that could be useful to perform more complete studies on interactions between different cell types within the mammary tumor microenvironment (Koledova 2017). Choi and coworkers (2015) developed a fluidic microsystem to coculture mammary tumor spheroids with ductal epithelial cells and mammary fibroblasts to replicate the microarchitecture of ductal carcinomas in situ. Another group of researchers have developed a model of spheroids using MCF-7 and MDA-MB-231 cocultured with human endothelial and stromal cells in an ECM composed by proteins and collagen I, with the purpose of replicate the architecture of the tissue, chemical environment, and cellular organization of a growing and invasive tumor (Benton et al. 2015).

Even with certain limitations or disadvantages, which can be overcome in the near future, this type of organotypic culture presents new perspectives in breast cancer research and in the field of oncology. The improvement of minimally invasive methods to access tissue together with optimized culture and derivation methods can increase the use of such advanced preclinical platforms beyond the current set of patients (Pauli et al. 2017).

2.5 *Bioreactors 3D and Applications in Breast Cancer Research*

Bioreactors are culture systems that allow controlled and reproducible handling of large quantities of tissue-engineered 3D constructs in vitro with programmed and computer-controlled inputs and outputs to regulate environmental culture conditions. The main advantages of these systems are the monitoring of changes in scalability and methods that permit controlled changes in environment physicochemical parameters such as temperature, pH, nutrients, and oxygen, as well as mechanical stimulation such as flow rates and shear stress, while retaining aseptic conditions (Astashkina et al. 2012; Hickman et al. 2014; Portner and Giese 2007). Rotary cell culture systems (RCCSs) are examples of bioreactors which have been applied to 3D cancer cell culture for almost 40 years; indeed, the rotating wall vessel (RWV) bioreactor, horizontally rotating cylindrical culture vessels with no internal stirring mechanisms, was originally designed to recapitulate certain conditions that occur in the microgravity environment of space, and adapted for the culture of cancer cells, with the aim of study tumor formation and progression and tumor microenvironment.

Various breast tumor models have been successfully engineered using RWV-based bioreactors (Becker and Souza 2013; Hickman et al. 2014; Xu et al. 2014). Kaur et al. (2011) reported the utilization of a 3D rotary bioreactor to coculture human breast cancer cells and fibroblasts, generating structures that resembled breast cancer tissue. Krishnan et al. (2011) also applied this technology to investigate the breast cancer cell interactions with 3D multilayer osteoblastic tissue grown in a specialized bioreactor. In other study, a specialized bioreactor was utilized for the growth of osteoblastic tissue and its coculture with metastatic breast cancer cells. This system may be a relevant model for cancer colonization of bone since sequential stages of breast cancer interaction with osteoblasts, including breast cancer cell adhesion, penetration and colony formation, and osteoblastic tissue reorganization reflected some of the features, with considerable degree of physiological relevance, involved in breast cancer bone metastasis (Dhurjati et al. 2008; Xu et al. 2014).

The last-generation of microfluidic devices or microbioreactors is an efficient small-scale system with precise control of cell culture microenvironment. Microfluidic technology was introduced as a tool for biological analysis in the early 1990s (Hong and Quake 2003; Hickman et al. 2014). In one example of a cell-on-a-chip microfluidic platform for in vitro drug screening, breast tumor cells were encapsulated within alginate which was gelled in situ within one microfluidic–microchannel and then were cultured into a 3D tumor spheroid. Doxorubicin was then perfused through another microchannel and it was found that compared to a 2D assay of doxorubicin's effect on cancer cell viability, the drug displayed a reduced effectiveness in 3D culture (Chen et al. 2010; Tsui et al. 2013). Recent advances in microfluidic technology improved the development of "organ-on-chip," this approach aims to capture key structural and physiological characteristics of the target tissue. "Tumor-on-a-chip" (in vitro vascularized microtumors) incorporates human tumor and stromal cells that grow in a 3D extracellular matrix which depends for survival on nutrient delivery through living, perfused microvessels. A group of researchers reported that breast cancer cells grew vigorously in this platform and responded to standard therapies (Sobrino et al. 2016). In addition, microfluidic systems have been used to perform studies of metastases to bone. Bersini et al. (2014) developed a 3D microfluidic in vitro model to analyze the extravasation of highly metastatic breast cancer cells into an osteo-cell conditioned microenvironment. They observed that extravasated cancer cells can proliferate and generate micrometastases within this conditioned microenvironment. In another example, Buchanan et al. (2014) developed a microfluidic tumor vascular model system for coculture of tumor and endothelial cells under varying flow shear stress conditions. This study demonstrated that endothelial cells assemble a confluent endothelium on the microchannel lumen that maintains the channel integrity under physiological flow shear stresses. Also, the cancer cells cocultured with endothelial cells under low flow conditions significantly enhanced the expression of pro-angiogenic factors (Xu et al. 2014).

In addition to these models, Muraro et al. (2017) used a perfusion-based bioreactor to maintain the tumor microenvironment of freshly excised breast cancer tissue obtained from 27 breast cancer patients and used this platform to test the therapeutic effect of antiestrogens and checkpoint-inhibitors on the cancer cells. In other research, Goliwas et al. (2017) presented a tissue-engineered surrogates model, which consist of breast cancer epithelial cells and human cancer-associated fibroblasts within a 3D volume of ECM (comprised of fibrin, collagen type I, and basement membrane) continuously perfused with nutrients through five microchannels in a perfused bioreactor system. This group used several approaches to analyze cell growth and response to paclitaxel in tissue-engineered models of breast cancer. These models represent a valuable tool in breast cancer research because they are complex systems that allow incorporating other 3D models such as multicellular spheroids and organoids derived from patients and at the same time controlling the tumor microenvironment through computers, which ultimately allows recapitulating the tumor conditions as in vivo, and achieve more representative results toward the human condition.

2.6 Organotypic Culture of Human Breast Cancer Tissue Slices

The use of tissue slices in research was reported for the first time in the 20s by Otto Warburg. At that time, tissue slices were prepared by hand using scalpels; their applications increased from the 80s, with the development of precise-automated tissue slicers allowing the aseptic and rapid preparation of living slices with better quality (Bach et al. 1996; Krumdieck et at. 1980; Smith et al. 1985).

Organotypic culture of tumor-derived slices which retain the tumor microenvironment is gaining acceptation because results obtained using in vitro, in vivo, and even other 3D approaches do not allow extrapolation, by example, with what occurs in cancer patients in terms of physiology, metabolism, and pharmacokinetics of antineoplastic agents. For these reasons, to study breast cancer behavior it is very important to maintain or reconstitute an environment that is similar to that of the tumor in situ. One method to preserve the tissue architecture avoiding excessive manipulation is to cultivate breast cancer tissue slices as intact as possible in order to be considered as a suitable preclinical drug-testing model to individualize antineoplastic treatment, a growing concern in cancer therapy (Conde et al. 2012; Holliday et al. 2013; Grosso et al. 2013).

Considered by experts as "microtumors" that contain practically all tissue cells from which they are derived, breast cancer tumor slices retain the histological and 3D structure of the original tumor from which they are derived, while maintaining intercellular and extracellular interactions and elements of the cell matrix and, most interestingly, retaining their metabolic capacity and can be an alternative model to study multicellular processes (Mestres et al. 2006; de Graaf et al. 2010; Morrisey et al. 2015).

With this system, greater reproducibility is achieved by experiments performed with the same organ, and a great quantity of slices that can be cultured under controlled conditions can be obtained. An additional advantage is that a considerable reduction in the number of animals can be achieved. For its preparation, the Krumdieck and Brendel Vitron slicers and, more recently, the Leica VT1200 vibratome and other different vibrating microtomes can be used. For organotypic culture of breast cancer slices, it is important that the tumoral fragment to be used should be obtained during surgery, under aseptic conditions and selected by a pathologist to get a representative sample of the tumor. This sample must be transported to the laboratory and processed rapidly. It is recommended that less than 2 h should pass from surgical resection to the incubation of the slices. To preserve the viability of the tissue during the slicing process, physiological buffers are constantly aerated with carbogen gas. The slice thickness is also very important, depending on the experimental needs it varies between 250 and 400 μm (Fig. 2a).

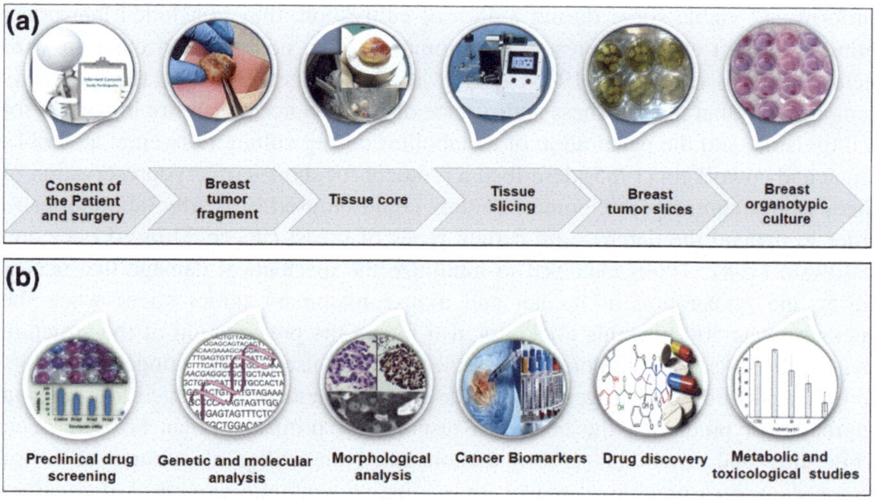

Fig. 2 Generation of breast cancer tissue slices for organotypic culture and its applications. **a** Schematic representation of tissue slices generation. After the consent from the patient is obtained, tumor samples are collected immediately after surgical resection and transported to the laboratory in ice-cold culture medium, with no more than 2 h from harvest to surgery, tissue cores of 5–8 mm diameter are generated under aseptic conditions, tissue cores are generated with coring tools designed to give sharp, well-formed cylindrical live tissue cores for use in the Krumdieck tissue slicer (Alabama Research and Development, Birmingham, AL); they are important to produce tissue slices with reproducible size. Tissue cores are placed into the tissue slicer in ice-cold physiological buffers continuously oxygenated and sliced at a thickness of 250–400 μm. Slices are collected and placed into six-well tissue culture dishes with complete culture media and used according to the design of the experiments. **b** Examples of applications of this 3D cultures include ex vivo preclinical studies, such as screening of molecules, which can be used as cancer biomarkers, genetic, molecular, metabolic and toxicological studies, morphological analysis and investigations on new anticancer drugs

2.7 Applications of Organotypic Cultures of Breast Cancer Slices

In the first scientific publications, in which human breast cancer tissue was cultivated, free-hand methods were used. Tannerberger and Bacigalupo (1970) and Tannerberger and Mohr (1973) reported pioneering work in the field of organ cultures using breast cancer tissue. They analyzed individual cytostatic therapy after pre-therapeutic sensitivity assays in vitro, and the biological characterization of human tumors by organ culture and individualized cancer treatment, respectively. In his report from 1973, Tannerberger mentions for the first time the concept of oncobiogram.

Afterward, Nissen and colleagues described a new technique where 1 mm^3 pieces of human mammary carcinomas tissue were cultivated, the breast cancer tissue was sliced with aid of a stereoscopic microscope and the procedure yield uniform and viable slices during a 48 h of cultivation, they concluded that organ cultures (slices) closely represent carcinomas in vivo compared to other in vitro techniques (Nissen et al. 1983). Using a similar methodology, these authors demonstrated that the thickness and surface of breast cancer slices are important for cell division, and the penetration of metabolites during culture (Nissen et al. 1984). Stone and coworkers (1985) described a protocol for short-term cryopreservation of slices from mammary carcinomas, so they could be used in nearby laboratories in order to expand the database on certain types of breast cancer. Mira-y-Lopez and Ossowski (1987, 1990) managed to minimize the mechanical damage that occurs during the preparation of human and mouse mammary tumor slices when the thickness was 500 μm; this thickness also favors the preservation of the structure and functioning of the tumor, and together with the use of plasminogen activators and quantification of estrogen receptor levels, this model represented a promising approach for predicting the hormonal response of primary human breast cancer, while Jing and coworkers (1996) demonstrated the utility of cultures of breast tumor slices as a preclinical model for evaluating synthetic vitamin A derivatives (retinoids), which are negative regulators of tumor growth in breast cancer.

In all the publications between 1983 and 1996, breast tumor slices were obtained using the Stadie-Riggs microtome. In the following reports from 2004 onward, the Krumdieck precision-automated tissue slicer and Leica vibratome were used. This is important to note because even with the aid of the Stadie-Riggs microtome, the tissue gets compressed and it is not easy to obtain reproducible data because the tissue slices had not uniform size or thickness, restricting the investigations to experiments lasting only minutes, or few hours (Bach et al. 1996).

In the first report using automated, more precise tissue slicers, breast cancer slices were used to investigate the effect of vitamin D3 on the regulation of markers associated with an invasive phenotype (Barbosa et al. 2004). Since then, and to date, several studies and applications have been published using organotypic cultures of breast cancer slices (Fig. 2b). In the next paragraphs and following a chronological order, the studies published to date are described. Stoff-Khalili and coworkers (2005) used adenoviral vectors as therapeutic alternatives to breast cancer and found that the use of the adenoviral promoter CXCR-4 is suitable for breast cancer virotherapy mediated by adenovirus. Mestres-Ventura and coworkers (2005) considered that due to the preservation of their internal histological structure, metabolic, and functional characteristics, breast cancer tumor slices act as "microtumors." This author reported a method that combines the use of mammary tumor slices on silicon chips equipped with electrochemical sensors to study the sensitivity of breast tumor slices incubated in the presence of taxol (Mestres et al. 2006). On the other side, Van der Kuip and coworkers (2006) incubated mammary tumor slices with taxol and found that tissue viability was inversely proportional to the concentration of taxol. With these results, they suggested that this technique can be a valid tool to investigate the resistance and effectiveness of antineoplastic agents. Sonnenberg and coworkers (2008) conducted toxicity assays using in vitro cultures of carcinoma-associated fibroblasts and organotypic cultures of lung cancer and breast cancer tissue slices exposed to cisplatin and paclitaxol. Their results suggest that carcinoma-associated fibroblasts are more resistant to cisplatin when they are within the tissue microenvironment, than when they were isolated by enzymatic digestion. Milani and coworkers (2010) found that the vitamin D signaling pathway is functional in breast cancer slices, and Pennington and coworkers (2010) demonstrated the usefulness of breast cancer slices to analyze the capacity of infection and replication of oncolytic adenoviruses as preclinical system for the evaluation of viral therapies. Rajendran and coworkers (2011) also used adenoviral vectors to evaluate different methods of administering genes for the localized treatment of breast cancer at preclinical level. They expanded the existing knowledge on the specificity of tumor cells and biological factors that determine the successful administration of viral genes. Conde and coworkers (2012) highlighted that the organotypic culture of breast cancer tumor slices is the best alternative for studies that seek to individualize cancer treatments. Holliday and coworkers (2013) analyzed the practical aspects of using organotypic cultures of breast cancer tumors as a preclinical model and validated its usefulness. Grosso and coworkers (2013) used organotypic cultures of breast cancer slices as an ex vivo preclinical tool to evaluate changes in rapamycin-mediated gene expression in individual patients. Milani and coworkers (2013) worked with mammary tumor slices from post-menopausal women to study the transcriptional effects of 1,25 dihydroxyvitamin D3 on genes involved in hormone-dependent signaling pathways. Naipal and coworkers (2014) developed an assay with cultures of mammary tumor slices to investigate homologous recombination due to mutations in BRCA1 or BRCA2 genes in hereditary breast cancer. Corben and coworkers (2014) determined that organotypic cultures represent a more accurate tool for evaluating the efficacy of drugs in the ex vivo treatment of primary tumors and/or associated metastases during the preclinical and clinical development of treatments for breast cancer. Faversani and coworkers (2014) examined the ex vivo response to doxorubicin in organotypic cultures of primary human breast tumors. Recently, our group used organotypic cultures of breast cancer tumor slices to study natural compounds with antineoplastic potential. We demonstrated that this system represents an alternative model to evaluate natural or synthetic compounds with antitumor activity, or synergistic properties with known antineoplastic drugs (Carranza-Torres et al. 2015). Naipal and coworkers (2016) used this model to evaluate the cytotoxic response of individual tumors to chemotherapy with 5-fluorouracil, adriamycin, and cyclophosphamide. With this assay it is possible to distinguish the response of individual tumors to anticancer drugs; therefore, the individualization of breast cancer treatment can be improved.

Other authors have used slices of mammary tumors obtained from transgenic mice for ex vivo molecular analysis and to study the 3D structure and stromal–epithelial interaction in tumor biology, as well as the response to drugs, cytokines, and extracellular environment under defined conditions. They concluded that this model represents a system for ex vivo studies that closely mimics the in vivo scenario (Parajuli and Doppler 2009). Similarly, Vesci and coworkers (2015) used xenotransplanted mice to obtain tumor slices which were incubated with trastuzumab or docetaxel. They found that breast tumor slices represent an experimental model of cancer that is more predictive than cell cultures and that allow the evaluation of the antitumor efficacy of large (trastuzumab) and small (docetaxel) molecules.

Limitations of this system are related to the challenge of standardizing the processing criteria; for example, the different intraoperative manipulation, ischemia times, optimal transport and processing solutions and the reduction of mechanical stress during slicing, the intratumoral heterogeneity, as well as the consistency of the tumors, which have an impact on the quality of the slices and therefore on their cultivation.

3 Conclusions

As the organotypic models for breast cancer research become more widely used, there is a strong need to standardize the conditions of testing, analysis, validation criteria, and acceptance of the obtained results. Organotypic cultures currently represent a relevant biological tool that fills the gap that exists between 2D in vitro cultures and animal tests; simultaneously, they are an alternative to experiments using animals and can help predict the assessment of risk to humans with a view to individualized therapy in breast cancer. Lisa Hutchinson, former Chief Editor of Nature Reviews Clinical Oncology, recently pointed that in the search for tools for the treatment of patients with cancer, a better understanding of the tumor microenvironment and the interactions that occur in the stroma may be the key to find answers about how cancer begins, and to learn more about the metastatic process. However, her best estimate of the time to reach a "biological cure" for cancer is measured in much more than decades (Wagstaff 2017). If this is true, our generation and the following ones must continue doing investigation on cancer to contribute, albeit a little, to achieve this objective.

References

Amin MB et al (eds) (2017) AJCC cancer staging manual, 8th edn. Springer, New York, NY

Arun G et al (2016) Differentiation of mammary tumors and reduction in metastasis upon Malat1 lncRNA loss. Genes Dev 30(1):34–51

Astashkina A, Mann B, Grainger DW (2012) A critical evaluation of in vitro cell culture models for high-throughput drug screening and toxicity. Pharmacol Ther 134(1):82–106

Bach PH et al (1996) The use of tissue slices for pharmacotoxicology studies. ATLA 24:893–923

Baker LA, Tiriac H, Clevers H, Tuveson DA (2016) Modeling pancreatic cancer with organoids. Trends Cancer 2(4):176–190

Barbosa EM et al (2004) Vitamin D3 modulation of plasminogen activator inhibitor type-1 in human breast carcinomas under organ culture. Virchows Arch 444(2):175–182

Bassa LM et al (2016) *Rhodiola crenulata* induces an early estrogenic response and reduces proliferation and tumorsphere formation over time in MCF7 breast cancer cells. Phytomedicine 23(1):87–94

Blaas L et al (2016) *Lgr6* labels a rare population of mammary gland progenitor cells that are able to originate luminal mammary tumours. Nat Cell Biol 18(12):1346–1356

Becker JL, Souza GR (2013) Using space-based investigations to inform cancer research on Earth. Nat Rev Cancer 13(5):315–327

Benton G, DeGray G, Kleinman HK, George J, Arnaoutova I (2015) In vitro microtumors provide a physiologically predictive tool for breast cancer therapeutic screening. PLoS ONE 10(4): e0123312

Bersini S et al (2014) A microfluidic 3D in vitro model for specificity of breast cancer metastasis to bone. Biomaterials 35(8):2454–2461

Bischel LL, Beebe DJ, Sung KE (2015) Microfluidic model of ductal carcinoma in situ with 3D, organotypic structure. BMC Cancer 15:12

Buchanan CF et al (2014) Three-dimensional microfluidic collagen hydrogels for investigating flow-mediated tumor-endothelial signaling and vascular organization. Tissue Eng Part C Methods 20(1):64–75

Caddeo S, Boffito M, Sartori S (2017) Tissue engineering approaches in the design of healthy and pathological in vitro tissue models. Front Bioeng Biotechnol 5:40

Campbell JJ, Davidenko N, Caffarel MM, Cameron RE, Watson CJ (2011) A multifunctional 3D co-culture system for studies of mammary tissue morphogenesis and stem cell biology. PLoS ONE 6(9):e25661

Carranza-Torres IE et al (2015) Organotypic culture of breast tumor explants as a multicellular system for the screening of natural compounds with antineoplastic potential. Biomed Res Int 2015:618021

Chanda A et al (2017) Identification of the SUMO E3 ligase PIAS1 as a potential survival biomarker in breast cancer. PLoS ONE 12(5):e0177639

Chandhoke AS, Chanda A, Karve K, Deng L, Bonni S (2017) The PIAS3-Smurf2 sumoylation pathway suppresses breast cancer organoid invasiveness. Oncotarget 8(13):21001–21014

Chen MC, Gupta M, Cheung KC (2010) Alginate-based microfluidic system for tumor spheroid formation and anticancer agent screening. Biomed Microdevices 12(4):647–654

Cheung KJ, Gabrielson E, Werb Z, Ewald AJ (2013) Collective invasion in breast cancer requires a conserved basal epithelial program. Cell 155(7):1639–1651

Choi Y et al (2015) A microengineered pathophysiological model of early-stage breast cancer. Lab Chip 15(16):3350–3357

Coates AS et al (2015) Tailoring therapies—improving the management of early breast cancer: St gallen international expert consensus on the primary therapy of early breast cancer 2015. Ann Oncol 26(8):1533–1546

Colditz GA, Kaphingst KA, Hankinson SE, Rosner B (2012) Family history and risk of breast cancer: nurses' health study. Breast Cancer Res Treat 133(3):1097–1104

Colomer R et al (2017) Biomarkers in breast cancer: a consensus statement by the Spanish Society of Medical Oncology and the Spanish Society of Pathology. Clin Transl Oncol. https://doi.org/10.1007/s12094-017-1800-5

Conde SJ, Luvizotto R de A, de Síbio MT, Noqueira CR (2012) Human breast tumor slices as an alternative approach to cell lines to individualize research for each patient. Eur J Cancer Prev 21(4):333

Corben AD et al (2014) Ex vivo treatment response of primary tumors and/or associated metastases for preclinical and clinical development of therapeutics. J Vis Exp 92:e52157

Dai X, Cheng H, Bai Z, Li J (2017) Breast cancer cell line classification and its relevance with breast tumor subtyping. J Cancer 8(16):3131–3141

Dallas S et al (2016) Generation and characterization of a breast cancer resistance protein humanized mouse model. Mol Pharmacol 89(5):492–504

de Boo J, Hendriksen C (2005) Reduction strategies in animal research: a review of scientific approaches at the intra-experimental, supra-experimental and extra-experimental levels. Altern Lab Animal 33(4):369–377

de Graaf IA et al (2010) Preparation and incubation of precision-cut liver and intestinal slices for application in drug metabolism and toxicity studies. Nat Protoc 5(9):1540–1551

DeRose YS et al (2011) Tumor grafts derived from women with breast cancer authentically reflect tumor pathology, growth, metastasis and disease outcomes. Nat Med 17(11):1514–1520
DeRose YS et al (2013) Patient-derived models of human breast cancer: protocols for in vitro and in vivo applications in tumor biology and translational medicine. Curr Protoc Pharmacol 14 (Unit14):23
Dhurjati R, Krishnan V, Shuman LA, Mastro AM, Vogler EA (2008) Metastatic breast cancer cells colonize and degrade three-dimensional osteoblastic tissue in vitro. Clin Exp Metastasis 25(7):741–752
Drife JO (1986) Breast development in puberty. Ann N Y Acad Sci 464:58–65
Faversani A et al (2014) Survivin family proteins as novel molecular determinants of doxorubicin resistance in organotypic human breast tumors. Breast Cancer Res 16(3):R55
Ferlay J et al (2013) GLOBOCAN 2012 v1.0, Cancer incidence and mortality worldwide: IARC Cancer Base No. 11. Lyon, France: IARC. Available from: http://globocan.iarc.fr. Accessed 20 Nov 2017
Ferlay J et al (2015) Cancer incidence and mortality worldwide: sources, methods and major patterns in GLOBOCAN 2012. Int J Cancer 136(5):E359–E386
Fong EL, Harrington DA, Farach-Carson MC, Yu H (2016) Heralding a new paradigm in 3D tumor modeling. Biomaterials 108:197–213
Francies HE, Barthorpe A, McLaren-Douglas A, Barendt WJ, Garnett MJ (2016) Drug sensitivity assays of human cancer organoid cultures. In: Methods in molecular biology. Humana Press. https://doi.org/10.1007/7651_2016_10
Gallardo Pérez JC, Espinosa Castilla M, Meléndez Zajgla J, Maldonado Lagunas V (2006) Esferoides tumorales para evaluar estrategias anticancerosas. REB 25(4):101–107
Gauck D, Keil S, Niggemann B, Zänker KS, Dittmar T (2017) Hybrid clone cells derived from human breast epithelial cells and human breast cancer cells exhibit properties of cancer stem/ initiating cells. BMC Cancer 17(1):515
Ge WQ, Pu JX, Zheng SY (2012) Clinical application of the adenosine triphosphate-based response assay in intravesical chemotherapy for superficial bladder cancer. Asian Pac J Cancer Prev 13(2):689–692
Gerlinger M et al (2012) Intratumor heterogeneity and branched evolution revealed by multiregion sequencing. N Engl J Med 366(10):883–892
Goldacre MJ, Abisgold JD, Yeates DG, Vessey MP (2010) Benign breast disease and subsequent breast cancer: English record linkage studies. J Public Health (Oxf) 32(4):565–571
Goliwas KF et al (2017) Methods to evaluate cell growth, viability, and response to treatment in a tissue engineered breast cancer model. Sci Rep 7(1):14167
Goodwin PJ et al (2012) Breast cancer prognosis in BRCA1 and BRCA2 mutation carriers: an International prospective breast cancer family registry population-based cohort study. J Clin Oncol 30(1):19–26
Gradishar WJ et al (2016) Invasive breast cancer version 1.2016, NCCN clinical practice guidelines in oncology. J Natl Compr Canc Netw 14(3):324–354
Grantab R, Sivananthan S, Tannock IF (2006) The penetration of anticancer drugs through tumor tissue as a function of cellular adhesion and packing density of tumor cells. Cancer Res 66(2):1033–1039
Griffin G, Clark JM, Zurlo J, Ritskes-Hoitinga M (2014) Scientific uses of animals: harm-benefit analysis and complementary approaches to implementing the three Rs. Rev Sci Tech 33(1):265–272
Grosso SH et al (2013) Breast cancer tissue slices as a model for evaluation of response to rapamycin. Cell Tissue Res 352(3):671–684
Hahn WC, Weinberg RA (2002) Modelling the molecular circuitry of cancer. Nat Rev Cancer 2(5):331–341
Ham SL, Joshi R, Luker GD, Tavana H (2016) Engineered breast cancer cell spheroids reproduce biologic properties of solid tumors. Adv Healthc Mater 5(21):2788–2798

Harris LN et al (2016) Use of biomarkers to guide decisions on adjuvant systemic therapy for women with early-stage invasive breast cancer: American Society of Clinical Oncology clinical practice guideline. J Clin Oncol 34(10):1134–1150

He M et al (2015) The Hedgehog signalling pathway mediates drug response of MCF-7 mammosphere cells in breast cancer patients. Clin Sci (Lond) 129(9):809–822

Hiatt RA, Brody JG (2018) Environmental determinants of breast cancer. Annu Rev Public Health 39:21.1–21.21

Hickman JA et al (2014) Three-dimensional models of cancer for pharmacology and cancer cell biology: capturing tumor complexity in vitro/ex vivo. Biotechnol J 9(9):1115–1128

Hidalgo M et al (2014) Patient-derived xenograft models: an emerging platform for translational cancer research. Cancer Discov 4(9):998–1013

Hoffman RM (2015) Patient-derived orthotopic xenografts: better mimic of metastasis than subcutaneous xenografts. Nat Rev Cancer 15(8):451–452

Holen I, Speirs V, Morrisey B, Blyth K (2017) In vivo models in breast cancer research: progress, challenges and future directions. Dis Model Mech 10(4):359–371

Holliday DL, Brouilette KT, Markert A, Gordon LA, Jones JL (2009) Novel multicellular organotypic models of normal and malignant breast: tools for dissecting the role of the microenvironment in breast cancer progression. Breast Cancer Res 11(1):R3

Holliday DL, Speirs V (2011) Choosing the right cell line for breast cancer research. Breast Cancer Res 13(4):215

Holliday DL et al (2013) The practicalities of using tissue slices as preclinical organotypic breast cancer models. J Clin Pathol 66(3):253–255

Holtfreter JA (1944) A study of the mechanics of gastrulation, Part II. J Exp Zool 95:171–212

Hong JW, Quake SR (2003) Integrated nanoliter systems. Nat Biotechnol 21(10):1179–1183

Inman JL, Robertson C, Mott JD, Bissell MJ (2015) Mammary gland development: cell fate specification, stem cells and the microenvironment. Development 142(6):1028–1042

Ishiguro T et al (2017) Tumor-derived spheroids: relevance to cancer stem cells and clinical applications. Cancer Sci 108(3):283–289

Ivascu A, Kubbies M (2007) Diversity of cell-mediated adhesions in breast cancer spheroids. Int J Oncol 31(6):1403–1413

Jaganathan H et al (2014) Three-dimensional in vitro co-culture model of breast tumor using magnetic levitation. Sci Rep 4:6468

Jing Y, Xu XC, Lotan R, Waxman S, Mira-y-Lopez R (1996) Human breast carcinoma slice cultures retain retinoic acid sensitivity. Braz J Med Biol Res 29(9):1105–1108

Joshi K et al (2011) Method for novel anti-cancer drug development using tumor explants of surgical specimens. J Vis Exp 53:2846

Junttila MR, de Sauvage FJ (2013) Influence of tumour micro-environment heterogeneity on therapeutic response. Nature 501(7467):346–354

Kabat GC et al (2010) A multi-center prospective cohort study of benign breast disease and risk of subsequent breast cancer. Cancer Causes Control 21(6):821–828

Kamb A (2005) What's wrong with our cancer models? Nat Rev Drug Discov 4(2):161–165

Kaur P et al (2011) Human breast cancer histoid: an in vitro 3-dimensional co-culture model that mimics breast cancer tissue. J Histochem Cytochem 59(12):1087–1100

Kenny PA et al (2007) The morphologies of breast cancer cell lines in three-dimensional assays correlate with their profiles of gene expression. Mol Oncol 1(1):84–96

Kim JB, Stein R, O'Hare MJ (2004) Three-dimensional in vitro tissue culture models of breast cancer—a review. Breast Cancer Res Treat 85(3):281–291

Kimlin LC, Casagrande G, Virador VM (2013) In vitro three-dimensional (3D) models in cancer research: an update. Mol Carcinog 52(3):167–182

Knight A (2008) Systematic reviews of animal experiments demonstrate poor contributions toward human healthcare. Rev Recent Clin Trials 3(2):89–96

Koledova Z (2017) 3D coculture of mammary organoids with fibrospheres: a model for studying epithelial-stromal interactions during mammary branching morphogenesis. In: Koledova Z (eds) 3D cell culture. Methods in molecular biology, vol 1612. Humana Press, New York, NY, pp 107–124

Koren S, Bentires-Alj M (2015) Breast tumor heterogeneity: source of fitness, hurdle for therapy. Mol Cell 60(4):537–546

Krishnan V et al (2011) Dynamic interaction between breast cancer cells and osteoblastic tissue: comparison of two- and three-dimensional cultures. J Cell Physiol 226(8):2150–2158

Krumdieck CL, dos Santos JE, Ho KJ (1980) A new instrument for the rapid preparation of tissue slices. Anal Biochem 104(1):118–123

Lacroix M et al (2004) Gene regulation by phorbol 12-myristate 13-acetate in MCF-7 and MDA-MB-231, two breast cancer cell lines exhibiting highly different phenotypes. Oncol Rep 12(4):701–707

Lakhani SR, Ellis IO, Schnitt SJ, Tan PH, van de Vijver MJ (eds) (2012) WHO Classification of tumours of the breast, 4th edn. IARC, Lyon, France

Langley GR et al (2017) Towards a 21st-century roadmap for biomedical research and drug discovery: consensus report and recommendations. Drug Discov Today 22(2):327–339

LaRue KE, Khalil M, Freyer JP (2004) Microenvironmental regulation of proliferation in multicellular spheroids is mediated through differential expression of cyclin-dependent kinase inhibitors. Cancer Res 64(5):1621–1631

Leighton J (1951) A sponge matrix method for tissue culture; formation of organized aggregates of cells in vitro. J Natl Cancer Inst 12(3):545–561

Levenson AS, Jordan VC (1997) MCF-7: the first hormone-responsive breast cancer cell line. Cancer Res 57(15):3071–3078

Li G, Zhao F, Cui Y (2013) Proteomics using mammospheres as a model system to identify proteins deregulated in breast cancer stem cells. Curr Mol Med 13(3):459–463

Lin RZ, Chang HY (2008) Recent advances in three-dimensional multicellular spheroid culture for biomedical research. Biotechnol J 3(9–10):1172–1184

Linnemann JR, Meixner LK, Miura H, Scheel CH (2017) An organotypic 3D assay for primary human mammary epithelial cells that recapitulates branching morphogenesis. In: Koledova Z (eds) 3D cell culture. Methods in molecular biology, vol 1612. Humana Press, New York, NY, pp 125–137

Mak IW, Evaniew N, Ghert M (2014) Lost in translation: animal models and clinical trials in cancer treatment. Am J Transl Res 6(2):114–118

Marquette C, Nabell L (2012) Chemotherapy-resistant metastatic breast cancer. Curr Treat Options Oncol 13(2):263–275

Mavaddat N et al (2012) Pathology of breast and ovarian cancers among BRCA1 and BRCA2 mutation carriers: results from the Consortium of Investigators of Modifiers of BRCA1/2 (CIMBA). Cancer Epidemiol Biomark Prev 21(1):134–147

Mestres P, Morguet A, Schmidt W, Kob A, Thedinga E (2006) A new method to assess drug sensitivity on breast tumor acute slices preparation. Ann N Y Acad Sci 1091:460–469

Mestres-Ventura P, Morguet A, Schofer A, Laue M, Schmidt W (2005) Application of silicon sensor technologies to tumor tissue in vitro. In: Blumenthal RD (eds) Chemosensitivity: volume II. Methods in molecular medicine™, vol 111. Humana Press, Totowa, NJ, pp 109–125

Milani C et al (2010) Human breast tumor slices: a model for identification of vitamin D regulated genes in the tumor microenvironment. J Steroid Biochem Mol Biol 121(1–2):151–155

Milani C et al (2013) Transcriptional effects of 1,25 dihydroxyvitamin D(3) physiological and supra-physiological concentrations in breast cancer organotypic culture. BMC Cancer 13:119

Mira-y-Lopez R, Ossowski L (1987) Hormonal modulation of plasminogen activator: an approach to prediction of human breast tumor responsiveness. Cancer Res 47(13):3558–3564

Mira-y-Lopez R, Ossowski L (1990) Preservation of steroid hormone receptors in organ cultures of human breast carcinomas. Cancer Res 50(1):78–83

Morrissey B et al (2015) SEARCHBreast Workshop Proceedings: 3D modelling of breast cancer. Altern Lab Anim 43(6):367–375

Moscona A (1952) Cell suspensions from organ rudiments of chick embryos. Exp Cell Res 3(3):535–539

Muraro MG et al (2017) Ex-vivo assessment of drug response on breast cancer primary tissue with preserved microenvironments. Oncoimmunology 6(7):e1331798

Nagelkerke A, Bussink J, Sweep FC, Span PN (2013) Generation of multicellular tumor spheroids of breast cancer cells: how to go three-dimensional. Anal Biochem 437(1):17–19

Naipal KA et al (2014) Functional ex vivo assay to select homologous recombination-deficient breast tumors for PARP inhibitor treatment. Clin Cancer Res 20(18):4816–4826

Naipal KA et al (2016) Tumor slice culture system to assess drug response of primary breast cancer. BMC Cancer 16:78

Nallanthighal S, Elmaliki KM, Reliene R (2017) Pomegranate extract alters breast cancer stem cell properties in association with inhibition of epithelial-to-mesenchymal transition. Nutr Cancer 69(7):1088–1098

Netti PA, Berk DA, Swartz MA, Grodzinsky AJ, Jain RK (2000) Role of extracellular matrix assembly in interstitial transport in solid tumors. Cancer Res 60(9):2497–2503

Neve RM et al (2006) A collection of breast cancer cell lines for the study of functionally distinct cancer subtypes. Cancer Cell 10(6):515–527

Nguyen-Ngoc KV et al (2012) ECM microenvironment regulates collective migration and local dissemination in normal and malignant mammary epithelium. Proc Natl Acad Sci USA 109(39):E2595–E2604

Nissen E, Tanneberger S, Weiss H, Bender E (1983) In vitro cultivation of vital tissue slices: a new variation of organ culture technics. Biomed Biochim Acta 42(7–8):907–916

Nissen E, Tanneberger S, Weiss H, Arnold W (1984) In vitro cell and tissue cultures as a model for studying human tumors. Arch Geschwulstforsch 54(2):173–180

Oudar O (2000) Spheroids: relation between tumour and endothelial cells. Crit Rev Oncol Hematol 36(2–3):99–106

Pampaloni F, Steltzer EH, Masoti A (2009) Three-dimensional tissue models for drug discovery and toxicology. Recent Pat Biotechnol 3(2):103–117

Parajuli N, Doppler W (2009) Precision-cut slice cultures of tumors from MMTV-neu mice for the study of the *ex vivo* response to cytokines and cytotoxic drugs. Vitro Cell Dev Biol Anim 45(8):442–450

Pauli C et al (2017) Personalized in vitro and in vivo cancer models to guide precision medicine. Cancer Discov 7(5):462–477

Pennington K, Chu QD, Curiel DT, Li BD, Mathis JM (2010) The utility of a tissue slice model system to determine breast cancer infectivity by oncolytic adenoviruses. J Surg Res 163(2):270–275

Pound P, Bracken MB (2014) Is animal research sufficiently evidence based to be a cornerstone of biomedical research? BMJ 348:g3387

Pörtner R, Giese C (2007) An overview on bioreactor design, prototyping and process control for reproducible three-dimensional tissue culture. In: Marx U, Sandig V (eds) Drug testing in vitro: breakthroughs and trends in cell culture technology, Wiley-VCH Verlag GmbH & Co. KGaA, Weinheim, Germany, pp 53–78

Rajendran S et al (2011) Preclinical evaluation of gene delivery methods for the treatment of loco-regional disease in breast cancer. Exp Biol Med (Maywood) 236(4):423–434

Ravi M, Paramesh V, Kaviya SR, Anuradha E, Solomon FD (2015) 3D cell culture systems: advantages and applications. J Cell Physiol 230(1):16–26

Razzaghi H et al (2012) Mammographic density and breast cancer risk in white and african american women. Breast Cancer Res Treat 135(2):571–580

Reid BG et al (2014) Live multicellular tumor spheroid models for high-content imaging and screening in cancer drug discovery. Curr Chem Genom Transl Med 8(Suppl 1):27–35

Riffle S, Hegde RS (2017) Modeling tumor cell adaptations to hypoxia in multicellular tumor spheroids. J Exp Clin Cancer Res 36(1):102

Ritte R et al (2013) Height, age at menarche and risk of hormone receptor-positive and -negative breast cancer: a cohort study. Int J Cancer 132(11):2619–2629

Roberts TG Jr et al (2004) Trends in the risks and benefits to patients with cancer participating in phase 1 clinical trials. JAMA 292(17):2130–2140
Rodríguez CE, Reidel SI, de Kier Bal, Joffé ED, Jasnis MA, Fiszman GL (2015) Autophagy protects from trastuzumab-induced cytotoxicity in HER2 overexpressing breast tumor spheroids. PLoS ONE 10(9):e0137920
Ross JS et al (2003) Breast cancer biomarkers and molecular medicine. Expert Rev Mol Diagn 3(5):573–585
Sato T et al (2009) Single Lgr5 stem cells build crypt-villus structures in vitro without a mesenchymal niche. Nature 459(7244):262–265
Sato T et al (2011) Long-term expansion of epithelial organoids from human colon, adenoma, adenocarcinoma, and Barrett's epithelium. Gastroenterology 141(5):1762–1772
Schnitt SJ, Lakhani SR (2014) Breast cancer. In: Stewart BW, Wild CP (eds) World cancer report 2014, 1st edn. IARC, Lyon, France, pp 362–373
Senkus E et al (2015) Primary breast cancer: ESMO Clinical practice Guidelines for diagnosis, treatment and follow up. Ann Oncol 26(Suppl 5):v8–v30
Shekhar MP, Werdell J, Santner SJ, Pauley RJ, Tait L (2001) Breast stroma plays a dominant regulatory role in breast epithelial growth and differentiation: implications for tumor development and progression. Cancer Res 61(4):1320–1326
Sherman-Baust CA et al (2003) Remodeling of the extracellular matrix through overexpression of collagen VI contributes to cisplatin resistance in ovarian cancer cells. Cancer Cell 3(4): 377–386
Siegel RL, Miller KD, Jemal A (2017) Cancer statistics. CA Cancer J Clin 67(1):7–30
Smith PF et al (1985) Dynamic organ culture of precision liver slices for in vitro toxicology. Life Sci 36(1):1367–1375
Sobrino A et al (2016) 3D microtumors in vitro supported by perfused vascular networks. Sci Rep 6:31589
Sokol ES et al (2016) Growth of human breast tissues from patient cells in 3D hydrogel scaffolds. Breast Cancer Res 18(1):19
Song JW et al (2009) Microfluidic endothelium for studying the intravascular adhesion of metastatic breast cancer cells. PLoS ONE 4(6):e5756
Sonnenberg M et al (2008) Highly variable response to cytotoxic chemotherapy in carcinoma-associated fibroblasts (CAFs) from lung and breast. BMC Cancer 8:364
Speirs V et al (1998) Short-term primary culture of epithelial cells derived from human breast tumours. Br J Cancer 78(11):1421–1429
Stoff-Khalili MA et al (2005) Preclinical evaluation of transcriptional targeting strategies for carcinoma of the breast in a tissue slice model system. Breast Cancer Res 7(6):R1141–R1152
Stone KR, Craig RB, Palmer JO, Rivkin SE, McDivitt RW (1985) Short-term cryopreservation of human breast carcinoma cells for flow cytometry. Cytometry 6(4):357–361
Sung KE et al (2011) Transition to invasion in breast cancer: a microfluidic in vitro model enables examination of spatial and temporal effects. Integr Biol (Camb) 3(4):439–450
Sutherland RM (1988) Cell and environment interactions in tumor microregions: the multicell spheroid model. Science 240(4849):177–184
Sutherland RM, McCredie JA, Inch WR (1971) Growth of multicell spheroids in tissue culture as a model of nodular carcinomas. J Natl Cancer Inst 46(1):113–120
Tanneberger S, Bacigalupo G (1970) Experiences with individual cytostatic therapy of malignant tumors following pretherapeutic sensitivity test to cytostatics in vitro (oncobiogram). Arch Geschwulstforsch 35(1):44–53
Tannerberger S, Mohr A (1973) Biological characterization of human tumours by means of organ culture and individualized cytostatic cancer treatment. Arch Geschwulstforsch 42(4):307–315
Tsui JH, Lee W, Pun SH, Kim J, Kim DH (2013) Microfluidics-assisted in vitro drug screening and carrier production. Adv Drug Deliv Rev 65(11–12):1575–1588
Van der Kuip H et al (2006) Short term culture of breast cancer tissues to study the activity of the anticancer drug taxol in an intact tumor environment. BMC Cancer 6:86

Vesci L et al (2015) Trastuzumab and docetaxel in a preclinical organotypic breast cancer model using tissue slices from mammary fat pad: Translational relevance. Oncol Rep 34(3): 1146–1152

Visvader JE, Stingl J (2014) Mammary stem cells and the differentiation hierarchy: current status and perspectives. Genes Dev 28(11):1143–1158

Walenta S, Doetsch J, Mueller-Klieser W, Kunz-Schughart LA (2000) Metabolic imaging in multicellular spheroids of oncogene-transfected fibroblasts. J Histochem Cytochem 48(4): 509–522

Walsh AJ et al (2014) Quantitative optical imaging of primary tumor organoid metabolism predicts drug response in breast cancer. Cancer Res 74(18):5184–5194

Walsh AJ, Cook RS, Sanders ME, Arteaga CL, Skala MC (2016) Drug response in organoids generated from frozen primary tumor tissues. Sci Rep 6:18889

Wagstaff A (2017) Lisa Hutchinson: teasing out the signal from the noise. In: Cancer World Magazine (79). European School of Oncology, pp 26–28. http://cancerworld.net/wp-content/uploads/2017/08/CW79_-Profile.pdf. Accessed 25 Nov 2017

Weeber F, Ooft SN, Dijkstra KK, Voest EE (2017) Tumor organoids as a pre-clinical cancer model for drug discovery. Cell Chem Biol 24(9):1092–1100

Whittle JR, Lewis MT, Lindeman GJ, Visvader JE (2015) Patient derived xenograft models of breast cancer and their predictive power. Breast Cancer Res 17:17

Williams SA, Anderson WC, Santaguida MT, Dylla SJ (2013) Patient-derived xenografts, the cancer stem cell paradigm, and cancer pathobiology in the 21st century. Lab Invest 93(9): 970–982

World Health Organization (2017) Cancer fact sheet. http://www.who.int/mediacentre/factsheets/fs297/en/. Accessed 23 Nov 2017

Xu X, Farach-Carson MC, Jia X (2014) Three-dimensional in vitro tumor models for cancer research and drug evaluation. Biotechnol Adv 32(7):1256–1268

Zheng MJ et al (2015) Human mammary microenvironment better regulates the biology of human breast cancer in humanized mouse model. Med Oncol 32(2):8

Zheng MJ et al (2016) Screening and analysis of breast cancer genes regulated by the human mammary microenvironment in a humanized mouse model. Oncol Lett 2(6):5261–5268

Preclinical Organotypic Models for the Assessment of Novel Cancer Therapeutics and Treatment

Carol Ward, James Meehan, Mark Gray, Ian H. Kunkler, Simon P. Langdon, Alan Murray and David Argyle

Contents

C. Ward (✉) · M. Gray · D. Argyle
The Royal (Dick) School of Veterinary Studies and Roslin Institute, University of Edinburgh, Easter Bush, Roslin, Midlothian, EH25 9RG Edinburgh, UK
e-mail: C.Ward@ed.ac.uk

M. Gray
e-mail: s9900757@sms.ed.ac.uk

D. Argyle
e-mail: david.argyle@roslin.ed.ac.uk

C. Ward · J. Meehan · M. Gray · I. H. Kunkler · S. P. Langdon
Cancer Research UK Edinburgh Centre and Division of Pathology Laboratories, Institute of Genetics and Molecular Medicine, University of Edinburgh, Crewe Road South, EH4 2XU Edinburgh, UK
e-mail: j.meehan@hw.ac.uk

I. H. Kunkler
e-mail: iankunkler@yahoo.com

S. P. Langdon
e-mail: simon.langdon@ed.ac.uk

J. Meehan
School of Engineering and Physical Sciences, Institute of Sensors, Signals and Systems, Heriot-Watt University, EH14 4AS Edinburgh, UK

A. Murray
School of Engineering, Faraday Building, The King's Buildings, Mayfield Road, EH9 3JL Edinburgh, UK
e-mail: Alan.Murray@ed.ac.uk

Current Topics in Microbiology and Immunology (2021) 430: 225–246
https://doi.org/10.1007/82_2019_159

Published Online: 12 March 2019

Abstract The immense costs in both financial terms and preclinical research effort that occur in the development of anticancer drugs are unfortunately not matched by a substantial increase in improved clinical therapies due to the high rate of failure during clinical trials. This may be due to issues with toxicity or lack of clinical effectiveness when the drug is evaluated in patients. Currently, much cancer research is driven by the need to develop therapies that can exploit cancer cell adaptations to conditions in the tumor microenvironment such as acidosis and hypoxia, the requirement for more-specific, targeted treatments, or the exploitation of 'precision medicine' that can target known genomic changes in patient DNA. The high attrition rate for novel anticancer therapies suggests that the preclinical methods used in screening anticancer drugs need improvement. This chapter considers the advantages and disadvantages of 3D organotypic models in both cancer research and cancer drug screening, particularly in the areas of targeted drugs and the exploitation of genomic changes that can be used for therapeutic advantage in precision medicine.

1 Introduction

The simple definition of 'organoid' is 'resembling an organ,' implying that it should contain more than one cell type that is typically found in the organ of its origin, demonstrate an organ specific function and that the cells should show some degree of organization. However, given that in this case the 'organ' being modeled is a tumor, most if not all 3D models meet these criteria. 3D culture can encompass various types of model such as spheroids and organoids (terms which are often used interchangeably), or explants. Arguably, xenografts are also a type of 3D culture that fits the tumor organoid definition. For the sake of clarity, the various 3D systems meeting these criteria will be discussed here separately. For more information on the definition and historical perspective of organoid development, see Simian and Bissell 2017.

Although extensive funding supports cancer research each year, the majority of novel drugs and therapies fail to translate into clinical practice due to lack of success in clinical trials, either because of deficiencies in clinical efficacy, or issues with toxicity (Petsko 2010; Arrowsmith 2011). Approximately one drug in 5000–10,000 tested will gain FDA approval, with attrition rates in oncology particularly high with only 5% of anticancer therapies successfully undergoing phase I/II clinical trials (Zamboni et al. 2012; Ocana et al. 2011). This is despite strong and significant in vitro and in vivo preclinical data, suggesting that better preclinical methodologies are required to identify new therapeutics (Kamb 2005; Caponigro et al. 2011; Singh and Ferrara 2012; Siolas and Hannon 2013). The assessment of novel effective anticancer compounds is a compromise between the use of high-throughput, low-cost testing strategies, and high-cost, low-throughput methods which may give better translational results.

Most preclinical studies into the efficacy of anticancer drugs use cancer cell lines to assess various parameters such as cell survival, proliferation, cell death, migration, and invasion. 2D cultures are one of the most cost-effective strategies for high-throughput assays and are used to test large numbers of compounds and select those for further examination. These cell lines can be exploited to investigate the role of specific pathways, genes, or molecules using technologies such as CRISPR (Sander and Joung 2014; Ledford 2015), siRNA (Moore et al. 2010), or inducible transgenics (Pajic et al. 2000; Belteki et al. 2005; Saunders 2011; Xie et al. 2014). However, these systems lack the heterogeneity of the original tumor, since cell culture selects those clones that best proliferate and survive in culture conditions (Daniel et al. 2009; Hensley et al. 2016). Using a panel of cell lines of the cancer of interest, or from different tumors such as the NCI-60 panel, or the de novo expansion of cell lines from tumor tissue, can partially overcome this problem (Burdall et al. 2003; Langdon 2004), but 2D culture cannot replicate the complex stromal elements of tumors, interactions with immune cells, inflammatory mediators, growth factors, or the acidic/hypoxic conditions found in the tumor microenvironment (TME) (Karar and Maity 2009; Hanahan and Weinberg 2011; Ward et al. 2013). Targeting the adaptations, cancer cells adopt to survive in the tumor microenvironment, and it is currently a novel approach to the treatment of solid tumors, and more appropriate models are required to support this research (McAllister and Weinberg 2010; Pettersen et al. 2015).

2 3D in Vivo Models

(a) ***Mouse xenografts***

Following assessment in 2D model systems, candidate compounds move into in vivo studies: generally, mouse xenografts growing human tumor cell lines in strains of immunocompromised mice. These have shown mixed efficacy in translating novel drugs into clinical use, and often reproducibility of results is poor (Sausville and Burger 2006; Boedigheimer et al. 2013), but they do allow in vivo monitoring of tumor growth, drug toxicity, efficacy, and pharmacokinetic studies and will reflect the oxygen and pH gradients found in solid tumors. However, some cell lines do not form xenografts, metastatic models are limited, and immune responses, which affect tumor growth are compromised. Although xenografts incorporate some stromal elements, they are murine rather than human (Daniel et al. 2009). Xenografts are selected for rapidity of growth, while human tumors can take years to progress clinically; therefore, these models do not replicate the heterogeneity of the original tumor. Many xenografts are implanted subcutaneously and therefore do not reproduce the specific microenvironment of the tumor under investigation (Polilti and Pao 2011). Orthotopic xenografts enable tumor development in the organ of origin; this can allow evaluation in a preclinical tumor model which more closely mimics the disease process in humans. However, these models can be more difficult to establish compared with subcutaneous implantation.

(b) ***Genetically engineered mouse models (GEMMs)***

Spontaneous or carcinogen-generated tumors reflect the stromal interactions and angiogenesis found with tumor progression, but in a murine system with functioning immune responses. In GEMMs, tumors originate in a group of cells as a consequence of germline or somatic mutations and share the genetic heterogeneity and histopathology of human tumors. They allow expression of oncogenes in a tissue-specific manner as well as conditional expression or deletion of oncogenes and tumor suppressors, allowing many cancers to be modeled (Sharpless and Depinho 2006; Politi and Pao 2011). GEMMs are also useful to model treatment resistance, chemotherapy enhancement, co-treatments, and biomarker validation (Pietras and Hanahan 2005; Bell-McGuinn et al. 2007; Faca et al. 2008; Pitteri et al. 2009; Bellmunt et al. 2010; Polti et al. 2010). Several studies show that GEMMs model both the TME and stroma more accurately than xenografts (Olive et al. 2009; Graves et al. 2010) and better replicate the sensitivity of cancer to systemic therapy in clinical trials (Singh et al. 2010; Chen et al. 2012). However, the model is still murine.

(c) ***Patient-derived xenografts (PDXs)***

PDXs, involving the transfer of patient-derived tumor tissue into immunocompromised mice, (Kopetz et al. 2012; Julien et al. 2012; Huang et al. 2014), show pathology, growth, and metastatic outcomes that closely correspond to the original tumor and sustain the original histological markers, gene expression profiles, and proliferation indices (DeRose et al. 2011; Dean et al. 2012; Tentler et al. 2012; Stewart et al. 2017). This means that experimental data can be complemented with clinical information, such as response to treatment in the original patient. The use of PDXs is increasing in preclinical studies (Hidalgo et al. 2011; Tentler et al. 2012), in conjunction with 'co-clinical trials' which run preclinical PDX and clinical trials at the same time (Nardella et al. 2011). For example, one study demonstrated that PDXs developed from individual patient tumors could be used to select treatment options with an 88% response rate (Hidalgo et al. 2011). PDXs have also been used to model human phase II trials that can examine differences in therapy responses across different cancers (Townsend et al. 2016; Byrne et al. 2017). PDX models have now superseded the NCI-60 panel as the gold standard for cancer research because of the preservation of heterogeneity (Leford 2016).

While this system is more robust, PDXs do have some shortcomings. Passaging is time consuming, and some human stromal tumor elements are progressively lost over time. The rate of engraftment is poor for some cancer types such as prostate and ER^+ breast cancer (DeRose et al. 2011; Lawrence et al. 2013), and tissue from needle biopsies is often inadequate for growth. There is still deficiency in immune responses and increased growth rate of tumors compared with human cancers (Rayal et al. 2012; Huang et al. 2014), although tumor growth and drug validation in these models can take several months (Hildalgo et al. 2011; Gao and Chen 2015).

The same tumor sample can produce PDXs which may differ in genetics, or cell-surface markers, depending on the mouse strains used for tumor growth (Klco et al. 2014). The ratio of cancer cells to stromal cells can also drift; for example, pancreatic ductal adenocarcinomas (PDAs) contain high numbers of stromal cells (Guillaumond et al. 2013; Beloribi-Djefaflia et al. 2015; Delitto et al. 2015), but because cancer cells multiply at a greater rate than stromal cells, stromal/cancer cell interactions change as cell ratios change (Martinez-Garcia et al. 2014). A recent study of 1,100 PDX models across 24 types of cancer showed sizable phenotypic and genomic changes between the progression of PDX models and human cancers, with PDXs developing both chromosome and gene copy number alterations during passages (Ben-David et al. 2017), which could influence the response to drug treatment. Metabolic reprogramming has been identified as a key hallmark of cancer (Hanahan and Weinberg 2011), leading to increased interest in targeting specific enzymes as a novel therapeutic strategy in this disease. However, studies using human lung cancer tissue show that both glycolysis and glutaminolysis are increased when human lung tumor tissue is cultured in PDX models (with murine stroma) when compared with the same tissue cultured with human stroma (using tissue slices), implying that PDX models may not be predictive of patient responses in the context of drugs which target tumor metabolism (Lane et al. 2016).

PDXs are not suitable for high-throughput screening, or genetic manipulation (Gao and Chen 2015), but are extremely useful in situations where primary tumor material is scarce or rare, allowing long-term culture of much more diverse and heterogeneous tumor samples (Zhang et al. 2013). They can be developed from treatment-resistant or recurrent tumor tissue (Aparicio et al. 2015; Townsend et al. 2016; Stewart et al. 2017), allowing the mechanisms involved to be explored. Various depositories of diverse treated/untreated tumor PDXs have been produced or are being developed, (EuroPDX consortium, the Public Repository of Xenografts, and the National Cancer Institute—Patient-derived Models Repository) some of which are open source, such as that provided by Stewart et al., encompassing pediatric solid tumors, from 168 patients and 12 cancer types (Klco et al. 2014; Gao et al. 2015; Stewart et al. 2017; Townsend et al. 2016; Bruna et al. 2016).

(d) ***Improving animal models***

Animal models have been improved by 'humanizing' some aspects of the host, introducing human stromal elements or immune cells and thus humanizing the TME to some extent and allowing new immunotherapies to be explored (Proia 2006; Vudattu et al. 2014; Morton et al. 2016). Such methods have been used to engineer the primary tumor environment or mimic metastatic niches in breast and prostate cancers (Wang et al. 2010; Thibaudeau et al. 2014; Hesami et al. 2014; Holzapfel et al. 2014). One group has reported humanizing immunocompromised mice using aspirated hematopoietic stem and progenitor cells from bone marrow biopsies taken from cancer patients (Werner-Klein et al. 2014). However, the main obstacles to the use of animal models remain high cost and low throughput. The pros and cons of these animal models are outlined in Table 1.

Table 1 Listing the advantages and disadvantages of in vivo tumor models

	Strengths	Weaknesses
Mouse Xenografts	– Allow in vivo monitoring of tumor growth, drug toxicity, and efficacy – Reflect in vivo pH and oxygen gradients	– Limited metastatic models – Compromised immune responses – Murine stromal elements
GEMMs	– Share the genetic heterogeneity and histopathology of human tumors – Replicate both the TME and stroma more accurately than xenografts	– Model is still murine
PDXs	– Pathology, growth, and metastatic outcomes closely correspond to the original tumor – Preserve the heterogeneity of different cancers – Useful in cases where primary material is rare	– Availability/access to freshtumor tissue is a limiting factor – Passaging is time consuming – Stromal tumor elements lost over time – Rate of engraftment is poor for some cancer types – Deficiency in immune response

3 3D In Vitro Models

Xenograft models can have long latency periods; PDXs are maintained in a murine host, and the availability and access to fresh human tumor tissue is a limiting factor for many researchers. Further, the trend in ethical research is to try to limit the use of animals wherever possible. There is therefore, a strong need for low-cost, high-throughput, clinically and physiologically relevant models for cancer research. Several studies have illustrated that 3D gene signatures can predict prognosis in several independent datasets (Fournier et al. 2006; Martin et al. 2008), suggesting that 3D in vitro 'organoid' models may help bridge the gap between preclinical translation and clinical trials.

Cancer stem cells (CSCs) form a subpopulation of cancer cells within a tumor that have been attributed with the ability to initiate primary, metastatic, and recurrent tumor growth, and to be involved in treatment resistance (Mitra et al. 2015; Bomken et al. 2010). These cells have become an obvious focus for anti-cancer research. Tumor spheroids and organoids (see below) are both enriched for populations of CSCs (Lancaster and Knoblich 2014; Ishiguro et al. 2017; Capodanno et al. 2018). This allows for fast expansion of both organoid and spheroid culture systems, once the initial population of these 3D structures has been established. Further, both of these models can be used to harvest CSCs for research purposes (Ishiguro et al. 2017; Wang et al. 2018), an improvement on the traditional methodology for isolation and enrichment of these cells. CSCs from other mammals, for example from canine cancers, can also be obtained from this approach aiding studies in comparative and veterinary oncology (Capodanno et al; 2018).

(a) ***Spheroids***

Many cancer cell lines can be grown as small avascular spheroids using 3D scaffolds, hydrogels/collagen gels, or suspension culture as illustrated in Fig. 1 (Frederich et al. 2009; Li and Lu 2011; Verjans et al. 2017). These generate a more physiological model for cancer drug testing, displaying increased heterogeneity, cell–cell and cell–extracellular matrix (ECM) interactions, and differential gene/protein expression in comparison with monolayer cultures (Yamada and Cukierman 2007; Loessner et al. 2010). Drug and cell signaling responses can differ when 3D and 2D methodologies are compared (dit Faute et al. 2002; Yamada and Cukierman 2007; Loessner et al. 2010; Ravi et al. 2014). For example, in SKBR-3 breast cancer cells, the HER2 and HER3 receptors heterodimerize in 2D cultures, but HER2 forms homodimers in spheroids, with an increased response to HER2 inhibition (Pickl and Ries 2009). 3D cultures of human pancreatic, ovarian, liver, and lung cancer cells exhibit increased chemoresistance in contrast to 2D culture via mechanisms that involve interactions with the ECM/TME (Sethi et al. 1999; Loessner et al. 2010; Longati et al. 2013; Ekert et al. 2014).

In solid tumors, rapid cell growth induces gradients of hypoxia and acidosis to develop, affecting sensitivity to radiation and chemotherapy (Gatenby et al. 2007; Correia and Bissell 2012; Ward et al. 2013). Although spheroids are grown under atmospheric oxygen tensions, they develop O_2, CO_2, pH, and metabolite gradients as they reach approximately 400–600 uM in size, modeling the biology of solid tumors as illustrated in Fig. 1a and b. (Minchinton and Tannock 2006; Pouyssegur

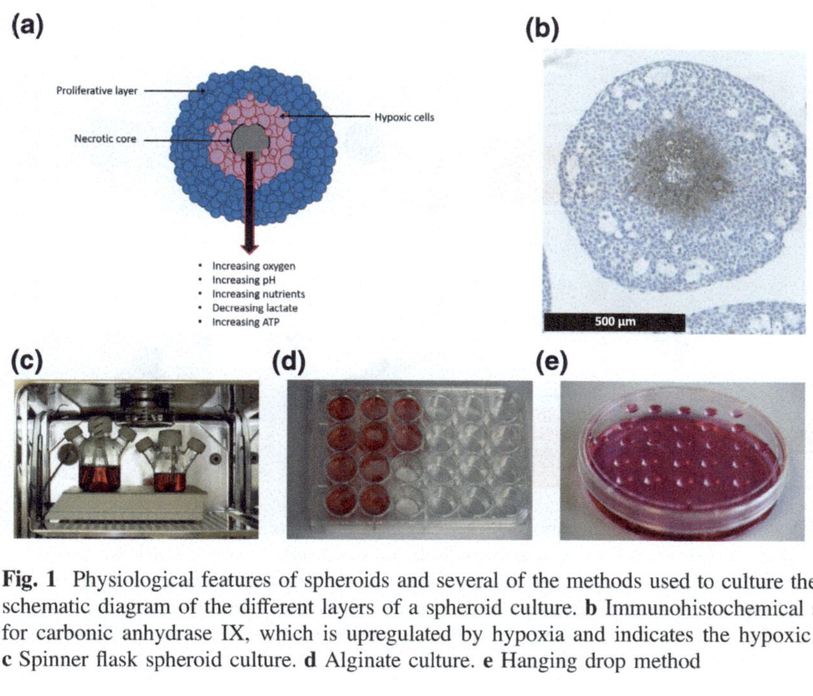

Fig. 1 Physiological features of spheroids and several of the methods used to culture them. **a** A schematic diagram of the different layers of a spheroid culture. **b** Immunohistochemical staining for carbonic anhydrase IX, which is upregulated by hypoxia and indicates the hypoxic region. **c** Spinner flask spheroid culture. **d** Alginate culture. **e** Hanging drop method

et al. 2006; Friedrich et al. 2009). These 3D models are now a necessity, particularly since novel treatment approaches targeting both pH regulation and/or glycolysis in hypoxic tumors are expected to be useful in combination with systemic and/or radiation therapy (Riffle et al. 2017). One reason for translational drug failure may be the lack of exploration of the effects of new drugs in conditions found in the TME.

Spheroids can be produced from cancer cells adapted by CRISPR, siRNA, shRNA, or inducible transgenics and can incorporate other cell types such as fibroblasts or be embedded in ECM proteins to examine cell–cell and cell–matrix interactions, migration, and invasion (Schmeichel and Bissell 2003; Heneweer et al. 2005; Hsiao et al. 2009; Ingram et al. 2010; Herrmann et al. 2014; McCracken et al. 2014; Ward et al. 2015) (Fig. 2a). For example, the leading edge of pancreatic cancer cells contains RhoA-dependent projections which can be targeted by anti-metastatic treatments; these were first identified in 3D culture models (Timpson et al. 2011). Tumor-derived human and murine breast cancer spheroids embedded into collagen showed that cells expressing basal epithelial markers lead to collective invasion, both in 3D culture and in in vivo studies across breast cancer subtypes (Cheung et al. 2013).

Co-culture of spheroids and cancer-associated fibroblasts (CAFs) demonstrated that CAFs influence treatment resistance and invasion (Crawford et al. 2009; Li and Lu 2011; Horie et al. 2012; Straussman et al. 2012; Clark et al. 2013; Jaganathan et al. 2014; Kim et al. 2015). Co-culture systems have been used to replicate the metastatic niche of prostate cancer cells in bone using co-culture with osteoblasts

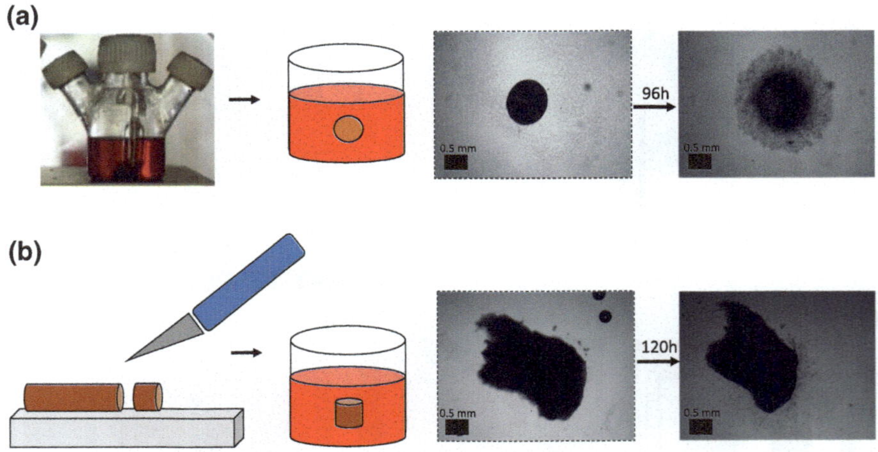

Fig. 2 3D invasion assays using human breast cancer cell spheroids and human breast cancer tumor explants. **a** Spheroids grown in spinner flasks and embedded in collagen I. Images were obtained at 0 h and 96 h of culture. **b** Tumor explants prepared by trimming fat from biopsy samples and cutting into pieces approximately 1 mm^3, before embedding in collagen I. Images were obtained at 0 and 120 h of culture. Images were acquired using phase contrast microscopy. Original magnification x25

and endothelial cells (Hsiao et al. 2009). Other 3D co-culture systems, using niches from lung or bone marrow stroma, demonstrated the role of TSP1 in tumor dormancy (Ghajar et al. 2013), while the relationship between brain endothelial cells and cancer stem cells has been examined in glioblastoma using 3D models of the perivascular niche (Infanger et al. 2013). A 3D co-culture of pancreatic stellate cells with PDA cells has produced a model that could be used to investigate PDA–stroma interactions in high-throughput assays for drug screening (Ware et al. 2016). Spheroid generation using a defined size microwell assay with non-adherent hydrogels can generate uniform spheroids from cervical, breast, and head and neck carcinoma cells within 3–6 days (Singh et al. 2015, 2016), making this system useful for high-throughput assays.

(b) ***Organoids***

Organoids are created by culturing tissue samples with factors to stimulate the replication of pluripotent stem cells. It can take 3 weeks for organoids to grow to approximately 1 mm^3 at which stage they contain the major cell types of the original tissue (Lancaster and Knoblich 2014). Tumor organoids can be developed from tumor tissue or from isolated CSCs and replicate many of the features of the primary tumor. For example, glioblastoma organoids from patient-derived CSCs demonstrated heterogeneity, a proliferative outer rim surrounding a hypoxic interior of senescent cells with dispersed quiescent radioresistant CSCs. Tumors generated from these patient-derived organoids showed the histological features of the parental tumor (Hubert et al. 2016).

Organoids can also be developed from murine tumor models. These have been used to explore the role of β-catenin and PI3 K in colorectal cancer; a screen of inhibitors using this system associated 4EPB1 and Akt with cellular survival and motility in disease progression (Riemer et al. 2017). They have been developed from resected tissue and biopsies from colorectal (CRC) and prostate cancer, breast cancer, PDA, liver cancer, metaplastic epithelia from Barrett's esophagus, and also from circulating tumor cells (Sato et al. 2011; Gao et al. 2014; Boj et al. 2015; Weeber et al. 2015; Broutier et al. 2017; Schutte et al. 2017, Sachs et al. 2018).

Tumor organoids contain differentiated cell types, are biologically stable, and can be frozen for later use and expanded once established (Sato et al. 2011; Sachs and Clevers 2014). Organoids allow the culture of tumors from individual patients, while maintaining the genetic and morphological diversity found within the corresponding patient tumor (van de Wetering et al. 2015; Weeber et al. 2015; Broutier et al. 2017; Pauli et al. 2017); for example, CRC organoids displayed a high degree of genetic similarity with the tumor tissue of origin (Schutte et al. 2017). PDA organoids generated from different stages of the cancer accurately replicated tumor progression based on transcriptomic and proteomic analysis (Boj et al. 2015) and demonstrated little change in histology, morphology, cytology, or expression of differentiation markers from the originating tumor after 16 days in culture (Huang et al. 2015). It remains to be established whether these similarities are sustained long term. Although several studies suggest that mutation patterns are maintained during multiple passages (van de Wetering et al. 2015; Weeber et al. 2015; Fujii et al. 2016; Pauli et al. 2017; Schutte et al. 2017), one study using fluorescent markers showed that one color predominated after 30–40 days in culture, suggesting clonal drift over time (Fujii et al. 2016). Organoids can also provide tissue for the study of rare cancers or those where cell line models are scarce. A further advantage is that control organoids from the same patient can be developed from tumor-adjacent tissue. Several groups have reported the culturing of matched normal and tumor organoids from treatment naïve surgical resections or biopsies (Boj et al. 2015; van de Wetering et al. 2015).

Organoid cultures still lack tumor stroma, vasculature, and immune cell interactions, and those derived from early-stage cancers appear to grow more quickly; therefore, more malignant tumors may need optimized conditions for growth. They will not mirror the accumulated genetic mutations found in later stages of progression if developed from early-stage tumors, but drug treatments should allow generation of treatment-resistant models. This methodology may lead to personalization of drug testing for a plethora of cancers (Sachs and Clevers 2014), but the length of time for growth can vary depending on the cancer type. Costs may be also an issue, and it is still unclear whether this method is applicable to all epithelial cancers.

Some of these issues are overcome by marrying organoid and PDX methodology. In one PDA model, tumor organoids were grown in orthotopic sites in mice, where they formed early-grade tumors that could develop and progress from local invasion to metastasis, while maintaining the histology of the original tumor tissue (Boj et al. 2016). This methodology was used to characterize specific alterations between premalignant and malignant states using transcriptomic and proteomic studies. Organoids grown from normal tissue can be genetically engineered, using technologies such as CRISPR, and therefore manipulated to study specific cancer mutations and cancer development and metastasis (Dekkers et al. 2013; Schwank et al. 2013; Matano et al. 2015; Fujii et al. 2016; Drost et al. 2017; Roper et al. 2017). As with spheroids, other cell types can be cultured with organoids. For example, co-culture of intraepithelial lymphocytes with murine intestinal organoids showed lymphocytes entering and leaving organoids (Nozaki et al. 2016), but better tumor models will be achieved with the co-culture of stromal elements, particularly in tumors such as PDA, where the stroma is particularly important.

Organoids have been extremely useful in the study of some types of cancer. For example, de novo prostate cancer cell line culture is difficult, with only seven widely available cell lines that lack expression of many of the genetic lesions involved in this cancer. Using organoids from prostate cancer patients showed that they developed similarities to the histology of the original tumor and contained many of the mutations in *SPOP*, and *FOXA1* as well as *CHD1* loss, which are known to be involved in prostate cancer (Gao et al. 2014). Further these organoids demonstrated differences in responses to drugs and after 3 months culture, the organoid lines were shown to still share matching somatic mutations and transcriptomes with the original tumor. Organoids have also been established from rare CRC tumors including neuroendocrine cancers and serrated adenomas and grown from biopsies from metastatic disease and from circulating tumor cells (Gao et al. 2014; Yu et al. 2014, Fujii et al. 2016).

CRC organoids responded to drug treatment based on the molecular status of the original tumor (Gao et al. 2014; van de Wetering et al. 2015). Screening of 83 drugs in these organoids confirmed that *TP53* loss of function mutants did not respond to the MDM2 inhibitor nutlin 3a, nor *RAS* mutants to EGFR inhibitors and demonstrated responsiveness of *RNF43* mutants to inhibitors of Wnt secretion. Organoids generated from liver metastases were more metastatic in xenograft culture even though the gene expression and mutational status were found to be comparable with the original tumor (van de Wetering et al. 2015). Comparisons between a primary gastric cancer and its ovarian metastatic growth found that the metastasis lost amplification of the *FGFR2* gene, but had gained alteration in transforming growth factor-β receptor 2 (Nadauld et al. 2014). Other researches illustrated that combination treatment using pan-HER and MEK inhibitors caused cell-cycle arrest in $Kras^{G12D}$ mutant CRC organoids, but cell death in *KRAS* wild-type organoids; combining this treatment with BCL-2 inhibition surmounted this resistance, a result counter to those seen in cell lines (Verissimo et al. 2016). Screens of single compounds or co-treatments in CRC, endometrial, and uterine cancer organoids have uncovered several possible original treatment strategies (Pauli et al. 2017).

A large multicenter, prospective trial (TUMOROID—NL49002.031.14) is assessing whether the drug responses of organoids can be used to evaluate treatment responses in metastatic CRC, breast, or NSCLC patients. The use of organoids in patient selection for clinical trials is under examination in the SENSOR study (NL50400.031.14 EudraCT 2014-003811-13), using biopsies from metastatic CRC and NSCLC. This is of importance since it is likely that potential novel treatments could fail if evaluated in an inappropriate patient group which is of real concern in the development of specific, targeted therapies. There are several initiatives to develop both cancer and normal organoids as resources for researchers. For example, the Hubrecht Organoid Technology (HUB) biobank, which forms a section of the Human Cancer Models Initiative, aims to generate around 1000 models in collaboration with The National Cancer Research Institute, Cancer Research UK, and the Wellcome Trust Sanger Institute. This initiative develops and analyzes tumor organoids from pancreatic, prostate, breast, CRC, and lung tumors, collating patient data and drug sensitivity. It was recently reported that over 100 breast cancer organoids from primary and metastatic tumors have also been cultivated (Sachs et al. 2018).

(c) ***Tumor Explants***

Tumor explants are a type of organoid derived from small tumor fragments often cultured on a semisolid support (Fig. 2b). Using primary tumor tissue in ex vivo culture maintains the stromal components of the tumor and the tumor architecture (van der Kuip et al. 2006; Gu et al. 2013a; Witkiewicz et al. 2015; Karekla et al. 2017).

Cultured explants offer a route toward precision medicine because multiple drug responses can be identified and monitored within days (Ward et al. 2015), a far more acceptable timescale in comparison with PDXs or organoids. Ex vivo tumor explants using naive biopsy tissue could provide analogous insights to neoadjuvant clinical trials, to verify responsiveness of the primary tumor to specific therapies, and could be used to examine novel therapies, co-treatments, and treatment scheduling. Explants can survive for over a month in culture, (Leeper et al. 2011; 2012; Katz et al. 2012; Ward et al. 2015); they reflect the heterogeneity of the tumor structure, are a good model to target microenvironmental adaptations and can allow comprehensive testing of preclinical models in heterogeneous tumor material subtypes. If tumor growth recurs, post-treatment resistance in individual patients can be investigated if tumor tissue is available. One of the difficulties in modeling breast cancer is the heterogeneity of the disease as observed in the clinic, since breast cancer subtypes often show differential responses to therapies. Ex vivo culture of unselected breast tumor explants can overcome this difficulty. Studies show that explanted breast tumor tissue conserved ER, PR, and HER2 status and proliferation rates in comparison with the original primary tumor tissue, which demonstrates the preservation of many characteristics of the original tumor in the ex vivo model (Dean et al. 2012). Tumor explants have also been used to examine novel carbonic anhydrase inhibitors in breast and topical chemotherapy in urothelial carcinoma (Bolenz et al. 2009; Ward et al. 2015).

Studies using human breast tumor explants illustrated that inhibitors of Rac GTPases (EHT 1864) and STAT3 (Stattic) could prevent spread of breast cancers and cell proliferation at the invading edge of a tumor (Katz et al. 2012). In other work, treatment of breast cancer explants with the CDK4/6 inhibitor PD-0332991 showed suppression of proliferation as measured by Ki67 staining, in all but a small subset of explants resistant to treatment. In these resistant tumors, lymph node metastases were also unaffected by treatment, and further research demonstrated that this was linked to the lack of the retinoblastoma (RB) tumor suppressor protein (Dean et al. 2012).

PD-0332991 has also been examined in PDA tumor explants (Witkiewicz et al. 2015). The CDK4/6 inhibitor p16ink4a is commonly lost in PDA, but in most models and in xenografts, inhibition of CDK4/6 is a fairly ineffective treatment and resistance quickly develops. In the explant model, however, CDK4/6 inhibition suppressed proliferation in all explants tested, except for one demonstrating loss of RB; a result mirrored in PDX of the same tumors, modeled in parallel (Witkiewicz et al. 2015), suggesting differential drug responses across systems. A study in prostate cancer explants using the Jak2 inhibitor AZD 1480 (Gu et al. 2013b) demonstrated that responsive explants underwent apoptosis when nuclear Stat5a/b decreased in response to the inhibitor, while in resistant explants, it did not. Prostate tumor explants have also been used to test the effectiveness of novel Stat5a/b inhibitory compounds (Liao et al. 2015).

Explants have been used to examine the links between metabolism, hypoxia, and inflammation in Barrett's esophagus, a pre-neoplastic lesion which can progress to esophageal adenocarcinoma (Phelan et al. 2016), and to examine the role of bile acids in gastroduodenal reflux which causes increased expression of MUC1, associated with early carcinogenic changes in this condition (Mariette et al. 2008). They have also been used in head and neck cancer to study the effectiveness of a phytochemical treatment, Lupeol (Bhattacharyya et al. 2017). Results showed a good response to Lupeol, and using explants from a cisplatin-resistant tumor, it was found that this molecule could resensitize the tumors to cisplatin (Bhattacharyya et al. 2017). In non-small cell lung carcinoma, tumor explants could be used to predict patient response to therapy and to monitor clinically relevant biomarkers (Karekla et al. 2017). Explants have been used to confirm and investigate changes between low- and high-grade tumors in several settings such as endometrial cancer (Cornel et al. 2012). Analysis of sarcoma development suggests that while the p53 tumor suppressor pathway may be intact, the p53 ubiquitin ligase, *MDM2*, is often amplified or overexpressed (Toledo and Wahl 2006). Although preclinical studies in cell lines with amplification of *MDM2* using the MDM2 inhibitor Nutlin-3a produced positive results, only 1 patient in 20 showed a partial response in clinical trials in sarcoma patients (Muller et al. 2007; Ray-Coquard et al. 2012). Pishas et al. used sarcoma tumor explants from a range of subtypes and showed that there was no correlation between *MDM2* amplification and response to Nutlin-3a (Pishas et al. 2013) and that non-responsiveness to the drug was conferred by hypermethylation of the p53 target gene *GADD45A*. Explants have also been used in radiation studies. For example, the effectiveness of inhibiting TGF-β as a treatment or possible radiosensitizer of high-grade glioma has been examined in tumor explants (Bayin et al. 2016). Results showed that most, but not all high-grade gliomas, were responsive, and the authors suggest that tumor explants could provide a rapid platform to determine patient response to this therapy and stratify treatment.

Currently, the main in vivo methodology for studying metastasis generally involves the use of mouse models. Culture of tumor explants enables invasive changes in an appropriate TME to be assessed and novel therapeutics to be analyzed in patient samples in a fast and quantitative format (Ward et al. 2015) (Fig. 2b). The effect of different ECM components and stromal cells can also be evaluated and monitored in this system. The main benefits of explants in these assays are that growth/response/invasion can be monitored continuously and appraised at any time over the course of an experiment, and after use, the explant tissue can be fixed and analyzed.

Normal skin explants are being used to investigate tumor immunology and possible anti-tumor vaccination development. One strategy currently used is to load dendritic cells with tumor associated antigens, but the cost of generating these cells and treating them is extremely high. The use of skin tissue explants to load resident dendritic cells with tumor antigens in situ has been successfully explored as an alternative method (Ruben et al. 2014). Immune checkpoint inhibition has demonstrated promise in several malignancies, but successful treatment is dependent on infiltration of $CD8^+$ T cells. Prostate tumor explants were successfully treated to increase numbers of $CD8^+$ T cells while decreasing numbers of immune suppressor cells (Muthuswamy et al. 2016), illustrating their use in investigation of immunotherapeutic strategies. At present, one drawback of tumor explants is the lack of vascularization which prevents study of angiogenesis and its inhibitors. However, novel research into vascularization of tumor explants has been recently published (Bazou et al. 2016). This group has developed a method of inducing vasculature formation into xenografts of human breast cancer and melanoma, ex vivo, and considered that it should be possible to adapt this methodology for vascularization of human tumor explants. The advantages and disadvantages of 3D in vitro organoid systems are listed in Table 2.

Table 2 Listing the advantages and disadvantages of 3D in vitro tumor models

	Setup Methods	Strengths	Weaknesses
Spheroids	Spheroids are set up using cancer cell lines from 2D culture (various methods are illustrated in Fig. 1). These can be grown from genetically modified cells and can incorporate other cell types	– Exhibit cell–cell and cell–ECM interactions – Mimic in vivo oxygen, pH, and metabolic gradients – Useful in high-throughput drug screening assays	– Use cancer cell lines, which can lose the biological traits of the cancer from which they derived
Organoids	Organoids are developed from tissue samples/ cancer stem cells and contain the cell types found in the tissue of origin. These cells can be genetically modified and cultured with other cell types	– Biologically stable – can be frozen for later use – Allow the culture of tumors from individual patients – Can provide tissues for the study of rare cancers or where cell line models are scarce – Useful in high-throughput drug screening assay	– Clonal drift may occur over time – High costs – Lack tumor stroma, vasculature, and immune cell interactions
Explants	Explants are derived from human tumor tissue – usually biopsies. Genetic modification is difficult, but they can be cultured with other cell types	– Maintain the stromal components and tissue architecture of the primary tumor – Good model to target microenvironmental adaptations – Maintain histology and proliferation rates of the original tumor – Growth/response/ invasion can be monitored continuously	– Lack of vascularization prevents the study of angiogenesis and its inhibition

4 Conclusions

It has been estimated that over 90% of drugs that make it through to clinical trials fail because of lack of efficacy and concerns over safety (Townsend et al. 2016; Ledford 2011). Given the particularly high attrition rate of novel anticancer therapies during progression from preclinical to clinical trials, it is obvious that current testing methodologies are problematic. Better, more physiological systems are clearly needed for (a) drug testing and (b) patient participation in an era of more targeted treatments and personalized medicine. The 'ideal' tumor model would be able to mimic the development of malignant cells, angiogenesis, stromal development, immune responses, metastasis, therapeutic efficacy, and progression to resistant disease. This chapter shows that there is not one specific methodology that can fulfill these criteria. Therefore, more than one model system will be required to provide better translational results.

Cell lines are a suitable model for drug screening, but lead compounds need to be tested in more intricate animal models that allow for pharmacokinetic and toxicity studies that can monitor off target effects. These in vivo models are useful, and GEMMs and PDXs can reflect many of the features of the tumor from which they were developed. But the time taken for tumor growth is not compatible with use as a clinical decision-making tool in precision medicine. However, in vitro models specific to an individual patient (such as organoids or tumor explants) can be used to test drugs and guide treatment in a more specialized manner. Results from these systems are available within days and could thus prevent the use of ineffective drugs for a given patient and limit access to more expensive drugs to those likely to benefit.

Access to human tumor tissue is problematic for many scientists, but because organoids can be stored they can be used as a research resource and passed to other researchers. Organoids could also be developed from GEMM mouse models and used in cancer research. They would be particularly useful in the study of the preclinical stages of cancer, and in the study of cancer resurgence from residual disease. This methodology has been used to study recurrence in breast cancer and showed that these residual cells develop to become transcriptionally divergent from both the original tumor and normal epithelium (Havas et al. 2017). Organoids can be used for high-throughput drug screening assays that meet industry standards (Huang et al. 2015; van de Wetering et al. 2015; Boehnke et al. 2016).

The most significant factor that is likely to govern the rate of success in oncology clinical trials is matching the experimental models to the clinical malignancy. The pros and cons of experimental systems suggest that new therapeutics should be examined using several methods.

Acknowledgements The authors MG, JM, CW, IK, AM, and DA are supported by funding from the UK Engineering and Physical Sciences Research Council, through the Implantable Microsystems for Personalised Anti-Cancer Therapy (IMPACT) program grant (EP/K-34510/1).

References

Aparicio S, Hidalgo M, Kung AL (2015) Examining the utility of patient-derived xenograft mouse models. Nat Rev Cancer 15:311–316

Arrowsmith J (2011) Trial watch: phase II failures: 2008–2010. Nat Rev Drug Discov 10:328–329

Bayin NS et al (2016) Patient-specific screening using high-grade glioma explants to determine potential radiosensitization by a TGF-β small molecule inhibitor. Neoplasia 18:795–805

Bazou D, Maimon N, Gruionu G, Munn LL (2016) Self-assembly of vascularized tissue to support tumor explants in vitro. Integr Biol 8:1301–1311

Bell-McGuinn KM, Garfall AL, Bogyo M, Hanahan D, Joyce JA (2007) Inhibition of cysteine cathepsin protease activity enhances chemotherapy regimens by decreasing tumor growth and invasiveness in a mouse model of multistage cancer. Cancer Res 67:7378–7385

Bellmunt J et al (2010) Activity of a multitargeted chemo-switch regimen (sorafenib, gemcitabine and metronomic capecitabine) in metastatic renal-cell carcinoma: a phase 2 study (ISOGUG-02-06). Lancet Oncol 11:350–357

Beloribi-Djefaflia S, Siret C, Lombardo D (2015) Exosomal lipids induce human pancreatic tumoral MiaPaCa-2 cells resistance through the CXCR4-SDF-1α signaling axis. Oncoscience 2:15–30

Belteki G et al (2005) Conditional and inducible transgene expression in mice through the combinatorial use of Cre-mediated recombination and tetracycline induction. Nucl Acids Res 33:e51

Ben-David U et al (2017) Patient-derived xenografts undergo mouse-specific tumor evolution. Nat Genet 49:1567–1575

Bhattacharyya S et al (2017) CDKN2A-p53 mediated antitumor effect of Lupeol in head and neck cancer. Cell Oncol 40:145–155

Boedigheimer MJ, Freeman DJ, Kiaei P, Damore MA, Radinsky R (2013) Gene expression profiles can predict panitumumab monoterapy responsiveness in human tumor xenograft models. Neoplasia 15:125–132

Boehnke K et al (2016) Assay establishment and validation of a high-throughput screening platform for three-dimensional patient-derived colon cancer organoid cultures. J Biomol Screen 21:931–941

Boj SF et al (2015) Organoid models of human and mouse ductal pancreatic cancer. Cell 160:324–338

Boj SF et al (2016) Model organoids provide new research opportunities for ductal pancreatic cancer. Mol Cell Oncology 3(1):e1014757

Bolenz C et al (2009) Topical chemotherapy in human urothelial carcinoma explants: a novel translational tool for preclinical evaluation of experimental intravesical therapies. Eur Urol 56:504–511

Bomken S, Fiser K, Heidenreich O, Vormoor J (2010) Understanding the cancer stem cell. Br J Cancer 103:439–445

Broutier et al (2017) Human primary liver cancer-derived organoid cultures for disease modeling and drug screening. Nature Med 23:1424–1435

Bruna A et al (2016) A biobank of breast cancer explants with preserved intra-tumor heterogeneity to screen anticancer compounds. Cell 167:260–274

Burdall SE, Hanby AM, Lansdown MR, Speirs V (2003) Breast cancer cell lines: friend or foe? Breast Cancer Res 5:89–95

Byrne AT et al (2017) Interrogating open issues in cancer precision medicine with patient-derived xenografts. Nature Rev Cancer 17:254–268

Capodanno Y et al (2018) Notch pathway inhibition targets chemoresistant insulinoma cancer stem cells. Endocr Relat Cancer 25:131–144

Caponigro G, Sellers WR (2011) Advances in the preclinical testing of cancer therapeutic hypotheses. Nat Rev Drug Discov 10:179–187

Chen Z et al (2012) A murine lung cancer co-clinical trial identifies genetic modifiers of therapeutic response. Nature 485:613–617

Cheung KJ, Gabrielson E, Werb Z, Ewald AJ (2013) Collective invasion in breast cancer requires a conserved basal epithelial program. Cell 155:1639–1651

Clark AK et al (2013) A bioengineered microenvironment to quantitatively measure the tumorigenic properties of cancer-associated fibroblasts in human prostate cancer. Biomaterials 34:4777–4785

Cornel KMC et al (2012) Overexpression of 17β-hydroxysteroid dehydrogenase type 1 increases the exposure of endometrial cancer to 17β-estradiol. J Clin Endocrinol Metab 97:E591–E601

Correia AL, Bissell MJ (2012) The tumor microenvironment is a dominant force in multidrug resistance. Drug Resist Update 15:39–49

Crawford Y et al (2009) PDGF-C mediates the angiogenic and tumorigenic properties of fibroblasts associated with tumors refractory to anti-VEGF treatment. Cancer Cell 15:21–34

Daniel VC et al (2009) A primary xenograft model of small-cell lung cancer reveals irreversible changes in gene expression imposed by culture in vitro. Cancer Res 69:3364–3373

Dean JL et al (2012) Therapeutic response to CDK4/6 inhibition in breast cancer defined by ex vivo analyses of human tumors. Cell Cycle 11:2756–2761

Dekkers JF et al (2013) A functional CFTR assay using primary cystic fibrosis intestinal organoids. Nat Med 19:939–945

Delitto D et al (2015) Downstream mediators of the intratumoral interferon response suppress antitumor immunity, induce gemcitabine resistance and associate with poor survival in human pancreatic cancer. Cancer Immunol Immunother 64:1553–1563

DeRose YS et al (2011) Tumor grafts derived from women with breast cancer authentically reflect tumor pathology, growth, metastasis and disease outcomes. Nat Med 17:1514–1520

dit Faute MA et al (2002) Distinctive alterations of invasiveness, drug resistance and cell-cell organization in 3D-cultures of MCF-7, a human breast cancer cell line, and its multidrug resistant variant. Clin Exp Metastasis 19:161–168

Drost J et al (2017) Use of CRISPR-modified human stem cell organoids to study the origin of mutational signatures in cancer. Science 358:234–238

Ekert JE et al (2014) Three-dimensional lung tumor microenvironment modulates therapeutic compound responsiveness in vitro—implication for drug development. PLoS ONE 9(3): e92248. https://doi.org/10.1372/journal.pone.0092248

Faca VM et al (2008) A mouse to human search for plasma proteome changes associated with pancreatic tumor development. PLoS Med 5:e123

Fournier MV et al (2006) Gene expression signature in organized and growth-arrested mammary acini predicts good outcome in breast cancer. Cancer Res 66:7095–7102

Freidrich J, Seidel C, Ebner R, Kunz-Schughart LA (2009) Spheroid-based drug screen: considerations and practical approach. Nat Protoc 4:309–324

Fujii M et al (2016) A colorectal tumor organoid library demonstrates progressive loss of niche factor requirements during tumorigenesis. Cell Stem Cell 18:827–838

Gao D et al (2014) Organoid cultures derived from patients with advanced prostate cancer. Cell 159:176–187

Gao D, Chen Y (2015) Organoid development in cancer genome discovery. Curr Opin Gen Develop 30:42–48

Gao H et al (2015) High-throughput screening using patient-derived tumor xenografts to predict clinical trial drug response. Nat Med 21:1318–1325

Gatenby RA et al (2007) Cellular adaptations to hypoxia and acidosis during somatic evolution of breast cancer. Br J Cancer 97:646–653

Ghajar CM et al (2013) The perivascular niche regulates breast tumour dormancy. Nat Cell Biol 15:807–817

Graves EE et al (2010) Hypoxia in models of lung cancer: implications for targeted therapeutics. Clin Cancer Res 16:4843–4852

Gu L, Liao Z, McCue P, Trabulsi EJ, Nevalainen MT (2013a) Ex vivo prostate cancer explant organ culture model system for targeted drug development in prostate cancer. J Clin Oncol Abstract https://doi.org/10.1200/jco.2013.31.6-suppl.110.a

Gu L et al (2013b) Pharmacologic inhibition of Jak2-Stat5 signaling by Jak2 inhibitor AZD1480 potently suppresses growth of both primary and castrate-resistant prostate cancer. Clin Cancer Res 19:5658–5674

Guillaumond F et al (2013) Strengthened glycolysis under hypoxia supports tumor symbiosis and hexosamine biosynthesis in pancreatic adenocarcinoma. PNAS USA 110:3919–3924

Hanahan D, Weinberg RA (2011) Hallmarks of cancer: the next generation. Cell 144:646–674

Havas KM et al (2017) Metabolic shifts in residual breast cancer drive tumor recurrence. J Clin Invest 127:2091–2105

Heneweer M et al (2005) Co-culture of primary human mammary fibroblasts and MCF-7 cells as an in vitro breast cancer model. Toxicol Sci 83:257–263

Hensley CT et al (2016) Metabolic heterogeneity in human lung tumors. Cell 164:681–694

Herrmann D et al (2014) Three-dimensional cancer models mimic cell-matrix interactions in the tumour microenvironment. Carcinogenesis 35:1671–1679

Hesami P et al (2014) A humanized tissue-engineered in vivo model to dissect interactions between human prostate cancer cells and human bone. Clin Exp Metastasis 31:435–446

Hidalgo M et al (2011) A pilot clinical study of treatment guided by personalized tumorgrafts in patients with advanced cancer. Mol Cancer Therapeut 10:1311–1316

Holzapfel BM et al (2014) Species-specific homing mechanisms of human prostate cancer metastasis in tissue engineered bone. Biomaterials 35:4108–4115

Horie M et al (2012) Characterization of human lung cancer-associated fibroblasts in three-dimensional in vitro co-culture model. Biochem Biophys Res Commun 423:158–163

Hsiao AY et al (2009) Microfluidic system for formation of PC-3 prostate cancer co-culture spheroids. Biomaterials 30:3020–3027

Huang L et al (2015) Ductal pancreatic cancer modeling and drug screening using human pluripotent stem cell- and patient-derived tumor organoids. Nat Med 21:364–1371

Huang M, Shen A, Ding J, Geng M (2014) Molecularly targeted cancer therapy: some lessons from the past decade. Trends Pharmacol Sci 35:41–50

Hubert CG et al (2016) A three-dimensional organoid culture system derived from human glioblastomas recapitulates the hypoxic gradients and cancer stem cell heterogeneity of tumors found in vivo. Cancer Res 76:2465–2477

Infanger DW et al (2013) Glioblastoma stem cells are regulated by interleukin-8 signaling in a tumoral perivascular niche. Cancer Res 73:7079–7089

Ingram M et al (2010) Tissue engineered tumor models. Biotech Histochem 85:213–229

Ishiguro T (2017) Tumor-derived spheroids: relevance to cancer stem cells and clinical applications. Cancer Sci 108:283–289

Jaganathan H et al (2014) Three-dimensional in vitro co-culture model of breast tumor using magnetic levitation. Sci Rep 4:6468. https://doi.org/10.1038/srep06468

Julien S et al (2012) Characterization of a large panel of patient-derived tumor xenografts representing the clinical heterogeneity of human colorectal cancer. Clin Cancer Res 18:5314–5328

Kamb A (2005) What's wrong with our cancer models? Nat Rev Drug Dis 4:161–165

Karar J, Maity A (2009) Modulating the tumor microenvironment to increase radiation responsiveness. Cancer Biol Ther 8:1994–2001

Karekla E et al (2017) Ex vivo explant cultures of non-small cell lung carcinoma enable evaluation of primary tumor responses to anticancer therapy. Cancer Res 77:2029–2039

Katz E et al (2012) Targeting of Rac GTPases blocks the spread of intact human breast cancer. Oncotarget 3:608–613

Kim S, Lee EK, Kuh H (2015) Co-culture of 3D tumor spheroids with fibroblasts as a model for epithelial-mesenchymal transition in vitro. Exper Cell Res 335:187–196

Klco JM et al (2014) Functional heterogeneity of genetically defined subclones in acute myeloid leukemia. Cancer Cell 25:379–392

Kopetz S, Lemos R, Powis G (2012) The promise of patient-derived xenografts: the best laid plans of mice and men. Clin Cancer Res 18:5160–5162

Lancaster MA, Knoblich JA (2014) Organogenesis in a dish: Modeling development and disease using organoid technologies. Science 345. https://doi.org/10.1126/science.1247125

Lane AN, Higashi RM, Fan TW (2016) Preclincal models for interrogating drug action in human cancers using stable isotope resolved metabolomics (SIRM). Metabolomics 12:118. https://doi.org/10.1007/s11306-016-1065-y

Langdon SP (2004) Isolation and culture of ovarian cancer cell lines. Method Mol Med 88:133–139

Lawrence MG et al (2013) A preclinical xenograft model of prostate cancer using human tumors. Nat Protoc 8:836–848

Ledford H (2011) 4 ways to fix the clinical trial. Nature 477:526–528

Ledford H (2015) CRISPR, the disruptor. Nature 522:20–24

Ledford H (2016) US cancer institute to overhaul tumour cell lines. Nature 530:391

Leeper AD et al (2011) Long-term culture of human breast cancer specimens and their analysis using optical projection tomography. J Vis Exp pii3085

Leeper AD et al (2012) Determining tamoxifen sensitivity using primary breast cancer tissue in collagen-based three-dimensional culture. Biomaterials 33:907–915

Li L, Lu Y (2011) Optimizing a 3D culture system to study the interaction between epithelial breast cancer and its surrounding fibroblasts. J Cancer 2:458–466

Liao Z et al (2015) Structure-based screen identifies a potent small molecule inhibitor of Stat5a/b with therapeutic potential for prostate cancer and chronic myeloid leukaemia. Mol Cancer Ther 14:1777–1793

Loessner D et al (2010) Bioengineered 3D platform to explore cell–ECM interactions and drug resistance of epithelial ovarian cancer cells. Biomaterials 31:8494–8506

Longati P et al (2013) 3D pancreatic carcinoma spheroids induce a matrix-rich, chemoresistant phenotype offering a better model for drug testing. BMC Cancer 13:95

Mariette C et al (2008) Activation of MUC1 mucin expression by bile acids in human esophaeal adenocarcinomatous cells and tissues in mediated by the phosphatidylinositol 3-kinase. Surgery 143:58–71

Martin JK, Patrick DR, Bissell MJ, Fournier MV (2008) Prognostic breast cancer signature identified from 3D culture model accurately predictes clinical outcome across independent datasets. PLoS ONE 3:e2994. https://doi.org/10.1371/journal.pone.0002994

Martinez-Garcia R et al (2014) Transcriptional dissection of pancreatic tumors engrafted in mice. Genome Medicine 6:27

Matano M et al (2015) Modeling colorectal cancer using CRISPR-Cas9-mediated engineering of human intestinal organoids. Nat Med 21:256–262

McAllister SS, Weinberg RA (2010) Tumor-host interactions: a far-reaching relationship. J Clin Oncol 28:4022–4028

McCracken KW et al (2014) Modelling human development and disease in pluripotent stem-cell-derived gastric organoids. Nature 516:400–404

Minchinton AI, Tannock IF (2006) Drug penetration in solid tumours. Nat Rev Cancer 6:583–592

Mitra A, Mishra L, Li S (2015) EMT, CTCs and CSCs in tumor relapse and drug-resistance. Oncotarget 6:10697–10711

Moore CB, Guthrie EH, Huang MTH, Taxman DJ (2010) Short hairpin RNA (shRNA): design, delivery, and assessment of gene knockdown. Methods Mol Biol 629:141–158

Morton JJ, Bird G, Refaeli Y, Jimeno A (2016) Humanized mouse xenograft models: narrowing the tumor-microenvironment gap. Cancer Res 76:6153–6158

Muller CR et al (2007) Potential for treatment of liposarcomas with the MDM2 antagonist Nutlin-3A. Int J Cancer 121:199–205

Muthuswamy R, Corman JM, Dahl K, Chatta GS, Kalinski P (2016) Functional reprogramming of human prostate cancer to promote local attraction of effector CD8(+) T cells. Prostate 76:1095–1105

Nadauld LD et al (2014) Metastatic tumor evolution and organoid modeling implicate *TGFBR2* as a cancer driver in diffuse gastric cancer. Genome Biol 428 http://genomebiology.com/2014/15/8/428

Nardella C, Lunardi A, Patnaik A, Cantley LC, Pandolfi PP (2011) The APL paradigm and the "co-clinical trial" project. Cancer Dis 1:108–116

Nozaki K et al (2016) Co-culture with intestinal epithelial organoids allows efficient expansion and motility analysis of intraepithelial lymphocytes. J Gastroenterol 51:206–213
Ocana A, Pandiella A, Siu LL, Tannock IF (2011) Preclinical development of molecular-targeted agents for cancer. Nat Rev Clin Oncol 8:200–209
Olive KP et al (2009) Inhibition of Hedgehog signalling enhances delivery of chemotherapy in a mouse model of pancreatic cancer. Science 324:1457–1461
Pajic A et al (2000) Cell cycle activation by c-myc in a burkitt lymphoma model cell line. Int J Cancer 87:787–793
Pauli C et al (2017) Personalized in vitro and in vivo cancer models to guide precision medicine. Cancer Discov 7:462–477
Petsko GA (2010) When failure should be the option. BMC Biol 8:61
Pettersen EO et al (2015) Targeting tumour hypoxia to prevent cancer metastasis. From biology, biosensing and technology to drug development: the METOXIA consortium. J Enzyme Inhib Med Chem 30(5):689–721
Phelan JJ et al (2016) Examining the connectivity between different cellular processes in the Barrett tissue microenvironment. Cancer Lett 371:334–346
Pickl M, Ries CH (2009) Comparison of 3D and 2D tumor models reveals enhanced HER2 activation in 3D associated with an increased response to trastuzumab. Oncogene 28:461–468
Pietras K, Hanahan D (2005) A multitargeted, metronomic, and maximum-tolerated dose "chemoswitch" regimen is antiangeogenic, producing objective responses and survival benefit in a mouse model of cancer. J Clin Oncol 23:939–952
Pishas KI et al (2014) Nutlin-3a efficacy in sarcoma predicted by transcriptomic and epigenetic profiling. Cancer Res 74:921–931
Pitteri SJ et al (2009) Integrated proteomic analysis of human cancer cells and plasma from tumor bearing mice for ovarian cancer biomarker discovery. PLoS ONE 4:e7916
Politi K, Fan PD, Shen R, Zakowski M, Varmus H et al (2010) Erlotinib resistance in mouse models of epidermal growth factor receptor-induced lung adenocarcinoma. Dis Model Mech 3:111–119
Politi K, Pao W (2011) How genetically engineered mouse tumour models provide insights into human cancers. J Clin Oncol 29:2273–2281
Pouyssegur J, Dayan F, Mazure NM (2006) Hypoxia signalling in cancer and approaches to enforce tumour regression. Nature 441:437–443
Proia DA, Kuperwasser C (2006) Reconstruction of human mammary tissues in a mouse model. Nat Protoc 1:206–214
Ravi M, Paramesh V, Kaviya SR, Anuradha E, Paul Solomon FDP (2014) 3D cell culture systems: advantages and applications. J Cell Physiol 230:16–26
Rayal F et al (2012) Molecular profiling of patient-derived breast cancer xenografts. Breast Cancer Res 14:R11
Ray-Coquard I et al (2012) Effect of the MDM2 antagonist R7112 on the P53 pathway in patients with MDM2-amplified, well-differentiated or dedifferentiated liposarcoma; an exploratory proof-of-mechanism study. Lancet Oncology 13:1133–1140
Riemer P et al (2017) Oncogenic β-catenin and PIK3CA instruct network states and cancer phenotypes in intestinal organoids. J Cell Biol 216:1567–1577
Riffle S, Naresh P, Albert M, Hegde RS (2017) Linking hypoxia, DNA damage and proliferation in multicellular tumor spheroids. BMC Cancer 17:338. https://doi.org/10.1186/s12885-017-3319-0
Roper J et al (2017) In vivo genome editing and organoid transplantation models of colorectal cancer and metastasis. Nature Biotechnol 35:569–576
Ruben JM et al (2014) In situ loading of skin dendritic cells with apoptotic bleb-derived antigen for the induction of tumor-directed immunity. Oncoimmunology 3:7e946360,https://doi.org/10.4161/21624011.2014.946360
Sachs N, Clevers H (2014) Organoid cultures for the analysis of cancer phenotypes. Curr Opin Gen Develop 24:68–73

Sachs N et al (2018) A living biobank of breast cancer organoids captures disease heterogeneity. Cell 172:1–14

Sander JD, Joung JK (2014) CRISPR-Cas systems for editing, regulating and targeting genomes. Nature Biotechnol 32:347–355

Sato T et al (2011) Long-term expansion of epithelial organoids from human colon, adenoma, adenocarcinoma and Barrett's epithelium. Gastroent 141:1762–1772

Saunders T (2011) Inducible transgenic mouse models. Methods Mol Biol 693:103–115

Sausville EA, Burger AM (2006) Contribution of human tumor xenografts to anticancer drug development. Cancer Res 66:3351–3354

Schmeichel KL, Bissell MJ (2003) Modeling tissue-specific signalling and organ function in three dimensions. J Cell Sci 116:2377–2388

Schutte M et al (2017) Molecular dissection of colorectal cancer in pre-clinical models identifies biomarkers predicting sensitivity to EGFR inhibitors. Nat Commun 8:14262

Schwank G et al (2013) Functional repair of CFTR by CRISPR/Cas9 in intestinal stem cell organoids of cystic fibrosis patients. Cell Stem Cell 13:653–658

Sethi T et al (1999) Extracellular matrix proteins protect small cell lung cancer cells against apoptosis: a mechanism for small cell lung cancer growth and drug resistance in vivo. Nat Med 5:662–668

Sharpless NE, Depinho RA (2006) The mighty mouse: genetically engineered mouse models in cancer drug development. Nature Rev Drug Discov 5:741–754

Simian M, Bissell MJ (2017) Organoids: a historical perspective of thinking in three dimensions. J Cell Biol 216:31–40

Singh M et al (2010) Assessing therapeutic responses in Kras mutant cancers using genetically engineered mouse models. Nat Biotechnol 28:585–593

Singh M, Ferrara N (2012) Modeling and predicting clinical efficacy for drugs targeting the tumor milieu. Nat Biotechnol 30:648–657

Singh M, Close DA, Mukundan S, Johnston PA, Sant S (2015) Production of uniform 3D microtumors in hydrogel microwell arrays for measurement of viability, morphology and signalling pathway activation. Assay Drug Dev Techol 13:570–583

Singh M, Mukundan S, Jaramillo M, Oesterreich S, Sant S (2016) Three-dimensional breast cancer models mimic hallmarks of size-induced tumor progression. Cancer Res 76:3732–3743

Siolas D, Hannon GJ (2013) Patient derived tumor xenografts: transforming clinical samples into mouse models. Cancer Res 73:5315–5319

Stewart E et al (2017) Orthotopic patient-derived xenografts of paediatric solid tumours. Nature 549:96–100

Straussman R et al (2012) Tumour micro-environment elicits innate resistance to RAF inhibitors through HGF secretion. Nature 487:500–504

Tentler JJ et al (2012) Patient-derived tumour xenografts as models for oncology drug development. Nat Rev Clin Oncol 9:338–350

Thibaudeau L et al (2014) A tissue-engineered humanized xenograft model of human breast cancer metastasis to bone. Dis Model Mech 7:299–309

Timpson P et al (2011) Spatial regulation of RhoA activity during pancreatic cancer cell invasion driven by mutant p53. Cancer Res 71:747–757

Toledo F, Wahl GM (2006) Regulating the p53 pathway: in vitro hypotheses, in vivo veritas. Nat Rev Cancer 6:909–923

Townsend EC et al (2016) The public repository of xenografts enables discovery and randomized phase II-like trials in mice. Cancer Cell 29:574–586

van der Kuip H et al (2006) Short term culture of breast cancer tissues to study the activity of the anticancer drug taxol in an intact tumor enviroment. BMC Cancer 6:86. https://doi.org/10.1186/1471-2407-6-86

van de Wetering M et al (2015) Prospective derivation of a 'Living Organoid Biobank' of colorectal cancer patients. Cell 161:933–945

Verissimo CS et al (2015) Targeting mutant RAS in patient-derived colorectal cancer organoids by combinatorial drug screening eLife, 5. https://doi.org/10.7554/elife.18489

Verjans E, Doijen J, Luyten W, Landuyt B, Schoofs L (2017) Three-dimensional cell culture models for anticancer drug screening: worth the effort? J Cell Physiol 1–11 doi.org/https://doi.org/10.1002/jcp26052

Vudattu NK et al (2014) Humanized mice as a model for aberrant responses in human T cell immunotherapy. J Immunol 193:587–596

Wang J et al (2010) A novel orthotopic and metastatic mouse model of breast cancer in human mammary microenvironment. Breast Cancer Res Treat 120:337–344

Wang X et al (2018) Enrichment of glioma stem cell-like cells on 3D porous scaffolds composed of different extracellular matrix. Biochem Biophys Res Comm 498:1052–1057

Ward C et al (2013) New strategies for targeting the hypoxic tumour microenvironment in breast cancer. Cancer Treat Rev 39:171–179

Ward C et al (2015) Evaluation of carbonic anhydrase IX as a therapeutic target for inhibition of breast cancer invasion and metastasis using a series of in vitro breast cancer models. Oncotarget 6:24856–2485670

Ware MJ et al (2016) Generation of an in vitro 3D PDAC stromal rich spheroid model. Biomaterials 108:129–142

Weeber F et al (2015) Preserved genomic diversity in organoids cultured from biopsies of colorectal cancer metastases PNAS USA 112:13308–13311

Werner-Klein M et al (2014) Immune humanization of immunodeficient mice using diagnostic bone marrow aspirates from carcinoma patients. PLoS ONE 9(e97860):6158

Witkiewicz AK et al (2015) Selective impact of CDK4/6 suppression of patient-derived models of pancreatic cancer. Oncotarget 6:15788–15801

Xie H et al (2014) Targeting lactate dehydrogenase—a inhibits tumorigenesis and tumor progression in mouse models of lung cancer and impacts tumor initiating cells. Cell Metabolism 19:795–780

Yamada KM, Cukierman E (2007) Modeling tissue morphogenesis and cancer in 3D. Cell 130:601–610

Yu M (2014) Ex vivo culture of circulating breast tumor cells for individualized testing of drug susceptibility. Science 345:216–220

Zamboni WC et al (2012) Best practices in cancer nanotechnology: perspective from NCI nanotechnology alliance. Clin Cancer Res 18:3229–3241

Zhang X et al (2013) A renewable tissue resource of phenotypically stable, biologically and ethnically diverse, patient-derived human breast cancer xenograft models. Cancer Res 73:4885–4897

Overview and Comparison of Intestinal Organotypic Models, Intestinal Cells, and Intestinal Explants Used for Toxicity Studies

Marc Maresca, Philippe Pinton, El Hassan Ajandouz, Sandrine Menard, Laurent Ferrier and Isabelle P. Oswald

Contents

Abstract The intestine is a complex organ formed of different types of cell distributed in different layers of tissue. To minimize animal experiments, for decades, researchers have been trying to develop in vitro/ex vivo systems able to mimic the cellular diversity naturally found in the gut. Such models not only help our understanding of the gut physiology but also of intestinal toxicity. This review describes the different systems used to evaluate the effects of drugs/contaminants on intestinal functions and compares their advantages and limitations. The comparison showed that the organotypic model is the best available model to perform intestinal toxicity studies, including on human tissues.

M. Maresca · E. H. Ajandouz
Aix Marseille Univ, CNRS, Centrale Marseille, iSm2, Marseille, France

P. Pinton · S. Menard · L. Ferrier · I. P. Oswald (✉)
Toxalim (Research Centre in Food Toxicology), Université de Toulouse, INRA, ENVT, INP-Purpan, UPS, Toulouse, France
e-mail: Isabelle.Oswald@inra.fr

Current Topics in Microbiology and Immunology (2021) 430: 247–264
https://doi.org/10.1007/82_2018_142

Published Online: 28 September 2018

1 Introduction

Humans are exposed to ingested contaminants, such as heavy metals, pesticides, fungal toxins (mycotoxins), and other natural or human-made toxins, every day. Although virtually all organs and tissues are exposed once the ingested toxins cross the intestinal wall, the gut is the first organ exposed and to the highest concentrations/doses of toxins. This is well documented in the case of mycotoxins, and many of these toxins are able to affect intestinal functions in both animals and humans (Grenier and Applegate 2013). For these reasons, toxicological studies focusing on the intestine, including toxicodynamic (effects/impact of toxins on the human gut) and toxicokinetic (metabolization and handling of the toxins by the human gut) aspects, are of primary importance.

Although some morphological variations can be observed in the general structure of the gastrointestinal tract, all parts of the human gut are formed by four functional layers: (i) the mucosa or epithelium layer that forms the most inner layer facing the gut lumen, (ii) the submucosa containing small blood vessels and immune cells, (iii) the muscularis propria responsible for intestinal peristaltism and containing the enteric nervous plexus, and (iv) the adventitia the outer layer containing the main blood vessels and nerves. Although all intestinal layers can be exposed to and affected by ingested toxins, the mucosal/epithelial layer is in direct contact with the intestinal lumen and is thus the most exposed.

The intestinal epithelium is formed by distinct cell types distributed along the crypt–villus axis. Although they all derived from a common stem cell progenitor located in the crypts, their morphology and roles differ. The main cell type (89–95% of total cells) comprises absorbent cells (i.e., enterocytes and colonocytes) that are mainly responsible for the terminal digestion of food constituents and the absorption/secretion of ions, nutrients, and water. These cells also have an immune function since they form the intestinal barrier and can produce pro-inflammatory cytokines and antimicrobial molecules such as nitric oxide and antimicrobial peptides. In addition, absorbent cells express enzymes and transporters that enable the metabolization and excretion of xenobiotics (chemical substance not naturally produced or expected to be present within the human or animal). These cells are thus the first line of defense against ingested toxins. Microfold cells (or M cells) lie on the gut-associated lymphoid tissue (GALT) of the Peyer's patches in the small intestine and represent a facilitating way for transcytosis. Goblet cells are the second most common type of cell and are responsible for the production/secretion of highly hydrated glycoproteins called mucins that form the mucus. The latter displays a mechanical and a chemical protective role in addition to providing board and lodging and a substrate for the commensal microbiota. Although they represent only 0.2–0.6% of epithelial cells, enteroendocrine cells play an important role as they release hormones that regulate intestinal function and the functions of other organs including the liver and pancreas. In the small intestine, Paneth cells are located close to stem cells, and the immune role they play is related to their ability to secrete antimicrobial agents such as antimicrobial peptides and lysozyme.

Finally, tuft cells, identified in 2011 (Gerbe et al. 2011), are the main source of endogenous intestinal opioids and provide defense against intestinal parasitic infections (Gerbe et al. 2016).

Although the cells that form the mucosal/epithelial layers play major roles in gut physiology and could be major targets of ingested toxins, it is now recognized that other cells are also important. For example, immune cells (lymphocytes, dendritic cells, macrophages) present in the mucosa and the submucosa are involved in normal and pathological inflammatory reactions of the gut. Similarly, the enteric nervous system, considered as the second brain due to the number of neurons located in the gut wall, controls gut motility and other functions; enteric glial cells are able, for example, to secrete molecules that act on intestinal barrier and prevent inflammation. Finally, the intestinal microbiota are now considered as a real organ since they play critical roles in almost all intestinal functions, including the degradation/metabolization of ingested toxins (Robert et al. 2017). The role of microbiota and the immune/nervous systems in maintaining the gut homeostasis is particularly well illustrated by the fact that perturbations of one or more of these systems are associated with functional (e.g., irritable bowel syndrome) and organic gastrointestinal disorders (e.g., Crohn’s disease).

This complex organ consisting of epithelial cells, immune cells, enteric nervous cells, microbiota each with its own function and role, is rendered even more complex by interconnections between the cells.

Because of the difficulty in accessing human tissue and to reduce animal experimentation, in vitro/ex vivo models aim to respond to the need to study the huge numbers of contaminants for intestinal toxicology (either toxicodynamic or toxicokinetic mechanisms). In this review, we describe the advantages and limitations of existing in vitro and ex vivo models used to mimic the human/animal gut, including primary cells or cell lines, intestinal explants and organoids. Our focus is on organoids as a promising emerging model.

2 Cell Models

Despite their inherent limitations, isolated cells have been extensively used in the field of toxicology. With respect to the gut, several cell models of both animal and human origin, from different parts of the gut, have provided scientists with useful tools to assess parameters of importance in toxicology, e.g., barrier function, metabolism, proliferation.

The main challenge using isolated cells is to maintain the organization of the epithelial lining in vitro. Classical approaches are represented by primary cell cultures, cell derived from healthy intestine and cell lines from tumor origin.

2.1 Primary Cell Cultures

Primary cells are isolated directly from animal or human intestinal tissue using mechanical and/or enzymatic dissociation. Thus, they closely represent a "real" tissue, with a low risk of mutation. However, their short lifespan limits their use to short-term toxicity studies. In 1984, Walton, Acton, and Stich (Walton et al. 1984) evaluated the toxicity of various molecules, including the mycotoxin aflatoxin B1, on freshly isolated cells from rainbow trout intestine, and observed a 10% mortality of untreated cells reached after 6 h. In 1997, Branka et al. (1997) described a promising technique, based on a new culture medium designed to maintain primary human normal colonocytes for 48 h. In 1998, Kaeffer and Briollais (1998) reported the use of a rotating bioreactor seeded with rat colon epithelial cells that allows consistent tissue reorganization up to 20 days with proliferative compartments and cells organized as monolayers. With this model, these authors confirmed the cytotoxic properties of seaweed-derived polysaccharides (Kaeffer et al. 1999), but the use of such systems is hard to standardize and they have rarely been used for toxicity studies.

2.2 Cells Derived from Healthy Intestine

Quaroni et al. (1979) established an intestinal cell line, derived from rat small intestine able to maintain an epithelioid morphology, establish tight junction, and exhibited a normal karyotype. The non-malignant of the cells was confirmed by their inability to induce tumors when injected in a naïve animal. These cells, referred to as IEC-6 cells, and other IEC-x cells described later have been used for toxicology studies. For example, Del Regno et al. (2015) showed that the mycotoxins, nivalenol, and deoxynivalenol exert pro-oxidant toxicity, an effect that involves NADPH oxidase and NFκB and NRf2-dependent pathways. IEC-6 has also been used to demonstrate that deoxynivalenol exacerbates the genotoxicity of gut microbiota (Payros et al. 2017). Earlier, Duizer et al. (1999) determined on IEC-18 cell line that cadmium alters epithelial barrier by modifying tight junction proteins such as zonula-occludens-1 and occludin. These studies demonstrated the advantage of using intestinal epithelial cell lines in toxicity studies as they make it possible to specifically investigate certain parameters such as epithelial barrier function, oxidative stress.

A few normal lines derived from human intestine are also commercially available, but mainly show a fibroblast phenotype, as CCD-112CoN cells. A normal epithelial cell line was described by Thompson et al. (1985) and is referred to as CCD-841CoN. However, these cells lack typical epithelial markers like cytokeratin and have not been used for toxicology studies so far. FHC cells isolated from the fetal human colon were described by Siddiqui and Chopra (1984), but these cells displayed karyotypic alterations and seemed to be tumorigenic rather than normal (Soucek et al. 2010).

In addition, two intestinal epithelial cell lines have been isolated from piglets. IPEC-J2 cells were isolated from the jejunum and IPEC-1 cells from a mixture of ileal and jejunal tissue of newborn piglets. These cell lines are able to form polarized monolayers with high transepithelial electrical resistance when cultured on pore-size filters. They have been used to investigate the toxic effect of metals such as zinc and cadmium (Zhou et al. 2017; Razzuoli et al. 2018), mycotoxins such as fumonisin, deoxynivalenol, and fusarenone-X (Loiseau et al. 2007; Pinton et al. 2012; Alassane-Kpembi et al. 2017a), as well as the ability of mycotoxin detoxifying agents to interact with drugs (Devreese et al. 2013).

2.3 Tumor Cell Models

Three main human intestinal epithelial cell lines have been used in laboratories: Caco-2, T84, and HT29, including many sub-clones. HT29 was the first to be established, in 1964, followed by Caco-2 in 1974, and T84 in 1984.

Cell lines obtained from human tumors exhibit variable degrees of differentiation that may be due to their origin. HT29 was obtained from a human colorectal adenocarcinoma of a 44-year-old female (Fogh and Trempe 1975), Caco-2 from a colorectal adenocarcinoma (Fogh et al. 1977) in a 72-year-old male and T84 from colorectal carcinoma derived from metastatic lung of a 72-year-old male (Dharmsathaphorn et al. 1984).

These human cell lines can be grown on culture inserts to polarize and differentiate. Their tight junctions are functional, and their transepithelial electrical resistance can be measured. Furthermore, they have apical microvilli, express brush border enzymes and both apical and basolateral specific transporters (Hidalgo et al. 1989). Caco-2 cells and T84 cells differentiate with the characteristics of mature absorptive cells, i.e., a polarized monolayer with an apical brush border, tight junctions, and brush border-associated hydrolases and/or transporters. The HT29 cell line presents two main subtypes: absorptive and mucus-secreting cells (HT29-MTX and HT29-16E clones respectively) (Augeron and Laboisse 1984). They are less differentiated and behave more like a pluripotent cell line with sparse microvilli and low or no brush border-enzyme activities (Rousset 1986; Bolte et al. 1997). Even though both Caco-2 and T84 appeared to be well-differentiated models of the intestinal epithelial monolayer, a recent study has highlighted their differences. They present distinct morphological, biochemical, and functional characteristics, since T84 cells maintain their original colonic characteristics throughout differentiation whereas Caco-2 cells acquire a mature small intestinal enterocytes signature (Devriese et al. 2017).

These human intestinal epithelial cell lines have been extensively used in toxicology studies. For example, Caco-2 cells have been used to investigate the cocktail effect of food contaminants such as mycotoxin and heavy metal. When these cells were exposed to deoxynivalenol and cadmium the interaction between these two contaminants ranged from nearly additive to antagonistic depending on their concentration (Le et al. 2018). This cell line was also used to compare the toxicity of the mycotoxin deoxynivalenol and several modified forms of this toxin (Pierron et al. 2016a, b) or to explain the mechanism of action of polychlorinated biphenyls on intestinal integrity (Choi et al. 2010). In the same way, human goblet cells (HT29-16E cells) were used to understand the mechanism of action of deoxynivalenol on mucus production. Deoxynivalenol was shown to activate the protein kinase R and the mitogen-activated protein kinase p38 ultimately leading to the inhibition of the expression of resistin-like molecule beta, a known positive regulator of mucin expression (Pinton et al. 2015).

2D cell culture on culture inserts has been extensively used as a screening tool to investigate and predict intestinal absorption of drugs, nutrients, and environmental toxics (Hidalgo et al. 1989; Artursson et al. 2001; Kobayashi et al. 2013; Sessa et al. 2014; Vázquez et al. 2015) as well as their potential deleterious effects on the intestine (Jensen-Jarolim et al. 1998; Okada et al. 2000; Hurley and McCormick 2003). Recently, combined analyses of the three cell lines (Caco-2, HT29, and T84) in monolayer were performed and showed that they were sensitive enough to distinguish between hazardous and non-hazardous proteins (Hurley et al. 2016). Testing compound toxicity on the three cell lines in parallel was shown to be more effective than testing it on only one cell.

Colon cancer cell lines have well-recognized limitations including numerous undefined DNA mutations, derive of the cell lines that sometime prevent reproducible results from one laboratory to another and the lack of epithelial cell diversity (for a review on Caco-2 see Sun et al. 2008).

Nevertheless, 2D monolayer of intestinal epithelial cells (either primary cultures or cell lines) is a useful tool as it maintains epithelial cell polarity and allows access to the apical and basolateral compartments. Furthermore, compared to animal experimentation, this technique is less expensive, easy to replicate, and allows a first set of screening trials, thus reducing the need for animal testing. Moreover, these models are excellent tools to decipher the direct interaction between toxics contaminants and intestinal epithelial cells, thus avoiding the complex interactions between epithelial and neighboring cells on the one hand, and between epithelial cells and microbes on the other. Tumoral cell lines therefore represent a widely used model for toxicity studies, and the literature shows some adaptations of these models. First, a recent study of co-culture that include intestinal, hepatic, and immune cells has nicely evidenced strong differences in the toxicity of two mycotoxins zearalenone and deoxynivalenol in individual cell lines compared with bi- or tri-culture system (Smith et al. 2018), highlighting the importance of interactions to determine toxicity of compounds in more complex systems. Second, methodological developments have allowed maintaining these models valuable to monitor deleterious effects of these contaminants on transepithelial resistance. As an example, Akbari et al. (2014) have observed that upon exposure to deoxynivalenol, intestinal cells show an initial drop in transepithelial resistance, followed by increase permeability to macromolecules. More recently, Flynn and Vohra have described an elegant coupling of permeability determination in Caco-2 cells with mass spectrometry to determine interactions of molecules in complex mixtures (Flynn and Vohra 2018). Thus, the authors have determined permeability of alkaloids from serpentine wood (*Rauwolfia serpentina*) across Caco-2 cells monolayers and have simultaneously measure metabolites, which may exhibit bioactivity and toxicity, in the basolateral compartment.

3 Explant Models

The explant (or organ culture) ex vivo model originally derives from the inverted intestinal sacs and ring models developed in the early 1950s to measure the absorption of nutrients (Wilson and Wiseman 1954). Organ culture was almost totally limited to embryonic tissues, until in 1959, Trowell cultured mature tissues for the first time (Trowell 1959). Later, using a modified version of Trowell's technique, Browning and Trier were able to successfully maintain adult human small intestinal mucosa in culture for up to 24 h (Browning and Trier 1969). This technique then expanded within the scientific and medical communities, and the concept was extended to other tissues, particularly bronchi.

The explant model was developed as an intermediate tool between animals and humans to study the homeostasis of the targeted organ and its behavior in physiological and pathological conditions. It is also an alternative to the use of isolated primary cells developed concomitantly (Kaminski 1953), which has its own limitations and advantages, as explained in the section dealing with cell models above.

Technically speaking, human intestinal explants are prepared from gut segments (either stomach, small intestine, or colon). Adventitia, muscularis, and submucosa may be removed, leaving only the epithelial layer, or not. Next, intestinal explants are cut off from the cleaned resections using (for example) surgical punches with different diameters. The human intestinal explants are then maintained in culture using different approaches including culture at the air–liquid interface, or immersion culture either in wells or after mounting in a Ussing chamber, a plastic device that allows specific access to either the apical/luminal or the basolateral/serosal compartments of the tissue (for details, see Randall et al. 2011).

Intestinal explants have the advantage of being a closer model to the human gut than cell models, since they harbor all the epithelial cell types normally present in the mucosa in the correct proportions. In addition to epithelial cells, intestinal explants also contain other cells such as immune cells, myofibroblasts, and enteric nervous cells that play important roles in gut physiology. The main limitation of this technique is the fact that the tissue undergoes progressive and rapid degeneration (within 12–48 h) after collection from patients, although improvements in culture conditions make it possible to maintain the tissue for up to 14 days in some cases (Randall et al. 2011).

Nowadays, although several thousands papers can be found in PubMed dealing with the use of animal or human intestinal explants, in-depth analysis of this literature revealed that only a limited number of articles deals with the use of intestinal explants in toxicology. A few studies have been performed on animal intestinal explants, mainly from rodents or pigs. Rat intestinal explants have been used for example to study the toxicity and metabolism of carcinogenic toxins, including the mycotoxin aflatoxin B1 (Autrup et al. 1978; Kolars et al. 1994). More recently, pig intestinal explants were used to study the toxicity of food-associated mycotoxins, especially type B trichothecenes. Upon exposure to these toxins, intestinal explants produced an inflammatory response (Pierron et al. 2016a; Alassane-Kpembi et al. 2017a) and activation of MAPKinases (Pinton et al. 2012; Garcia et al. 2018). A synergistic inflammatory response was observed between deoxynivalenol (DON) and nivalenol, particularly at low doses (Alassane-Kpembi et al. 2017b). Similarly, when tested on pig jejunal mucosal explants, the surfactant lauroyl-L-carnitine, 1-decanoyl-rac-glycerol, and nonaethylene glycol monododecyl ether at 2 mM caused profound change in the permeability and organization of the intestinal barrier (Danielsen and Hansen 2017). Lastly, it is worth noting that pigs fed DON contaminated feed and pig jejunal explants exposed to DON showed similar histological and signaling changes (Lucioli et al. 2013). In both models, DON led to histological lesions and an increase in the phosphorylation of ERK 1/2 and p38 but not in the phosphorylation of JNK.

Most studies dealing with the use of human intestinal explants in toxicology were conducted in the 1970s and 1980s and provided information on the toxicity and metabolic behavior of some natural and neoformed carcinogens in humans. Using radiolabeled molecules, Autrup et al. (1979) incubated aflatoxin B1 with colon explants and identified two DNA adducts derived from aflatoxin B1 (AFB1). The DNA binding level of AFB1 was lower than that of benzo(a)pyrene (BP), which was also tested in the study. Similarly, Autrup et al. (1980a, b) reported that the metabolic profiles of AFB1, BP, and 1,2-dimethylhydrazine (DMH) in colon explants from human and rat were similar, although the binding level of AFB and DMH to colonic DNA was higher in rat than in human. These authors also found a correlation between the DNA binding potential of BP and DMH in 42 individuals, suggesting that similar enzyme systems are involved in the metabolism of BP and DMH. Using human fetal explants, the same group found that the DNA binding level of AFB, BP, and certain N-nitrosamines is higher in the stomach than in the liver and the esophagus (Autrup et al. 1984). The profiles of the DNA adducts of fetal liver explants were identical to those observed in adult human tissues, suggesting that these carcinogens are metabolized at an early stage of age.

The usefulness of explants for toxicology is also illustrated by the study of Cohen et al. (1983) dealing with the oxidative metabolism of BP and the conjugative metabolism of 1-naphthol by healthy and cancerous colon explants. Several BP metabolites were identified, including monools, diols, triol, tetrols, and quinones, whereas in normal tissue, 1-naphthol (20 μM) was mainly conjugated to sulfate. Interestingly, the ratio of sulfate ester conjugates to glucuronic acid conjugates was significantly higher in the normal tissue than in the tumor tissue (Cohen et al. 1983).

Except for these studies, to our knowledge, human explants had not been used to investigate the toxicity or metabolism of ingested toxins until a recent study investigating the deacetylation of acetylated mycotoxins (3- and 15-acetyldeoxynivalenol) by human gut and liver explants (Ajandouz et al. 2016). This study identified the small intestine and the liver as major deacetylation sites and showed that 3-AcDON is more subjected to deacetylation than 15-AcDON, in agreement with the data obtained on cells and in animals.

In conclusion, it appears that the ex vivo explant model was underused in the last decades for the topic of the present review, since it early proved to be highly relevant to humans with respect to metabolism patterns of aflatoxin B1 and benzo (a)pyrene. One should argue that the model is fully functional only for a short time; it requires access to human tissues and that it is not really suitable for large-scale screening trials. Explants are undoubtedly a robust ex vivo model, ideally if used together with other in vitro models in integrative approaches, for the risk assessment of toxic contaminants in human food.

4 Organoids Models

Preserving tissue architecture and obtaining cellular differentiation is difficult in culture. However, maintaining highly organized structures, joining adjacent cells with asymmetrical distribution of membrane proteins and lipids in the two plasma membrane domains and tight junctions is particularly important for the study of intestinal absorption, metabolism, and toxicity (Rodriguez-Boulan and Nelson 1989). The development of intestinal organoid models has benefitted from the progress made in the characterization and culture of stem cells originating from the gut.

To perform long-term culture and provide the basic crypt–villus physiology in the absence of mesenchymal niche, Sato et al. (2009) combined the growth factor requirements of the intestinal epithelium observed in vivo. Gradients of specific factors were used for the culture including Wnt3a signaling, required for crypt proliferation, the Wnt agonist R-spondin that induces marked crypt hyperplasia in vivo, epidermal growth factor associated with intestinal proliferation and the transgenic expression of Noggin that induces an expansion of crypt numbers. To facilitate the growth and maintenance of intestinal organoids, cell lines can be engineered to secrete Wnt3a, R-spondin, and Noggin as shown with mouse fibroblast producing conditioned media used to grow both mouse and human intestinal crypt organoids (Powell and Behnke 2017). Organoids display the 3D architecture required to obtain the in vivo functional differentiation (Simian and Bissell 2017). Organoids rely on artificial extracellular matrices to facilitate their self-organization into 3D structures that resemble native tissue architecture. For this purpose, a semi-solid, laminin/collagen-rich material was used to support intestinal epithelial growth.

In this pioneer development of gut organoids, mouse crypt preparations as well as single-sorted $Lgr5^+$ cells were expanded for long-term culture. Single crypts underwent multiple crypt fission events and simultaneously generated villus-like epithelial domains. Single-sorted $Lgr5^+$cells also initiated these crypt–villus organoids. In these structures, the $Lgr5^+$ stem cells were able to divide and generate a self-renewing stem cell population and cells able to differentiate into enterocytes, enteroendocrine cells, and goblet cells. All these cells could exfoliate from the villus tip into the lumen, thus mimicking the physiological turnover of the adult intestinal epithelium. Since the seminal study of Sato et al. (2009), organoids have been produced from primary tissue, and from embryonic or induced pluripotent stem cells. Despite their diversity, they are used for the common goal of capturing complex biological interactions by re-creating physiologically relevant mechanical organ functions, cellular diversity, and spatial architecture (Hynds and Giangreco 2013).

Although intestinal organoids have no enteric nerve cells or immune cells, which limits their use in modeling inflammatory responses to infection or drugs (Fatehullah et al. 2016), the development of consistently stable cell culture conditions led to a broad array of applications. Various proof-of-concept investigations have shown that organoids grown from intestinal crypts can be used for controlled gene expression, used in various analytical procedures (microarrays, sequencing, immunohistochemistry, mass spectrometry, etc.) or cryopreserved for long-term storage. The 3D organoid model is a powerful tool for the study of a range of in vivo biological processes including stem cell functions and maintenance, candidate gene function in tissue physiology and carcinogenesis or tissue responses to drugs, virus or bacterial infections.

Healthy organoids represent a closed system with a clean, light appearance with clear epithelial integrity. When the number of dead or dying cells increases, the outer surface is disrupted and takes on an overall darker and less uniform appearance. To quantify cell death, it is possible to use traditional cell culture systems like MTT, which is converted to formazan through the NAD(P) H-dependent cellular oxidoreductase enzymes, the CellTiter-Glo® assay, which uses the luciferase reaction to quantify ATP levels, or the PrestoBlue® assay, which becomes fluorescent in the presence of the reducing cytosol associated with living cells. Nevertheless, for these assays, some modifications are needed to solubilize the semi-solid structure of the organoids (Young and Reed 2016). Damage to the epithelial cells can affect the intestinal barrier function. Recently, microinjection of fluorescent-labeled dextran and microscopy imaging were performed to assess the barrier permeability of organoids (Hill et al. 2017).

Large-scale proteome variations can be measured in organoids as demonstrated by Gonneaud et al. (2016). These authors used a stable isotope labeling by amino acids in cell culture (SILAC)-based quantitative proteomic approach to measure changes in protein expression changes in intestinal organoids under conditions inhibiting their growth and development. Organoids have been used to understand the role of the intercalating Paneth cells in crypt cell maintenance (Yin et al. 2014). They have also been used to demonstrate the role of the Toll-like Receptor 4 (TLR4) in intestinal cell proliferation (Neal et al. 2012) as well as in intestine-microbe/virus interactions (Leslie et al. 2015; Karve et al. 2017; Sigal et al. 2015; Foulke-Abel et al. 2014).

Regarding toxicology researches, the use of organoids is in its infancy and very few papers have been published to date. The impact of different potent inducers of intestinal epithelial cells apoptosis (TNF-α, X-ray irradiation and the chemotherapeutic drug, 5-Fluoro-Uracile) was assessed on intestinal organoids. The results concerning cell death and cell survival using microscopic and colorimetric approaches showed that organoids are useful and physiologically relevant models to test in vitro the damage caused by drugs and toxins (Grabinger et al. 2014).

As the intestine is exposed to various aryl hydrocarbon receptors (AhR) ligands (dioxins and dioxin-like, dibenzofurans, and biphenyls) and polycyclic aromatic hydrocarbon through contaminated food, organoids have been used to investigate whether AhR can affect their development and modulates signaling pathways associated with intestinal epithelial cell development. Thus, 2,3,7,8-tetrachlorodibenzo-p-dioxin and 6-formylindolo[3,2-b] carbazole, a tryptophan metabolite (FICZ) inhibited the development of organoids and FICZ also reduced the level of active β-catenin, suggesting that this metabolite interferes with the canonical Wnt signaling pathway (Park et al. 2016).

Similarly, the intestinal transport and toxicity of ricin was assessed using organoids. Results showed that ricin crosses the intestinal wall without apparent damage up to 24 h of exposure, whereas at 48 h, damage was visible and most cells had become rounded and detached from the scaffold. Importantly, in vivo results confirmed that no apparent histological damage was apparent in mice up to 16 h after ricin ingestion, thereby confirming the predictive value of organoids (Flora et al. 2013).

Recently, the effect of dietary nutrients on the growth of murine intestinal organoids was evaluated by measuring the surface area of organoid culture. Among these nutrients, caffeic acid inhibited organoid growth in a concentration-dependent manner, curcumin exhibited variable effectiveness, and vitamin C had no effect on organoids growth (Cai et al. 2018).

Organoids have also been shown to be an appropriate tool to study the function of different genes and enzymes involved in intestinal drug metabolism and toxicity. Crypt cultures were treated with different prototypical agonists of xenobiotic nuclear receptors (XNR) leading a notable induction of the representative downstream target genes, including cytochrome p450. To demonstrate that this induction was a consequence of ligand–receptor binding, organoids obtained from XNR-null mice were tested and failed to induce the specific Cyp450 genes (Lu et al. 2017).

Beyond studies of specific signaling pathways, organoids can be used to study the cellular mechanisms behind intestinal diseases through studies of gene expression or manipulation (Jung et al. 2011). Recent studies have shown that CRISPR/Cas9-mediated gene editing of healthy organoids is a potential tool to evaluate candidate gene function following exposure to drugs or toxic compounds (Drost et al. 2015). However, up to now, despite abundant knowledge and evidence for the successful implementation of 3D culture in fundamental biological research, these models for toxicity assessment has rarely been used (Astashkina and Grainger 2014).

Recently, Stappenbeck' group has adapted the intestinal epithelial organoids obtained from biopsies to established human primary epithelial cell monolayers from multiple regions of gastrointestinal tract from healthy subjects and inflammatory bowel disease patients (VanDussen et al. 2015). With this reproducible technique, organoids were grown for 2–3 weeks and then disrupted and seeded on culture inserts for 2–3 days to obtain monolayers. Like organoid culture, this technique makes it possible to establish and store cells from specific patients. This model is of particular interest to study pathogen adherence (VanDussen et al. 2015; Wang et al. 2017) and to access to apical compartment which is rather difficult with organoids (McCracken et al. 2014; Leslie et al. 2015). However, to our knowledge, the toxicological relevance of a toxic contaminant remains to be tested in this promising model.

5 Conclusion

As described in this chapter, different intestinal models, cell culture, explant, and organoids are used for toxicity studies. These models are increasingly complex, ranging from simple 2D models to 3D structures, including the structural and functional aspects of the mammalian intestinal epithelium. Intestinal cell cultures are easy to use and could be maintained virtually with no time limit, thereby enabling studies of long-term exposure. Although their survival time is limited, intestinal explants have the advantage of conserving the microenvironment of the gut mucosa. As they are currently not commercially available, they require access to human tissues obtained from surgical units. Organoids are promising models for intestinal toxicology. They represent an emerging system with high potential that recapitulates the lumen and crypt–villus structure of the intestine with cell types found in the intestinal epithelium. They can be obtained from stem cells pooled from different donors and are now available from different companies. This suggests that organoids are the more promising models for intestinal toxicology. One should nevertheless bear in mind the physiological complexity and the compensatory mechanisms of the intestinal mucosal barrier and also that the intestinal barrier is itself connected to other organs in well-controlled way. In addition, the recent development of organ-on-a-chip technologies that combined cell or organ culture with microfluidics allows a better mimicking of physiological regulations induced by shear stress or peristaltism. Given the tremendous development of intestinal in vitro/ex vivo models, especially organoids, significant advances in the ability of these models to approach animals and humans for toxicology studies one may expect in the near future.

Acknowledgements This research was partially supported by the French National Research Agency (projects CaDON ANR-15-CE21-0001, Fumolip ANR-16-CE21-0003, and ExpoMycoPig17-CARN-012-001). The authors thank D. Goodfellow for English editing.

Model	Advantages	Limitations
Primary cells	- Close to reality - Simple model enable easier identification of molecular events involved in toxicity	- Isolation of primary human cells requires access to human tissue - Short lifespan limits their use to very short-term studies - Lack of standardization
Immortalized cells	- Can be maintained for a long time in culture (allowing long-term exposure experiments) - Simple model enabling easier identification of molecular events involved in toxic mechanisms - Low variability of results obtained with this model	- Not all cell types present in the epithelium can be isolated and cultured - Immortalized cells are transformed
Tumor cell lines	- Commercially available - Can be maintained for a long time in culture (allowing long-term exposure experiments) - Enable separate evaluation of the effects of toxins on major cell types present in the epithelium when co-cultured (e.g. enterocytes and goblet cells)	- Display genetic alterations - Far from the physiological state
Explants	- Contain all cell types normally present in the epithelium, in the correct proportions - May also contain other cells types such as immune cells, myofibroblasts, and cells from the enteric nerve system (ENS) (neurons and enteric glial cells)	- Limited survival time (48 h) - Access to apical compartment only when mounted in a Ussing chamber - Require access to human intestinal tissue - Higher variability due to inter-individual differences in response to toxins
Organoids	- Contain most of the cell types normally present in the epithelium (but with no guarantee of them being in the correct proportion) - Can be maintained for a long time in culture, allowing long-term exposure to toxins - Commercially available (low variability) - Derived from healthy or diseased individuals	- Contain only epithelial cells (absence of immune or ENS cells except if co-cultured with other cells)

References

Ajandouz EH, Berdah S, Moutardier V et al (2016) Hydrolytic Fate of 3/15-acetyldeoxynivalenol in humans: specific deacetylation by the small intestine and liver revealed using in vitro and ex vivo approaches. Toxins 8:E232. https://doi.org/10.3390/toxins8080232

Akbari P, Braber S, Gremmels H, Koelink PJ et al (2014) Deoxynivalenol: a trigger for intestinal integrity breakdown. FASEB J 28:2414–2429. https://doi.org/10.1096/fj.13-238717

Alassane-Kpembi I, Gerez JR, Cossalter A-M et al (2017a) Intestinal toxicity of the type B trichothecene mycotoxin fusarenon-X: whole transcriptome profiling reveals new signaling pathways. Sci Rep 7:7530. https://doi.org/10.1038/s41598-017-07155-2

Alassane-Kpembi I, Puel O, Pinton P et al (2017b) Co-exposure to low doses of the food contaminants deoxynivalenol and nivalenol has a synergistic inflammatory effect on intestinal explants. Arch Toxicol 91:2677–2687. https://doi.org/10.1007/s00204-016-1902-9

Artursson P, Palm K, Luthman K (2001) Caco-2 monolayers in experimental and theoretical predictions of drug transport. Adv Drug Deliv Rev 46:27–43. https://doi.org/10.1016/S0169-409X(00)00128-9

Astashkina A, Grainger DW (2014) Critical analysis of 3-D organoid in vitro cell culture models for high-throughput drug candidate toxicity assessments. Adv Drug Deliv Rev 69–70:1–18. https://doi.org/10.1016/j.addr.2014.02.008

Augeron C, Laboisse CL (1984) Emergence of permanently differentiated cell clones in a human colonic cancer cell line in culture after treatment with sodium butyrate. Cancer Res 44:3961–3969

Autrup H, Barrett LA, Jackson FE et al (1978) Explant culture of human colon. Gastroenterology 74:1248–1257

Autrup H, Essigmann JM, Croy RG et al (1979) Metabolism of aflatoxin B1 and identification of the major aflatoxin B1-DNA adducts formed in cultured human bronchus and colon. Cancer Res 39:694–698

Autrup H, Harris CC, Schwartz RD et al (1980a) Metabolism of 1,2-dimethylhydrazine by cultured human colon. Carcinogenesis 1:375–380

Autrup H, Schwartz RD, Essigmann JM et al (1980b) Metabolism of aflatoxin B1, benzo[a]pyrene, and 1,2-dimethylhydrazine by cultured rat and human colon. Teratog Carcinog Mutagen 1:3–13

Autrup H, Harris CC, Wu SM et al (1984) Activation of chemical carcinogens by cultured human fetal liver, esophagus and stomach. Chem Biol Interact 50(1):15–25. https://doi.org/10.1016/0009-2797(84)90128-5

Bolte G, Beuermann K, Stern M (1997) The cell lines CACO-2, T84, and HT-29: models of enterocytic differenciation and function. J Pediatr Gastroenterol Nutr 24:473

Branka J, Vallette G, Jarry A et al (1997) Early functional effects of *Clostridium difficile* toxin A on human colonocytes. Gastroenterology 112:1887–1894. https://doi.org/10.1053/gast.1997.v112.pm9178681

Browning TH, Trier JS (1969) Organ culture of mucosal biopsies of human small intestine. J Clin Invest 48:1423–1432. https://doi.org/10.1172/JCI106108

Cai T, Qi Y, Jergens A et al (2018) Effects of six common dietary nutrients on murine intestinal organoid growth. PLoS ONE 13:e0191517. https://doi.org/10.1371/journal.pone.0191517

Choi YJ, Seelbach MJ, Pu H, Eum SY et al (2010) Polychlorinated biphenyls disrupt intestinal integrity via NADPH oxidase-induced alterations of tight junction protein expression. Environ Health Perspect 118:976–981. https://doi.org/10.1289/ehp.0901751

Cohen GM, Grafstrom RC et al (1983) Metabolism of benzo(a)pyrene and 1-naphthol in cultured human tumorous and non tumorous colon. Cancer Res 1983 43:1312-5

Danielsen EM, Hansen GH (2017) Intestinal surfactant permeation enhancers and their interaction with enterocyte cell membranes in a mucosal explant system. Tissue Barriers 3:e1361900. https://doi.org/10.1080/21688370.2017.1361900

Del Regno M, Adesso S, Popolo A et al (2015) Nivalenol induces oxidative stress and increases deoxynivalenol pro-oxidant effect in intestinal epithelial cells. Toxicol Appl Pharmacol 285:118–127. https://doi.org/10.1016/j.taap.2015.04.002

Devreese M, Pasmans F, De Backer P, Croubels S (2013) An in vitro model using the IPEC-J2 cell line for efficacy and drug interaction testing of mycotoxin detoxifying agents. Toxicol In Vitro 27:157–163. https://doi.org/10.1016/j.tiv.2012.09.020

Devriese S, Van den Bossche L, Van Welden S et al (2017) T84 monolayers are superior to Caco-2 as a model system of colonocytes. Histochem Cell Biol 148:85–93. https://doi.org/10.1007/s00418-017-1539-7

Dharmsathaphorn K, McRoberts JA, Mandel KG et al (1984) A human colonic tumor cell line that maintains vectorial electrolyte transport. Am J Physiol 246:G204–G208. https://doi.org/10.1152/ajpgi.1984.246.2.G204

Drost J, van Jaarsveld RH, Ponsioen B et al (2015) Sequential cancer mutations in cultured human intestinal stem cells. Nature 521:43–47. https://doi.org/10.1038/nature14415

Duizer E, Gilde AJ, Versantvoort CH, Groten JP (1999) Effects of cadmium chloride on the paracellular barrier function of intestinal epithelial cell lines. Toxicol Appl Pharmacol 155:117–126. https://doi.org/10.1006/taap.1998.8589

Fatehullah A, Tan SH, Barker N (2016) Organoids as an in vitro model of human development and disease. Nat Cell Biol 18:246–254. https://doi.org/10.1038/ncb3312

Flora AD, Teel LD, Smith MA, Sinclair JF, Melton-Celsa AR, O'Brien AD (2013) Ricin crosses polarized human intestinal cells and intestines of ricin-gavaged mice without evident damage and then disseminates to mouse kidneys. PLoS ONE 8:e69706. https://doi.org/10.1371/journal.pone.0069706

Flynn TJ, Vohra SN (2018) Simultaneous determination of intestinal permeability and potential drug interactions of complex mixtures using Caco-2 cells and high-resolution mass spectrometry: studies with *Rauwolfia serpentina* extract. Chem Biol Interact 290:37–43. https://doi.org/10.1016/j.cbi.2018.05.006

Fogh J, Fogh JM, Orfeo T (1977) One hundred and twenty-seven cultured human tumor cell lines producing tumors in nude mice. J Natl Cancer Inst 59:221–226

Fogh J, Trempe G (1975) New human tumor cell lines. In: Fogh J (ed) Human tumor cells in vitro, 1st edn. pp 115–159

Foulke-Abel J, In J, Kovbasnjuk O et al (2014) Human enteroids as an ex-vivo model of host-pathogen interactions in the gastrointestinal tract. Exp Biol Med Maywood NJ 239:1124–1134. https://doi.org/10.1177/1535370214529398

García GR, Payros D, Pinton P, Dogi CA (2018) Intestinal toxicity of deoxynivalenol is limited by *Lactobacillus rhamnosus* RC007 in pig jejunum explants. Arch Toxicol 92:983–993. https://doi.org/10.1007/s00204-017-2083-x

Gerbe F, van Es JH, Makrini L et al (2011) Distinct ATOH1 and Neurog3 requirements define tuft cells as a new secretory cell type in the intestinal epithelium. J Cell Biol 192:767–780. https://doi.org/10.1083/jcb.201010127

Gerbe F, Sidot E, Smyth DJ et al (2016) Intestinal epithelial tuft cells initiate type 2 mucosal immunity to helminth parasites. Nature 529:226–230. https://doi.org/10.1038/nature16527

Gonneaud A, Jones C, Turgeon N, et al (2016). A SILAC-based method for quantitative proteomic analysis of intestinal organoids. Sci Rep 6:38195. https://doi.org/10.1038/srep38195, 30 Nov 2016.

Grabinger T, Luks L, Kostadinova F et al (2014) Ex vivo culture of intestinal crypt organoids as a model system for assessing cell death induction in intestinal epithelial cells and enteropathy. Cell Death Dis 5:e1228. https://doi.org/10.1038/cddis.2014.183

Grenier B, Applegate TJ (2013) Modulation of intestinal functions following mycotoxin ingestion: meta-analysis of published experiments in animals. Toxins 5:396–430. https://doi.org/10.3390/toxins5020396

Hidalgo IJ, Raub TJ, Borchardt RT (1989) Characterization of the human colon carcinoma cell line (Caco-2) as a model system for intestinal epithelial permeability. Gastroenterology 96:736–749

Hill DR, Huang S, Tsai YH, Spence JR, Young VB (2017) Real-time measurement of epithelial barrier permeability in human intestinal organoids. J Vis Exp 130:e56960. https://doi.org/10.3791/56960

Hurley BP, McCormick BA (2003) Translating tissue culture results into animal models: the case of *Salmonella typhimurium*. Trends Microbiol 11:562–569. https://doi.org/10.1016/j.tim.2003.10.002

Hurley BP, Pirzai W, Eaton AD et al (2016) An experimental platform using human intestinal epithelial cell lines to differentiate between hazardous and non-hazardous proteins. Food Chem Toxicol Int J Publ Br Ind Biol Res Assoc 92:75–87. https://doi.org/10.1016/j.fct.2016.04.003

Hynds RE, Giangreco A (2013) Concise review: the relevance of human stem cell-derived organoid models for epithelial translational medicine. Stem Cells Dayt Ohio 31:417–422. https://doi.org/10.1002/stem.1290

Jensen-Jarolim E, Gajdzik L, Haberl I et al (1998) Hot spices influence permeability of human intestinal epithelial monolayers. J Nutr 128:577–581

Jung P, Sato T, Merlos-Suárez A et al (2011) Isolation and in vitro expansion of human colonic stem cells. Nat Med 17:1225–1227. https://doi.org/10.1038/nm.2470

Kaeffer B, Bénard C, Lahaye M et al (1999) Biological properties of ulvan, a new source of green seaweed sulfated polysaccharides, on cultured normal and cancerous colonic epithelial cells. Planta Med 65:527–531. https://doi.org/10.1055/s-1999-14009

Kaeffer B, Briollais S (1998) Primary culture of colonocytes in rotating bioreactor. In Vitro Cell Dev Biol—Anim 34:622–625. https://doi.org/10.1007/s11626-996-0008-8

Kaminski EJ (1953) A method for the preparation of isolated cells for use in high-resolution autoradiographic studies. Br J Radiol 26:378–380. https://doi.org/10.1259/0007-1285-26-307-378

Karve SS, Pradhan S, Ward DV, Weiss AA (2017) Intestinal organoids model human responses to infection by commensal and Shiga toxin producing *Escherichia coli*. PLoS ONE 12:e0178966. https://doi.org/10.1371/journal.pone.0178966

Kobayashi S, Shinohara M, Nagai T, Konishi Y (2013) Transport mechanisms for soy isoflavones and microbial metabolites dihydrogenistein and dihydrodaidzein across monolayers and membranes. Biosci Biotechnol Biochem 77:2210–2217. https://doi.org/10.1271/bbb.130404

Kolars JC, Benedict P, Schmiedlin-Ren P, Watkins PB (1994) Aflatoxin B1-adduct formation in rat and human small bowel enterocytes. Gastroenterology 106:433–439. https://doi.org/10.1016/0016-5085(94)90602-5

Le TH, Alassane-Kpembi I, Oswald IP, Pinton P (2018) Analysis of the interactions between environmental and food contaminants, cadmium and deoxynivalenol, in different target organs. Sci Total Environ 622–623:841–848. https://doi.org/10.1016/j.scitotenv.2017.12.014

Leslie JL, Huang S, Opp JS et al (2015) Persistence and toxin production by Clostridium difficile within human intestinal organoids result in disruption of epithelial paracellular barrier function. Infect Immun 83:138–145. https://doi.org/10.1128/IAI.02561-14

Loiseau N, Debrauwer L, Sambou T, Bouhet S et al (2007) Fumonisin B1 exposure and its selective effect on porcine jejunal segment: sphingolipids, glycolipids and trans-epithelial passage disturbance. Biochem Pharmacol 74:144–152. https://doi.org/10.1016/j.bcp.2007.03.031

Lu W, Rettenmeier E, Paszek M et al (2017) Crypt Organoid culture as an in vitro model in drug metabolism and cytotoxicity studies. Drug Metab Dispos Biol Fate Chem 45:748–754. https://doi.org/10.1124/dmd.117.075945

Lucioli J, Pinton P et al (2013) The food contaminant deoxynivalenol activates the mitogen activated protein kinases in the intestine: interest of ex vivo models as an alternative to in vivo experiments. Toxicon 66:31–36. https://doi.org/10.1016/j.toxicon.2013.01.024

McCracken KW, Catá EM, Crawford CM et al (2014) Modelling human development and disease in pluripotent stem-cell-derived gastric organoids. Nature 516:400–404. https://doi.org/10.1038/nature13863

Neal MD, Sodhi CP, Jia H et al (2012) Toll-like receptor 4 is expressed on intestinal stem cells and regulates their proliferation and apoptosis via the p53 up-regulated modulator of apoptosis. J Biol Chem 287:37296–37308. https://doi.org/10.1074/jbc.M112.375881

Okada T, Narai A, Matsunaga S et al (2000) Assessment of the marine toxins by monitoring the integrity of human intestinal Caco-2 cell monolayers. Toxicol Vitro Int J Publ Assoc BIBRA 14:219–226

Park J-H, Choi A-J, Kim S-J et al (2016) AhR activation by 6-formylindolo[3,2-b]carbazole and 2,3,7,8-tetrachlorodibenzo-p-dioxin inhibit the development of mouse intestinal epithelial cells. Environ Toxicol Pharmacol 43:44–53. https://doi.org/10.1016/j.etap.2016.02.007

Payros D, Dobrindt U, Martin P, Secher T, Bracarense AP, Boury M, Laffitte J, Pinton P, Oswald E, Oswald IP (2017) the food contaminant deoxynivalenol exacerbates the genotoxicity of gut microbiota. MBio 8:e00007–e00017. https://doi.org/10.1128/mBio.00007-17

Pierron A, Mimoun S, Murate LS et al (2016a) Intestinal toxicity of the masked mycotoxin deoxynivalenol-3-β-D-glucoside. Arch Toxicol 90:2037–2046. https://doi.org/10.1007/s00204-015-1592-8

Pierron A, Mimoun S, Murate LS, Loiseau N (2016b) Microbial biotransformation of DON: molecular basis for reduced toxicity. Sci Rep 6:29105. https://doi.org/10.1038/srep29105

Pinton P, Tsybulskyy D, Lucioli J et al (2012) Toxicity of deoxynivalenol and its acetylated derivatives on the intestine: differential effects on morphology, barrier function, tight junction proteins, and mitogen-activated protein kinases. Toxicol Sci 130:180–190. https://doi.org/10.1093/toxsci/kfs239

Pinton P, Graziani F, Pujol A, Nicoletti C et al (2015) Deoxynivalenol inhibits the expression by goblet cells of intestinal mucins through a PKR and MAP kinase dependent repression of the resistin-like molecule β. Mol Nutr Food Res 59:1076–1087. https://doi.org/10.1002/mnfr.201500005

Powell RH, Behnke MS (2017) WRN conditioned media is sufficient for in vitro propagation of intestinal organoids from large farm and small companion animals. Biol Open 6:698–705. https://doi.org/10.1242/bio.021717

Quaroni A, Wands J, Trelstad RL, Isselbacher KJ (1979) Epithelioid cell cultures from rat small intestine. Characterization by morphologic and immunologic criteria. J Cell Biol 80:248–265

Randall KJ, Turton J, Foster JR (2011) Explant culture of gastrointestinal tissue: a review of methods and applications. Cell Biol Toxicol 27:267–284. https://doi.org/10.1007/s10565-011-9187-5

Razzuoli E, Mignone G, Lazzara F, Vencia W et al (2018) Impact of cadmium exposure on swine enterocytes. Toxicol Lett 287:92–99. https://doi.org/10.1016/j.toxlet.2018.02.005

Robert H, Payros D, Pinton P, Théodorou V, Mercier-Bonin M, Oswald IP (2017) Impact of mycotoxins on the intestine: are mucus and microbiota new targets? J Toxicol Env Health: Part B: Crit Rev 20:249–275. https://doi.org/10.1080/10937404.2017.1326071

Rodriguez-Boulan E, Nelson WJ (1989) Morphogenesis of the polarized epithelial cell phenotype. Science 245:718–725

Rousset M (1986) The human colon carcinoma cell lines HT-29 and Caco-2: two in vitro models for the study of intestinal differentiation. Biochimie 68:1035–1040

Sato T, Vries RG, Snippert HJ et al (2009) Single Lgr5 stem cells build crypt-villus structures in vitro without a mesenchymal niche. Nature 459:262–265. https://doi.org/10.1038/nature07935

Sessa M, Balestrieri ML, Ferrari G et al (2014) Bioavailability of encapsulated resveratrol into nanoemulsion-based delivery systems. Food Chem 147:42–50. https://doi.org/10.1016/j.foodchem.2013.09.088

Siddiqui KM, Chopra DP (1984) Primary and long term epithelial cell cultures from human fetal normal colonic mucosa. In vitro 20:859–868

Sigal M, Rothenberg ME, Logan CY et al (2015) Helicobacter pylori activates and expands Lgr5 (+) stem cells through direct colonization of the gastric glands. Gastroenterology 148(1392–1404):e21. https://doi.org/10.1053/j.gastro.2015.02.049

Simian M, Bissell MJ (2017) Organoids: a historical perspective of thinking in three dimensions. J Cell Biol 216:31–40. https://doi.org/10.1083/jcb.201610056

Smith MC, Gheux A, Coton M, Madec S et al (2018) In vitro co-culture models to evaluate acute cytotoxicity of individual and combined mycotoxin exposures on Caco-2, THP-1 and HepaRG human cell lines. Chem Biol Interact 281:51–59. https://doi.org/10.1016/j.cbi.2017.12.004
Soucek K, Gajduskova P, Brazdova M, Hyzdalova M et al (2010) Fetal colon cell line FHC exhibits tumorigenic phenotype, complex karyotype, and TP53 gene mutation. Cancer Genet Cytogenet 197:107–116. https://doi.org/10.1016/j.cancergencyto.2009.11.009
Sun H, Chow EC, Liu S et al (2008) The Caco-2 cell monolayer: usefulness and limitations. Expert Opin Drug Metab Toxicol 4:395–411. https://doi.org/10.1517/17425255.4.4.395
Thompson AA, Dilworth S, Hay RJ (1985) Isolation and culture of colonic epithelial cells in serum-free medium. J Tissue Culture Meth 9:117–122. https://doi.org/10.1007/bf01797782
Trowell OA (1959) The culture of mature organs in a synthetic medium. Exp Cell Res 16:118–147. https://doi.org/10.1016/0014-4827(59)90201-0
VanDussen KL, Marinshaw JM, Shaikh N et al (2015) Development of an enhanced human gastrointestinal epithelial culture system to facilitate patient-based assays. Gut 64:911–920. https://doi.org/10.1136/gutjnl-2013-306651
Vázquez M, Devesa V, Vélez D (2015) Characterization of the intestinal absorption of inorganic mercury in Caco-2 cells. Toxicol Vitro Int J Publ Assoc BIBRA 29:93–102. https://doi.org/10.1016/j.tiv.2014.09.013
Walton DG, Acton AB, Stich HF (1984) DNA repair synthesis following exposure to chemical mutagens in primary liver, stomach, and intestinal cells isolated from rainbow trout. Cancer Res 44:1120–1121
Wang Y, DiSalvo M, Gunasekara DB et al (2017) Self-renewing monolayer of primary colonic or rectal epithelial cells. Cell Mol Gastroenterol Hepatol 4(165–182):e7. https://doi.org/10.1016/j.jcmgh.2017.02.011
Wilson TH, Wiseman G (1954) The use of sacs of everted small intestine for the study of the transference of substances from the mucosal to the serosal surface. J Physiol 123:116–125
Yin X, Farin HF, van Es JH et al (2014) Niche-independent high-purity cultures of $Lgr5^+$ intestinal stem cells and their progeny. Nat Methods 11:106–112. https://doi.org/10.1038/nmeth.2737
Young M, Reed KR (2016) Organoids as a model for colorectal cancer. Curr Colorectal Cancer Rep 2016(12):281–287
Zhou X, Li Y, Li C (2017) Autophagy plays a positive role in zinc-induced apoptosis in intestinal porcine epithelial cells. Toxicol in vitro 44:392–402. https://doi.org/10.1016/j.tiv.2017.08.006

Correction to: Human Organotypic Models for Anti-infective Research

Astrid Hendriks, Ana Rita Cruz, Elisabetta Soldaini, Andrea Guido Oreste Manetti and Fabio Bagnoli

Correction to: Chapter "Human Organotypic Models for Anti-infective Research" A. Hendriks et al., in: Current Topics in Microbiology and Immunology, https://doi.org/10.1007/82_2018_130

The original version of this chapter was published with the copyright holder "Springer Nature Switzerland AG" and without open access. This has now been changed to the copyright holder "The Author(s)" and open access under the terms of the Creative Commons Attribution 4.0 International License (http://creativecommons.org/licenses/by/4.0/). For further details see license information in the chapter.

The updated online version of this chapter can be found at https://doi.org/10.1007/82_2018_130

Current Topics in Microbiology and Immunology (2021) 430:265–265
https://doi.org/10.1007/82_2019_174

Published Online: 19 April 2019

The manufacturer's authorised representative in the EU is Springer Nature Customer Service Centre GmbH, Europaplatz 3, 69115 Heidelberg, Germany. If you have any concerns regarding our products, please contact ProductSafety@springernature.com

Printed and bound by CPI Group (UK) Ltd, Croydon, CR0 4YY
30/09/2026
02225689-0011